LANDSCAPE GRADING

For every element that we design in the landscape, there is a corresponding grading concept, and how these concepts are drawn together is what creates a site grading plan. This study guide explores these concepts in detail to help you learn how to grade with confidence in preparation for the Grading, Drainage and Construction Documentation section of the Landscape Architecture Registration Examination (LARE).

This updated second edition is designed as a textbook for the landscape architecture student, a study guide for the professional studying for the LARE, and a refresher for licensed landscape architects. New to this edition:

- Additional illustrations and explanations for grading plane surfaces and warped planes, swales, berms, retention ponds, and drain inlets;
- Additional illustrations and explanations for grading paths, ramp landings, ramp/stair combinations and retaining walls;
- A section on landscape and built element combinations, highlighting grading techniques for parking lots, culverts and sloping berms;
- A section on landscape grading standards, recognizing soil cut and fill, determining pipe cover, finding FFE, and horizontal and vertical curves;
- Updated information about the computer-based LARE test;
- All sections updated to comply with current ADA guidelines;
- An appendix highlighting metric standards and guidelines for accessibility design in Canada and the UK.

With 223 original illustrations to aid the reader in understanding the grading concepts, including 32 end-of-chapter exercises and solutions to practice the concepts introduced in each chapter, and 10 grading vignettes that combine different concepts into more robust exercises, mimicking the difficulty level of questions on the LARE, this book is your comprehensive guide to landscape grading.

Valerie Aymer is a licensed landscape architect and a member of the American Society of Landscape Architects. She received her Master's in Landscape Architecture in 2002 from Cornell University and her professional license in 2009. She has practiced in Florida, New Jersey, Pennsylvania, Virginia, and New York, most notably on the World Trade Center projects in Lower Manhattan. In 2015, she joined the faculty of the Department of Landscape Architecture at Cornell University where she teaches several technical courses including site engineering.

LANDSCAPE GRADING

A Study Guide for the LARE

Second Edition

Valerie E. Aymer, PLA, ASLA

Routledge
Taylor & Francis Group
LONDON AND NEW YORK

Second edition published 2020
by Routledge
2 Park Square, Milton Park, Abingdon, Oxon OX14 4RN

and by Routledge
52 Vanderbilt Avenue, New York, NY 10017

Routledge is an imprint of the Taylor & Francis Group, an informa business

First edition published by lulu.com 2011

British Library Cataloguing-in-Publication Data
A catalogue record for this book is available from the British Library

Library of Congress Cataloging-in-Publication Data
Names: Aymer, Valerie, author.
Title: Landscape grading : a study guide for the LARE / Valerie Aymer.
Description: 2. | Abingdon, Oxon ; New York, NY : Routledge, 2020. | Includes bibliographical references and index.
Identifiers: LCCN 2019057257 (print) | LCCN 2019057258 (ebook) | ISBN 9780367439057 (hardback) | ISBN 9780367439071 (paperback) | ISBN 9781003006404 (ebook)
Subjects: LCSH: Grading (Earthwork)–Examinations–Study guides. | Landscape construction–Examinations–Study guides. | Building sites–Examinations–Study guides.
Classification: LCC TH380 .A96 2020 (print) | LCC TH380 (ebook) | DDC 624.1/52–dc23
LC record available at https://lccn.loc.gov/2019057257
LC ebook record available at https://lccn.loc.gov/2019057258

ISBN: 978-0-367-43905-7 (hbk)
ISBN: 978-0-367-43907-1 (pbk)
ISBN: 978-1-003-00640-4 (ebk)

Typeset in Univers
by Wearset Ltd, Boldon, Tyne and Wear

For Product Safety Concerns and Information please contact our EU representative:
GPSR@taylorandfrancis.com
Taylor & Francis Verlag GmbH, Kaufingerstraße 24, 80331 München, Germany.

This text is dedicated to my study partners, my professors, my students, and all the future landscape architecture professionals.

Contents

Contents

Preface

As a student of landscape architecture in graduate school I was always fascinated by the technical aspects of the profession. How does a design come into being? How does what you draw on the paper become real? Grading gives structure to designs, it explains if your design is feasible and how it will ultimately change the existing landscape. It creates places from spaces. It helps designers work with, rather than against the existing topography.

Landscape grading is the key to a successful landscape design. For every element that we design in the landscape, there is a corresponding grading concept. How these concepts come together is what makes up a grading plan. It tells a Contractor how the design should interface with the existing landscape.

Whether it is at a small scale or a large scale the principles of landscape grading are the same. In this book, we will explore the basics of landscape grading through explanations and examples of how to solve simple problems. This book will show you where to begin, how to find your spot elevations, and how to connect your proposed contour lines. None of the concepts in this book are new, they are, however, presented in a way to help you make the leap from concept to application. In the subsequent years since the original book was published, I have used the techniques in this book on landscape architecture projects of varying size and scale. I have also refined the explanations and added depth where needed based on several years of teaching landscape grading. The edition is a collection of teaching notes with additional figures, exercises and vignettes that add clarity and provide additional practice.

This book is designed as a "how to solve a problem" guide rather than a "why do we solve the problem this way" guide. If you would like more detail into why we solve grading problems the way we do, there is a list of recommended reading in the back of the book. My goal, when writing this book, was to provide enough knowledge so that when presented with a grading vignette you know how to tackle it. Remember there are many ways to solve a grading problem, this book will add to your body of knowledge or give you a good starting point in your studying. I hope you find this study guide useful.

Although a lot of grading and drainage topics are covered in the book, it is not a comprehensive overview of all the topics covered in the LARE Section 4: Grading, Drainage and Construction Documentation. For a complete and comprehensive list of topics on the examination, check www.clarb.org. The information in this book is current and the subject matter has been reviewed by licensed landscape architects; any errors are unintentional. Additional research of topics, study and practice of vignettes should be undertaken for this examination. It is the responsibility of each person taking the examination to prepare for it thoroughly. Any similarity to existing LARE practice vignettes and current examination problems is unintentional and coincidental. To sign up for the examination or to check your eligibility to take the exam check with your state's licensing board and www.clarb.org.

I hope this book helps you better understand the concepts of site grading to move you forward in your professional life.

Acknowledgments

The author wishes to thank Professor Sherene Baugher, PhD, for her advice and guidance, Professors Emeritus Peter J. Trowbridge, FASLA, and Daniel W. Krall, FASLA, for their continued support and mentorship, Mr. Andrew Lavallee, FASLA, for his continued support, the peer reviewers for their welcome comments, my site engineering students, past and present, for their insights, and Marvin I. Adleman whose spirit lives on in these pages.

Section One
Basics of grading

1.1 What is grading?

Landscape grading is the artistic manipulation of land using basic mathematics and scientific principles. In construction, there are two stages of grading, rough or coarse grading, which moves large quantities of earth based on cut and fill calculations, and fine or finish grading, which is the art of meeting proposed spot elevations as shown on a civil and/or landscape architecture grading plan.

What then is a grading plan? It is the two-dimensional representation of the physical world as it exists and as we propose to change it based on a landscape design. The purpose of a grading plan is twofold. First, the grading plan shows how our two-dimensional drawing will be superimposed on the three-dimensional earth through contour line manipulation and spot elevations. The second and more crucial point of a grading plan is to show where surface water is flowing on a site, both existing and proposed surface runoff. We must account for water percolation and surface runoff by using basic grading techniques tied into grey and green infrastructure (e.g. stormwater system, surface drainage and low impact development systems, etc.) to determine where the water is and will need to be after we've designed a site.

Whether a site is designed for functionality or aesthetics, it is always the grading plan which tells the designer whether or not the design is feasible. The process of grading is fluid and there is more than one correct answer when grading a site. How you grade a site will be determined by existing site conditions, proposed drainage structures and ultimately the client's needs. In practice, landscape architects only need to provide spot elevations that contractors must meet, but on sites with lots of topography or on specialized designs, proposed contour grading is essential.

1.2 Understanding spot elevations

The following concepts, spot elevations, slopes, and interpolation are given as a brief overview. To begin grading anything, you need to understand and be comfortable with spot elevations and how to determine them. Spot elevations are identifying numerical marks on a site plan or topographical survey. They denote elevation, a given point above sea level or another predetermined vertical marker. In the US, civil engineers and landscape architects show spot elevations as decimal parts of a foot (0.00') and architects show spot elevations as feet and inches (0'–0"). Converting between engineering and architectural units is necessary to match the site design to both the architectural and civil elements on site. For example, because 1" = 1/12 of a foot, divide 1 by 12 to get the decimal foot equivalent 0.0833'. Repeat the process for each whole number. If, for example, you need to convert 5–1/2" into decimal feet. First, convert 5–1/2" to a decimal number, 5.5", then divide 5.5" by 12 to get the decimal foot equivalent 0.45833'. Round the decimal to the nearest hundredth and 0.46 is the decimal foot equivalent of 5–1/2". Spot elevations as written as either 1/100th of a foot or 1/10th of a foot. The reverse calculation is also possible. A spot elevation of 44.66' = 44' + 0.66'; 0.66' × 12 = 7.99' or rounded to the nearest whole number 8", therefore 44.66' = 44'-8". Figure 1.2–1 lists common decimal foot equivalents.

Spot elevations shown on a plan are usually preceded by a '+' for proposed or an 'x' for existing spot elevations, indicating the exact location of the elevation on the site. They do not always show units. For US grading plans, it is understood that (0.00) = decimal feet. A grading plan should have a graphic scale associated with it indicating the units of the plan. In tight spots, an arrow is used to pinpoint the location of the spot elevation without "muddying up" the drawing.

In addition to general spot elevations, there are several naming conventions used to highlight specific site elements. These elements have height at or above existing grade and are built elements or civil engineering structures. Abbreviations are used to identify the character of these elements. The following examples are the most commonly used: top and bottom elevation of wall (TW/BW); top and bottom elevation of curb (TC/BC); and top and bottom elevation of stairs (TS/BS). The abbreviations RIM EL, INV IN and INV OUT also precede spot elevations and indicate the height at the top of a drain inlet or rim elevation, and the bottom elevation that connects to a pipe, invert in and invert out. Providing these abbreviations in front of the spot elevations adds a level of clarity to a grading plan. A short list of symbol and abbreviation conventions follows in Section 1.6, Graphic conventions.

ARCHITECTURAL UNITS (FEET AND INCHES)	DECIMAL UNITS (FEET)
1/4"	$0.0208\overline{33}'$
1/2"	$0.041\overline{66}'$
1"	$0.08\overline{33}'$
2"	$0.1\overline{66}'$
3"	0.25'
4"	$0.\overline{33}'$
5"	$0.41\overline{66}'$
6"	0.5'
7"	$0.58\overline{33}'$
8"	$0.\overline{66}'$
9"	0.75'
10"	$0.8\overline{33}'$
11"	$0.91\overline{66}'$
12"	1.00'

SPOT ELEVATIONS REPRESENTATIONS	
× 10.25'	EXISTING (AS SHOWN ON SURVEYS)
+ 10.25'	PROPOSED ELEVATIONS ON HARDSCAPE
+ 10.5'	PROPOSED ELEVATIONS ON LANDSCAPE
+ 10' - 6"	PROPOSED ARCHITECTURAL SPOT ELEVATIONS

1.2–1
Common decimal foot equivalents and spot elevation nomenclature

1.3 Determining slopes

Spot elevations are created by surveyors and are measured in a grid format. They provide the most accurate vertical information about a site. Finding the slope between two spot elevations requires knowing the vertical distance or the change in height, and the horizontal distance between two spot elevations. In order to find the slope, divide the vertical distance by the horizontal distance. In Figure 1.3–1 for example, given Point A at elevation 45.3' and Point B at elevation 47.3' and a horizontal distance of 40', to find the slope first subtract Point A from Point B to find the vertical change between the two spot elevations, 47.3'−45.3' = 2'. Second, divide by the horizontal distance, 2'/40' = 0.05. Multiply the result by 100 and we know that the ground is sloping from Point B to Point A at 5%.

Slopes can be called out as fractions, decimals, ratios or percentages. It is necessary to be familiar and comfortable with the equivalent values as they could be expressed in any of these ways on the examination. The chart in Figure 1.3–2 highlights commonly used slope equivalents.

The slope formula is generalized as slope = vertical distance ÷ horizontal distance (S = V/H). As shown in Figure 1.3–3, it can be expressed in three ways, knowing any two variables of the equation will allow you to solve for the remaining one.

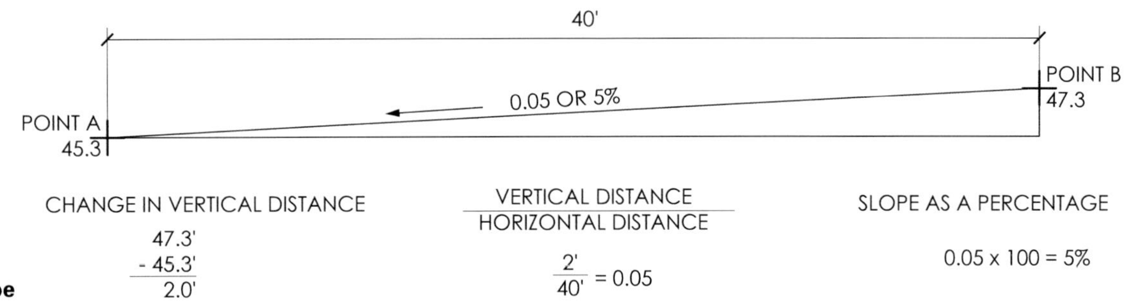

1.3–1
Calculating slope

DECIMAL	RATIO	PERCENTAGE	FRACTION
0.01	100:1	1%	1/100
0.02	50:1	2%	2/100
0.03	33:1	3%	3/100
0.04	25:1	4%	1/25
0.05	20:1	5%	1/20
$0.08\overline{33}$	12:1	$8.\overline{33}\%$	1/12
0.10	10:1	10%	1/10
0.20	5:1	20%	1/5
0.25	4:1	25%	1/4
0.33	3:1	33%	1/3
0.50	2:1	50%	1/2

1.3–2
Commonly used slope equivalents

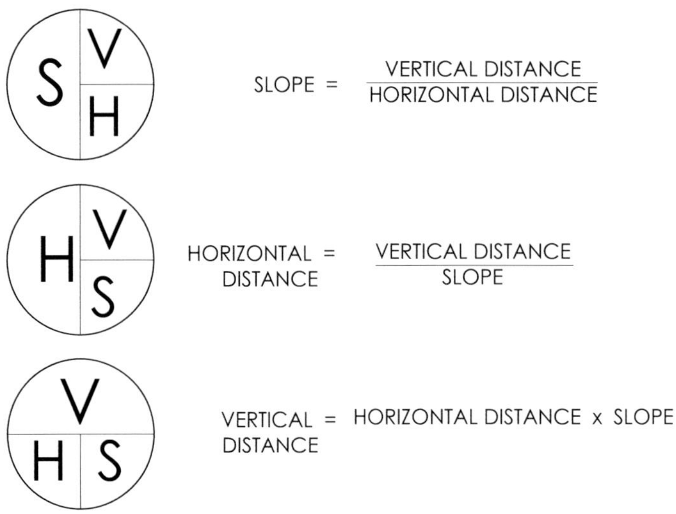

1.3–3
Representations of the slope formula

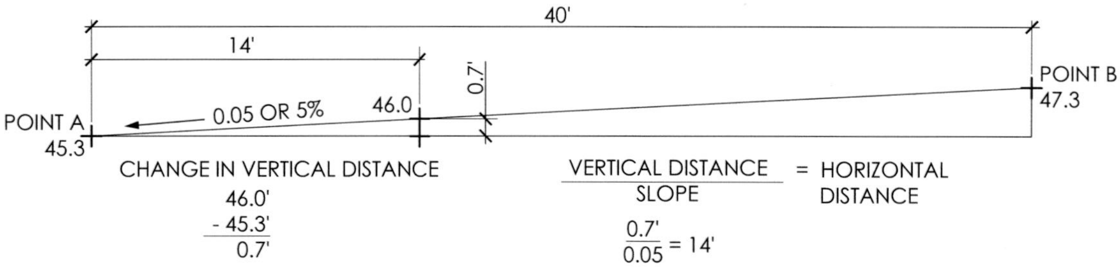

1.4–1
Using interpolation to find whole number spot elevation

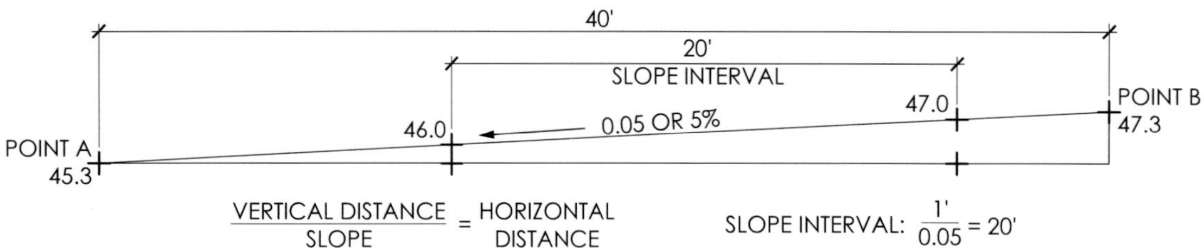

1.4–2
Finding the slope interval

1.4 Interpolation and slope intervals

Given any two spot elevations and the slope, we can determine where the whole number spot elevations occur between them using interpolation. Interpolation uses any combination of the slope formula to find the whole number spot elevation that is nearest to a known spot elevation. In Figure 1.4–1, for example, we would interpolate to find where the whole number spot elevation 46.0' occurs horizontally between 45.3' and 47.3'. First, subtract to find the vertical change, 46.0'−45.3'=0.7'. Second, divide by the slope, 0.05, to find the horizontal location, 0.7'/0.05 = 14'.

Using the slope, we can also determine the slope interval. The slope interval is the consistent horizontal distance between a uniformly-assigned vertical distance. In most cases, we use the slope interval to describe the consistent horizontal distance over a vertical change of one foot. To find the slope interval for a one foot vertical change, divide 1' by the slope. In Figure 1.4–2, the slope interval is 20 feet, 1'/0.05=20'. The next whole number spot elevation, 47.0', will be 20' away from the 46.0' spot elevation. Figure 1.4–3 highlights common slope intervals for a 1' foot change in elevation.

Interpolation finds the first whole number spot elevation and the slope interval automates finding any subsequent whole number spot elevations at a known slope. Slope intervals can be created for vertical intervals other than one foot. First, determine the slope interval for a one foot vertical distance and then multiply that number by the vertical distance.

By using interpolation and slope intervals in a grid format, we can determine whole number contour lines. Figure 1.4–4 shows a grid with spot elevations. First, find the slopes between the different spot elevations and then use interpolation and slope intervals to locate the whole number spot elevations.

SLOPE	SLOPE INTERVAL IN FEET
1%	100
2%	50
3%	$33.\overline{33}$
4%	25
5%	20
$8.\overline{33}$%	12
10%	10
20%	5
25%	4
33%	3
50%	2

1.4–3
Slope intervals for commonly used slopes

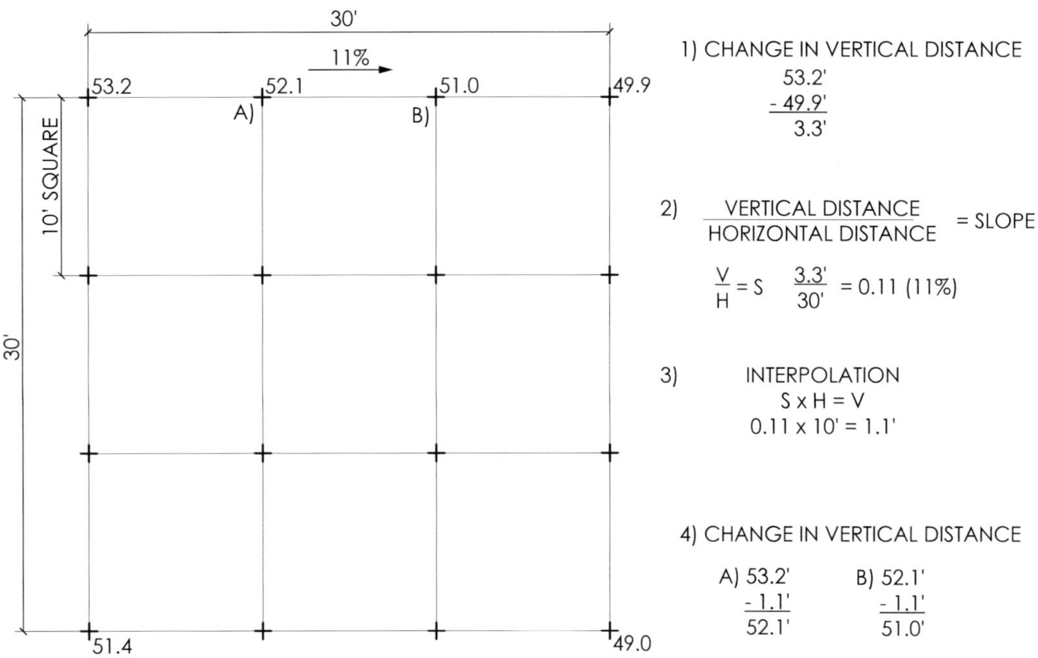

1.4–4
30×30 Grid showing interpolation of spot elevations

1.5 Contour grading

Contour lines are a representation of the whole number spot elevations on a plan. If a site is large enough or steep enough, the surveyed spot elevations are converted to contour lines for greater clarity of the site's topography. Contour lines are easier to read than a plan full of spot elevations, however, with contour lines some accuracy is lost. They show an average change in elevation over a specific distance. On a topographic plan or map, the change in elevation is a consistent, uniform, vertical distance. This is called the contour interval. Common contour intervals are 1', 2', 10', 50' and 100', other regular intervals are also used for larger scale plans such as USGS maps. Make sure to note the contour interval of a grading problem before starting the grading solution. Figure 1.5–1 shows the completed 30'×30' grid with the existing contour lines.

Contour lines follow several established conventions. Figure 1.5–2 illustrates some of these conventions. Contour lines are always labeled on the uphill side. Existing contour lines are shown as dashed lines and their corresponding numbers are shown in parentheses. Every fifth contour line, the contour index line, is darker. Proposed contour lines are solid lines; their corresponding numbers do not have parentheses. As with the existing contour lines, every fifth line is darker. They are connected to the existing contour lines by tick marks at both ends. Tick marks are drawn perpendicularly to the existing contour line. Proposed contour lines should create fluid, smooth connections to the existing contours; avoid jerky and excessively wavy lines.

Contour lines never merge with each other, but form closed rings. If two contour lines with the same value are adjacent they are separated by a spot elevation. Contour lines of different values will sometimes appear to merge; this indicates a vertical

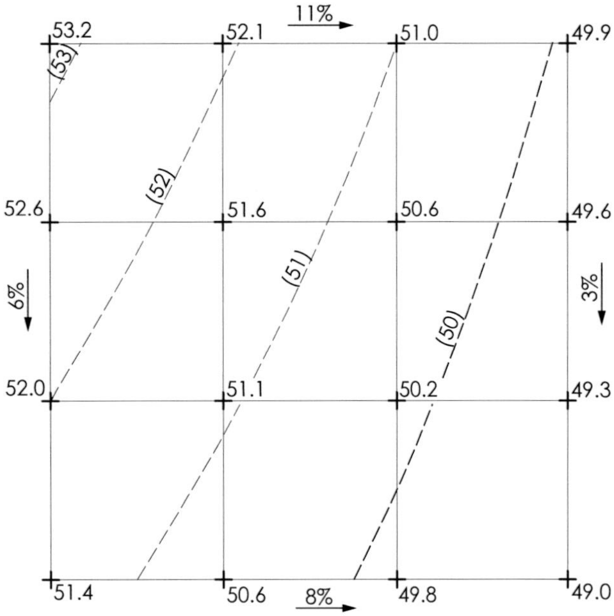

1.5–1
Spot elevations converted to contour lines

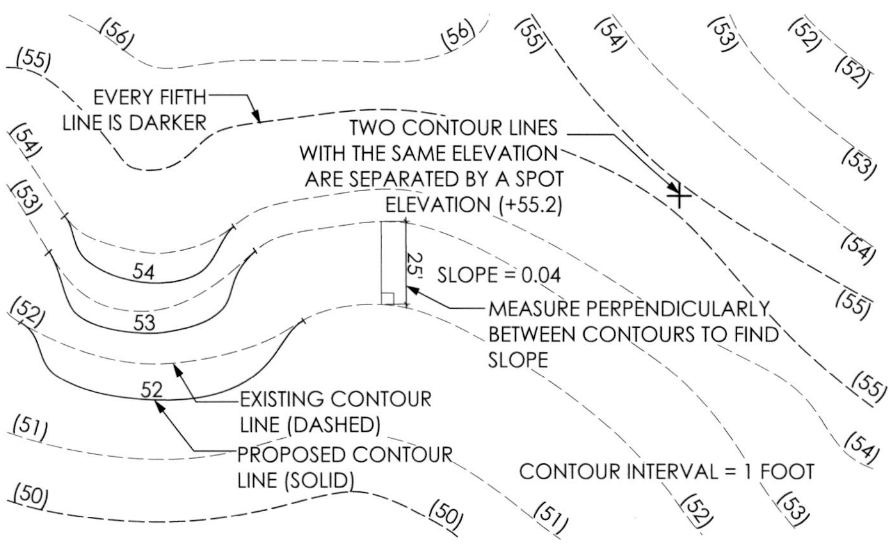

1.5–2
Proposed contours attached to existing contours and measuring slope on contour lines

structure such as a cliff or a wall; the closer together the contour lines the steeper the slope of the land, and the farther apart the shallower the slope of the land.

Determine the slope between two contour lines by measuring perpendicularly from one contour line to the other. We know the fixed vertical distance from the contour interval, and using the slope formula, dividing by the horizontal distance will provide the slope between the two contours. As illustrated in Figure 1.5–3, to find the slope between several existing contour lines, measure perpendicularly from the highest contour line to the lowest contour line and calculate the vertical change in elevation. Measure the horizontal distance of the perpendicular line. Using the slope formula,

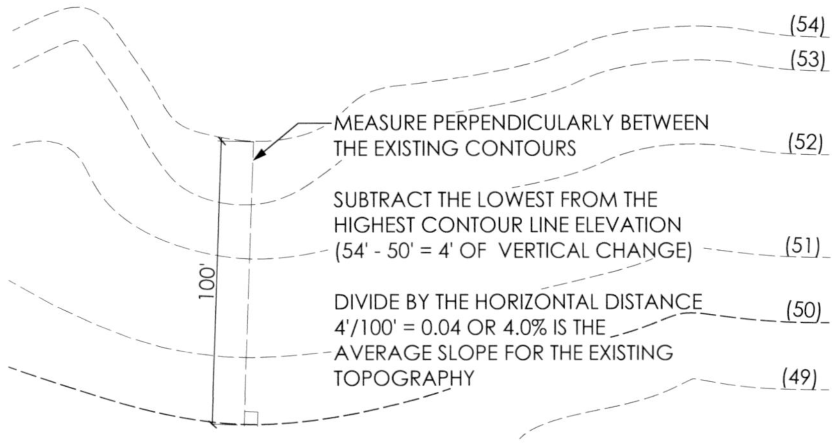

1.5–3
Estimating the total slope for the existing topography on a drawing

interpolate to find the average slope for the existing topography. This concept is very useful and should be done at the beginning of a grading problem to get an estimate of the total slope for the problem. The slope between each individual contour line may vary, but the total slope for the topography will be averaged out. It will give you a good estimate of the slope for the drawing, and a good slope to use for solving some of the larger drainage and built elements problems.

1.6 Graphic conventions

Knowing some simple graphic conventions is also useful. These conventions add clarity to your grading solution. There are three types of conventions to be aware of: symbols, abbreviations and line types. Figure 1.6–1 provides commonly used symbols, abbreviations and line types. Keep in mind that these symbols are not covered in the LARE Orientation Information booklet but are symbols that the Council of Landscape Architectural Registration Boards (CLARB) will assume that you know. CLARB will often define and label their symbols on the grading problem if they are different from those commonly used.

1.7 Summary

The process of site grading determines where surface water will flow on a site after a design is created. A grading plan is the two-dimensional representation of this process. The three basic math techniques needed for every grading problem are determining spot elevations as decimal foot equivalents, determining slope between two spot elevations; and interpolation, determining the whole number spot elevations between two spot elevations.

Contour lines are graphic representations of whole number spot elevations. They are an estimation of a site's topography. Contour lines show vertical change in consistent uniform intervals called a contour interval.

SYMBOLS, ABBREVIATIONS AND LINETYPES	
	DRAIN INLET
	CATCH BASIN
	ELEVATIONAL MARKER (USUALLY SHOWN IN SECTIONS)
℄	CENTERLINE
N	NORTH ARROW: ORIENTS THE DRAWING TO NORTH, IN THESE EXAMPLES NORTH IS UP
0 10 20 40FT	GRAPHIC SCALE: INDICATES HOW MUCH ONE FOOT HORIZONTALLY WILL BE ON A DRAWING. IN THIS EXAMPLE, ONE FOOT HORIZONTALLY EQUALS 20 FEET. USE AN ENGINEERING OR ARCHITECTURAL SCALE RULER TO DRAW A DESIGN SOLUTION BASED ON THE GRAPHIC SCALE
+ ×	SPOT ELEVATION: PROPOSED AND EXISTING
HP	HIGH POINT: INDICATES A HIGH POINT OR PEAK ON A SITE ACCOMPANIED BY A SPOT ELEVATION
LP	LOW POINT: INDICATES A LOW POINT OR DEPRESSION ON A SITE ACCOMPANIED BY A SPOT ELEVATION
HPS	HIGH POINT OF SWALE: INDICATES THE BEGINNING OR HIGHEST POINT ON A SWALE
TC/BC	TOP OF CURB/BOTTOM OF CURB: USED IN ROAD PROBLEMS WITH SPOT ELEVATIONS TO INDICATE CURB HEIGHT; PROVIDE SPOT ELEVATIONS FOR TOP AND BOTTOM OF CURB
TW/BW	TOP OF WALL/BOTTOM OF WALL: USED IN RETAINING WALL PROBLEMS WITH SPOT ELEVATIONS TO INDICATE HEIGHT OF WALL; PROVIDE SPOT ELEVATIONS FOR TOP AND BOTTOM OF WALL
FFE	FINISH FLOOR ELEVATION OR FINISHED FLOOR ELEVATION: THE ELEVATION OF THE FLOOR OF A BUILDING AS IT RELATES TO THE INTERIOR AND EXTERIOR OF A BUILDING. IT IS USUALLY THE FIRST FLOOR BUT IT COULD BE ANY FLOOR OF A BUILDING WHERE THE INTERIOR AND EXTERIOR MEET; ACCOMPANIED BY A SPOT ELEVATION.
	CENTERLINE: ALWAYS LABELED AT THE BEGINNING AND END WITH ℄
	FLOWLINE: CENTERLINE OF A VALLEY OR SWALE; SHOWS DIRECTION OF WATER FLOW
	RIDGELINE: CENTERLINE OF A RIDGE; USED TO DENOTE THE EDGE OF A WATERSHED
	WATER LINE: INDICATES THE EDGE OF A WATER BODY (E.G. STREAM, POND, ETC)
—W——W——W—	UTILITY LINE: IN THIS INSTANCE IT IS A PIPED WATER LINE
CLL	CONTRACT LIMIT LINE: LIMIT OF CONSTRUCTION WITHIN A PROPERTY LINE
	PROPERTY LINE: PROPERTY BOUNDARY

1.6–1
Graphic conventions used in landscape grading

1.8 Exercises (answers start on page 179)

1-1) Find the decimal foot equivalents for the following architectural measurements:

4' - 6"	53' - 9"
111' - 2"	40' - 3"
75' - 0 1/4"	32' - 1"

1-2) Find the architectural measurements for the following decimal foot equivalents:

$97.\overline{33}$'	$28.91\overline{66}$'
7.75'	$56.58\overline{33}$'
112.0104'	250.0208'

1-3) Find the VERTICAL distance for each of the following:

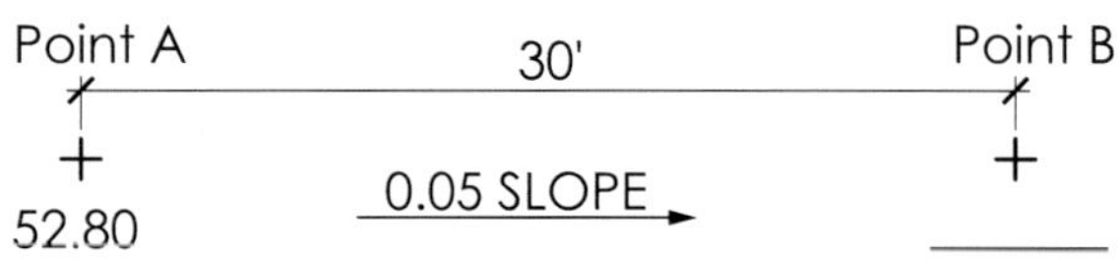

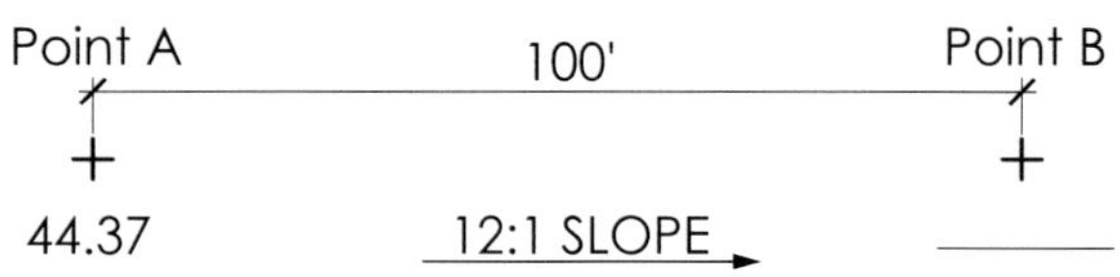

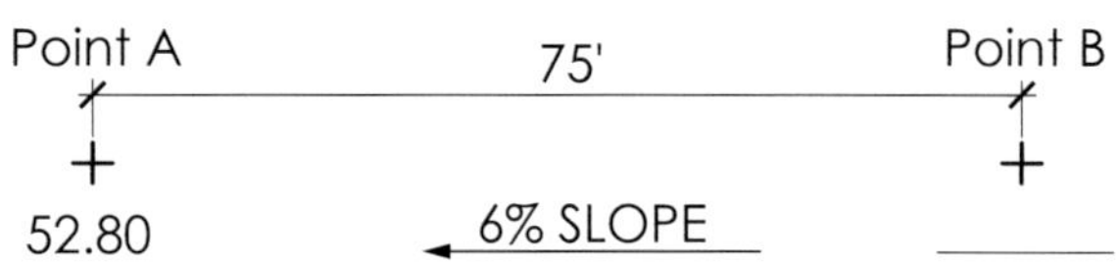

1-4) Find the SLOPE for each of the following:

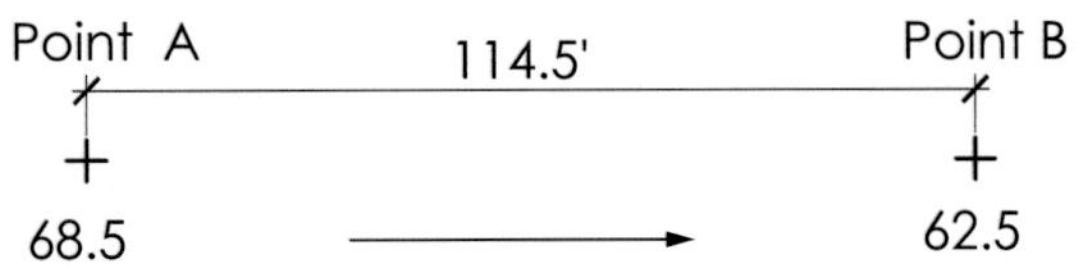

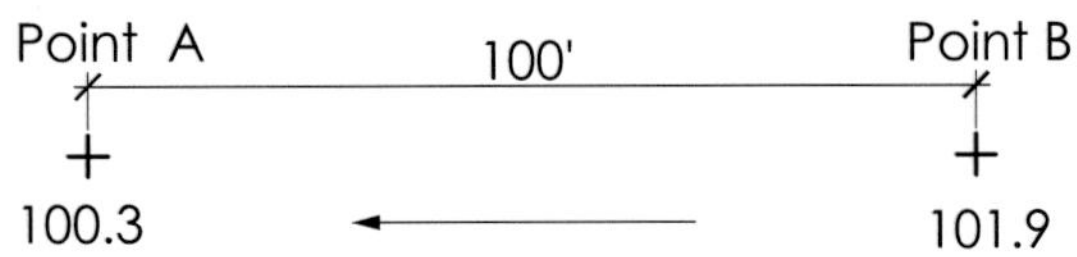

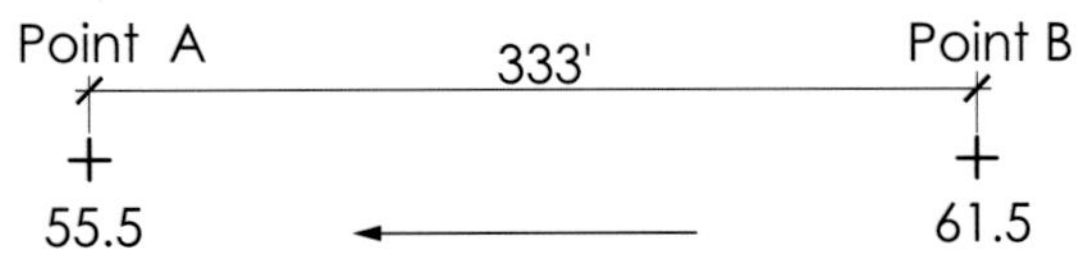

1-5) Find the HORIZONTAL distance for each of the following:

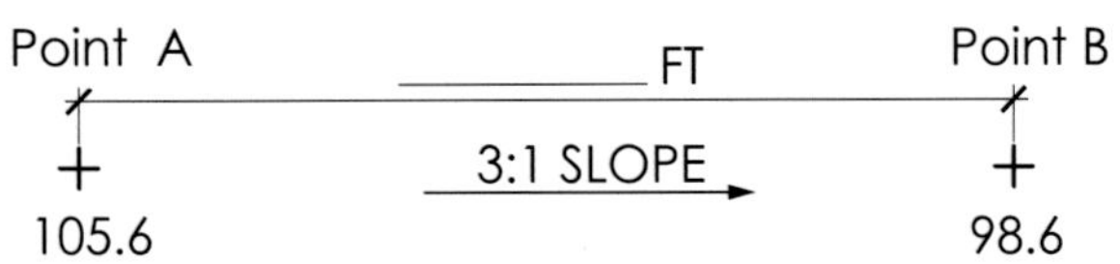

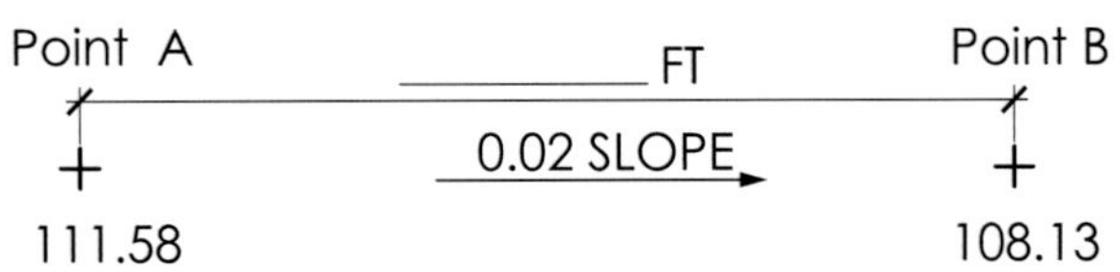

1-6) Complete the 10 x 10 grid to find the contour lines:

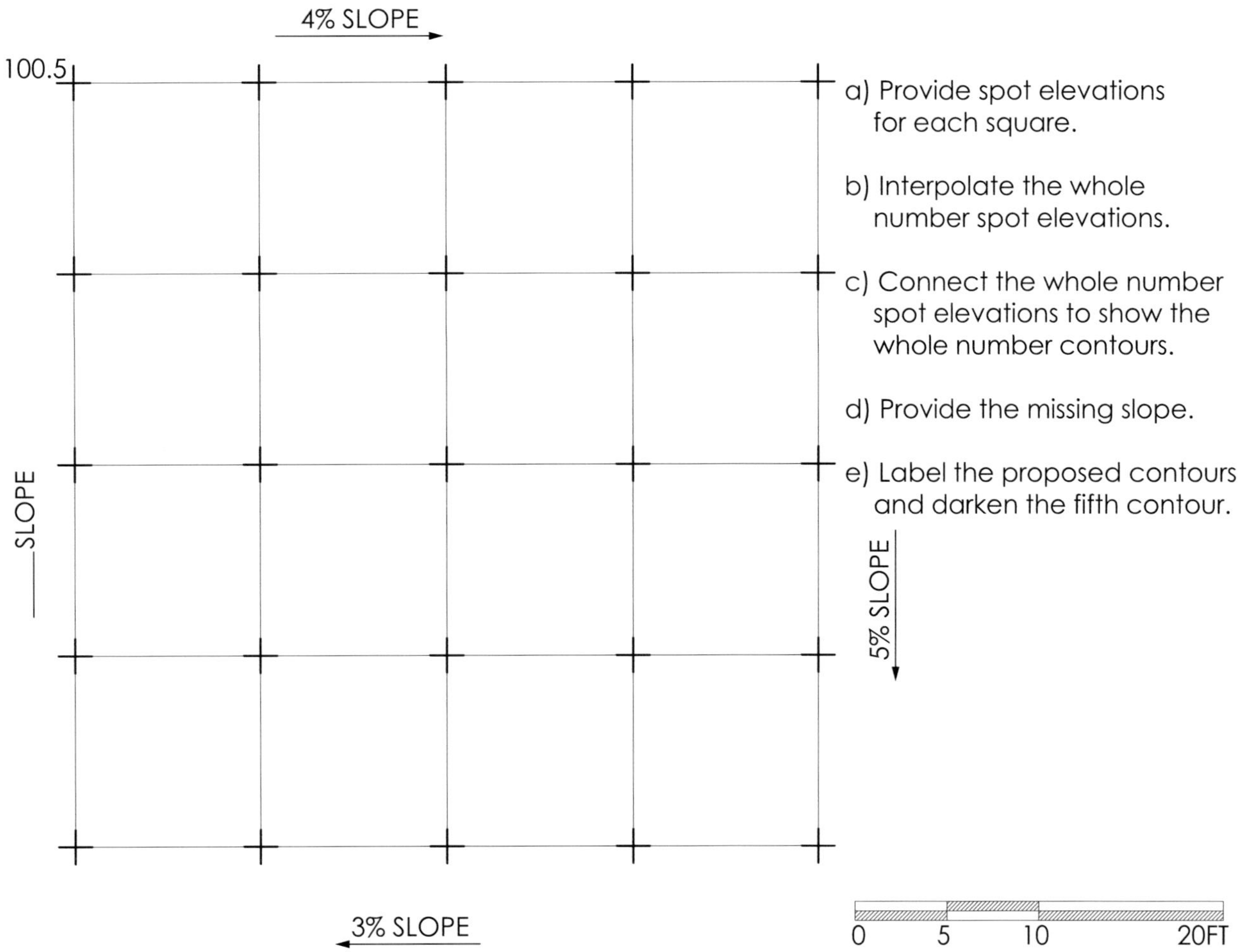

a) Provide spot elevations for each square.

b) Interpolate the whole number spot elevations.

c) Connect the whole number spot elevations to show the whole number contours.

d) Provide the missing slope.

e) Label the proposed contours and darken the fifth contour.

1.9 Literature cited

Council of Landscape Architectural Registration Boards. 2017. *L.A.R.E. Orientation Guide.* Fairfax, Virginia: CLARB. www.clarb.org/docs/default-source/take-the-exam/lareorientation guide.pdf.

Harris, Charles W. and Nicholas T. Dines. 1997. *Time-Saver Standards for Landscape Architecture. 2nd Ed.* New York: McGraw-Hill.

Hopper, Edward J. ed. 2006. *Landscape Architectural Graphic Standards.* New York: John Wiley & Sons, Inc.

Strom, Steven and Kurt Nathan. 1998. *Site Engineering for Landscape Architects. 3rd Ed.* New York: John Wiley & Sons, Inc.

Strom, Steven, Kurt Nathan and Jake Woland. 2013. *Site Engineering for Landscape Architects. 6th Ed.* New York: John Wiley & Sons, Inc.

Untermann, Richard K. 1973. *Grade Easy: An Introductory Course in The Principles and Practices of Grading and Drainage.* McLean, VA: Landscape Architecture Foundation.

Section Two
Landforms

2.1 What are landforms?

Landforms are the identifiable shapes that the earth naturally forms due to geological processes. As landscape architects, we mimic and manipulate these basic forms to create functional designs. The natural shapes we are most concerned with are plane surfaces, ridges, valleys, peaks and depressions. These are manifested differently throughout the world and are named by their vegetative and animal habitats. For example, a bog, a swamp, and a fen are different types of wetlands. The processes that form and sustain these different wetlands vary; and the types of bog, swamp, and fen vegetation varies, but the underlying shape they form is a depression, a low point in the topography. It is the underlying landform in all cases that we manipulate in landscape grading. These natural forms are most easily recognized on a larger scale (e.g. 100, 200, 500 scale). However, because these problems are at such a large scale, intricate grading is difficult. Identification of the landforms, knowing where to put the problem elements and calculating slope between contour lines becomes essential.

2.2 Watersheds

Together, plane surfaces, ridges, valleys, peaks, and depressions form watersheds. A watershed is a natural drainage basin for a particular area. It carries surface runoff, water from rainfall events, from the highest regions of the watershed downhill until it collects in streams, rivers, ponds, and eventually the ocean.

Watersheds can be divided into subdrainage basins where water flows to a particular stream or pond within a watershed or they can be linked together to form a larger watershed region. These larger watershed regions are not determined by state government boundaries, but often encompass several states or more than one country. Additional information about watershed processes can be found in the recommended reading. As illustrated in Figure 2.2–1, watersheds are defined by ridges and peaks at the high points and valleys and depressions at the low points.

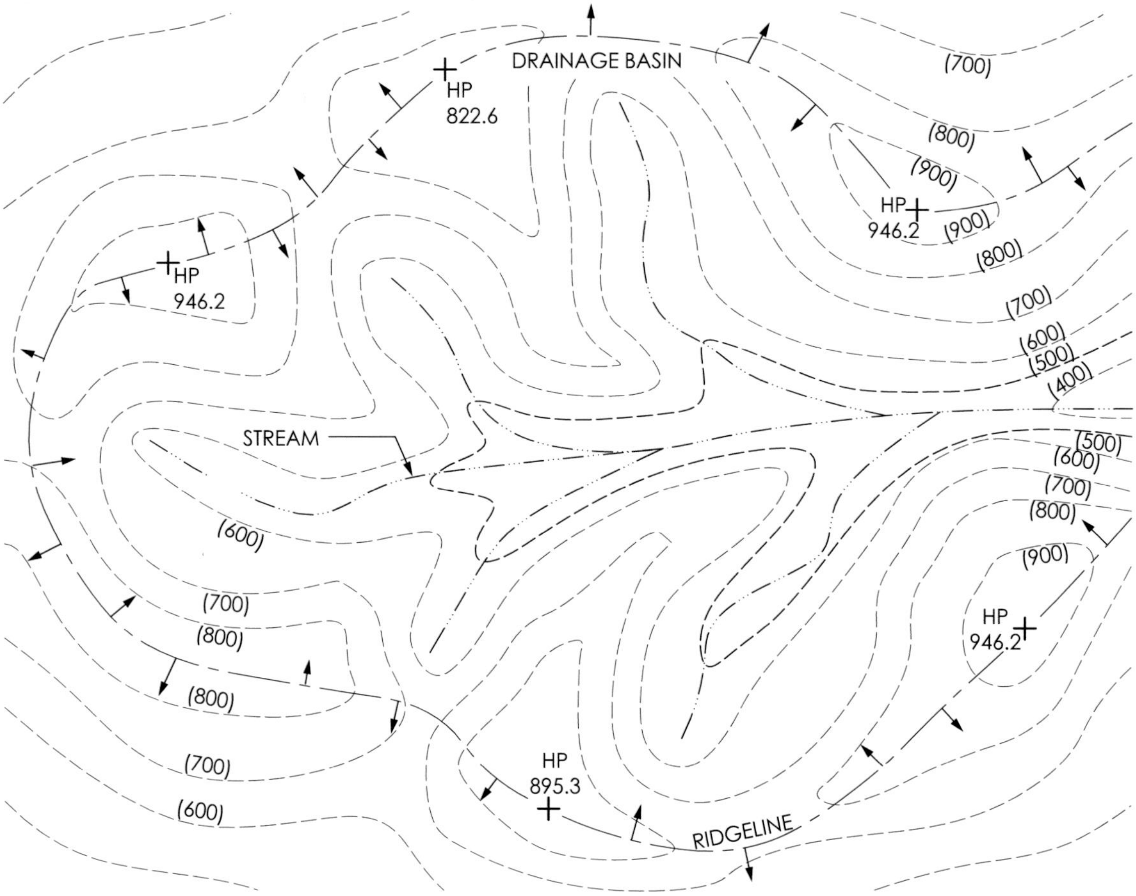

2.2–1
Watershed region

2.3 Ridges, peaks, and high points

Ridges are characterized by a shallow longitudinal slope that follows the ascending or descending topography and steeper cross slopes that drain water off of the sides of the ridge; the steeper the ridge, the faster the surface runoff. A ridgeline, as shown in Figure 2.3–1, is a continuous line through the center of the ridge with arrows pointing away from it showing the flow of water. It is a dividing line between watersheds, the point where surface runoff will flow downhill on either side of the line.

Topographically, ridges point downhill; they point to lower elevations. The steepness of a ridge is determined by how close together the contour lines are. To minimize or slow surface runoff, adjust the slope of a ridge by moving the contour lines farther apart. A ridge becomes a cliff, a shear vertical face, when the contour lines overlap or are very close together.

Ridges meet at peaks or high points. Peaks, as illustrated in Figure 2.3–2, occur when three or more ridges converge. A typical example of a peak is a mountain top. High points occur when two ridges meet. These high points are labeled with "HP" plus a spot elevation. There may be several high points in a watershed.

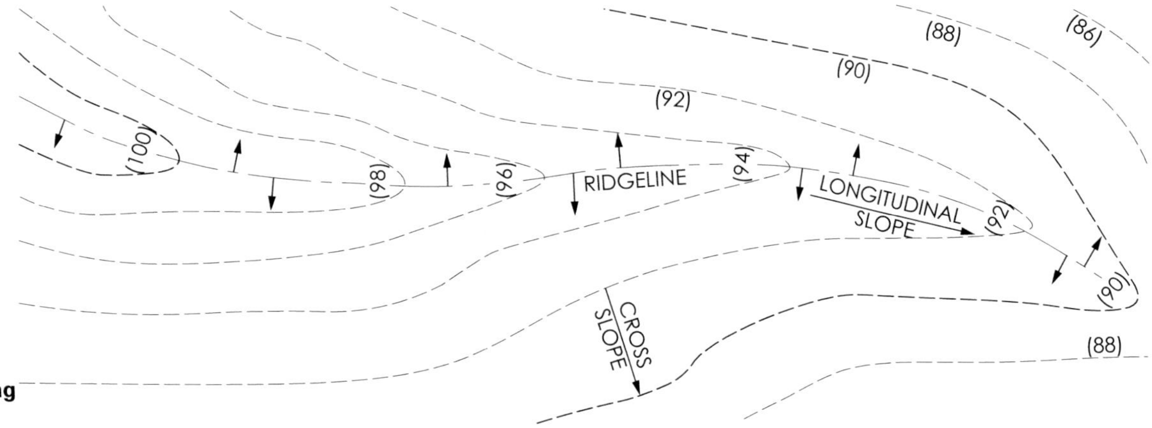

2.3–1
Ridge showing ridgeline

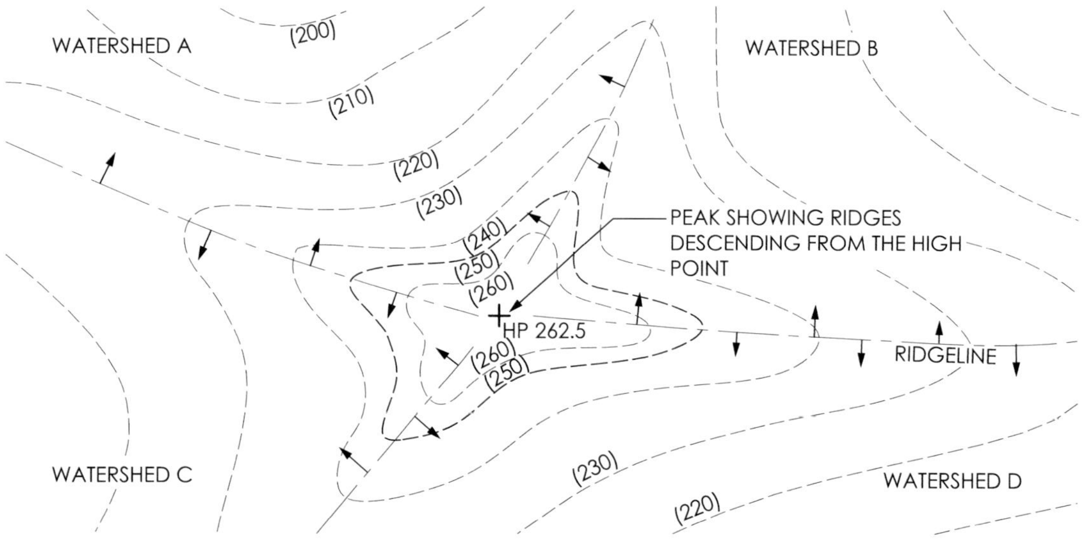

2.3–2
Ridgelines meeting at a peak

A point along a watershed boundary where two ridges meet may be the highest spot elevation, the starting point, for a valley as shown in Figure 2.3–3. Although it is a not necessarily the highest point on the ridgeline, it is still higher than much of the surrounding topography.

2.4 Valleys, ponds, and low points

Valleys are the collection areas for surface runoff that is flowing off of a ridge. Valleys, like ridges, have a shallow longitudinal slope and a steeper cross slope. As shown in Figure 2.4–1, the longitudinal slope of the valley follows the descending topography and the side or cross slope drains the surrounding area into the valley. The longitudinal slope starts at a high point, the point where two or more ridges meet. As with ridges, the steeper the longitudinal and cross slope the faster the water will drain through a

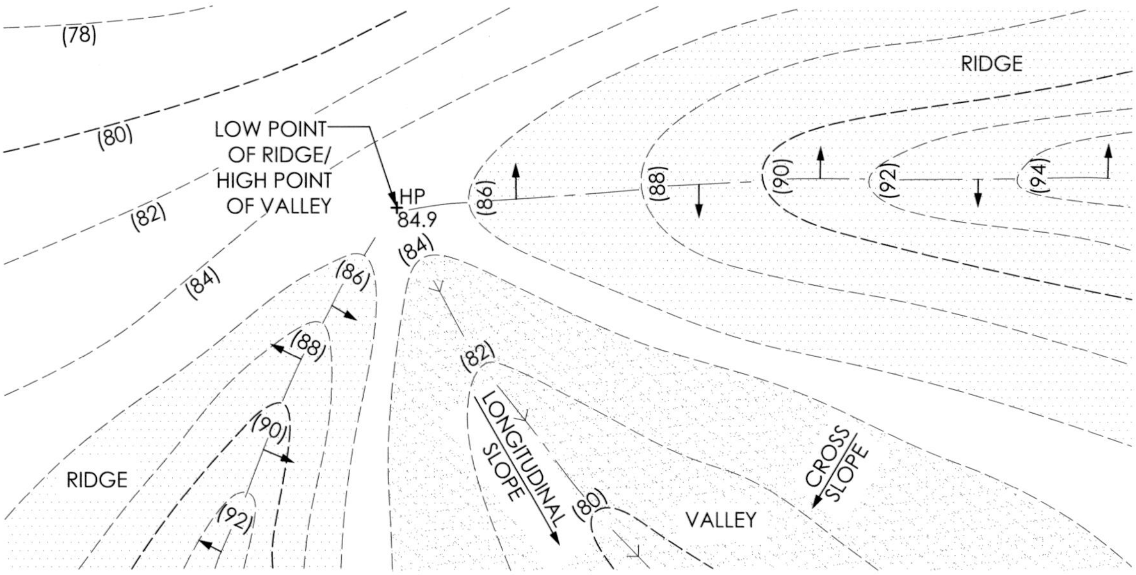

2.3–3
Ridgelines meeting at a high point and the subsequent valley

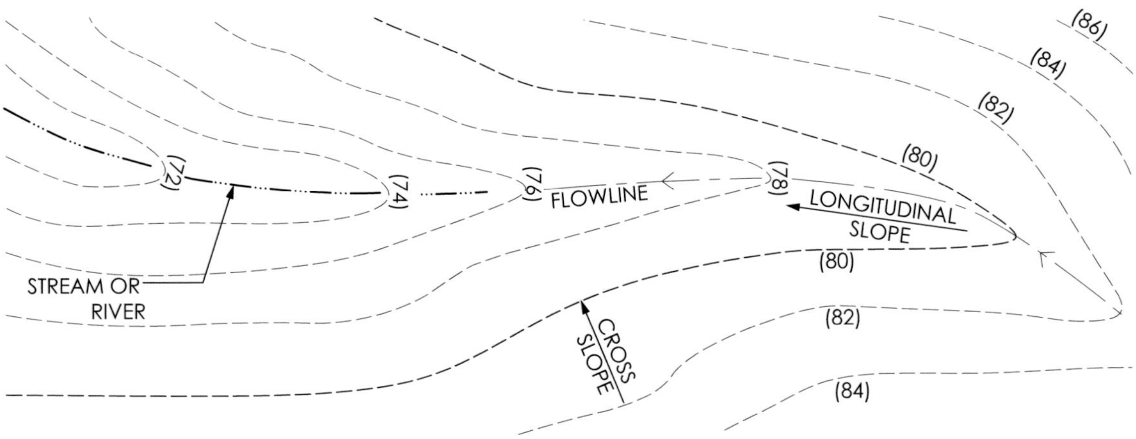

2.4–1
Valley with stream

valley. To minimize or slow the flow of surface runoff; adjust the slopes of the valley by moving the contour lines apart.

Topographically, valleys point uphill; they point to higher elevations. A valley is usually characterized by a stream or river carrying water towards a catchment basin or pond and will be surrounded by a flood plain of varying size and slope. A flowline is a centerline with an open arrow symbol showing the direction of water flow. A stream or river is shown as, or outlined with, a line with three dots and a dash. This is a helpful way of identifying some valleys on a watershed problem. An associated flood plain should be outlined based on the requirements of the individual exam problem.

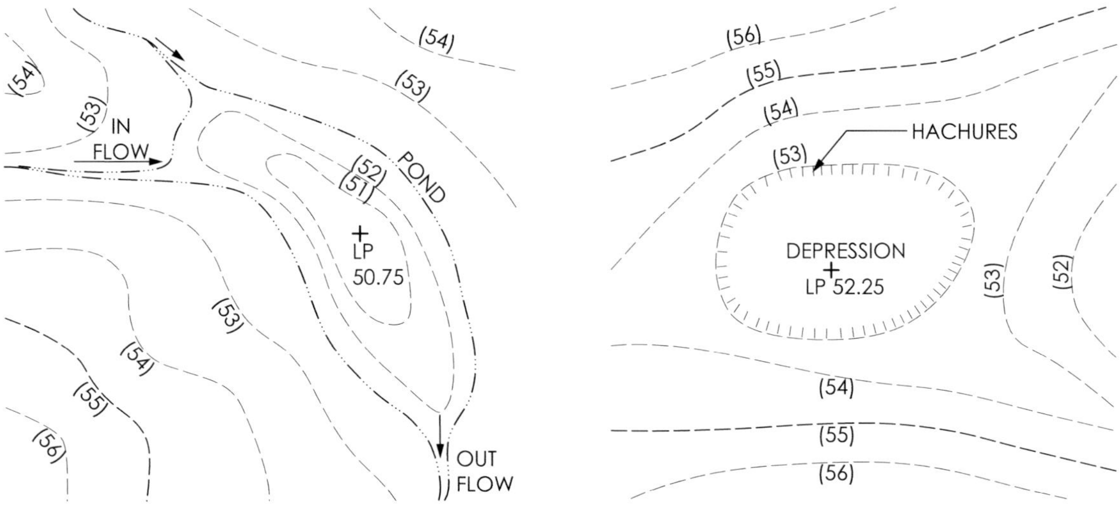

2.4–2
Two types of depression: the pond and the low point

Depressions may be continuously filled with water, e.g. a pond, or temporarily filled with water. Ponds are depressions and like rivers or streams are outlined by the identifying water line and are surrounded by a flood plain. Depressions that are not consistently filled with water are identified with hachure marks on the inside of the depression, as illustrated in Figure 2.4–2. Both types of depression are usually labeled with "LP" plus a spot elevation identifying the depth of the pond or depression.

2.5 Plane surfaces

Plane surfaces are not restricted to ridges or valleys and can occur throughout the watershed wherever there is a shallow slope. Plane surfaces are the areas within a watershed where siting structures is the easiest. Instead of leveling steep slopes, plane surfaces reduce the need for extensive grading and earth moving. The location of the plane surface in a watershed will determine the type of structure that should be built on it.

As illustrated in Figure 2.5–1, if the plane surface is located in the flood plain, athletic fields are well-suited since flooding will do minimal damage. If the plane surface is located at the top of a ridgeline, a road or a building may be a good choice. An examination problem may ask you to site one or more built elements on a shallow slope. Use the slope calculation to determine the steepness of the contours; it may not be readily apparent where the shallowest slopes exist. Make sure to highlight any required setbacks from any ponds, streams, or wetlands as required by the examination problem prior to starting the solution. Often times there appear to be more than one possible solution but adding setbacks will help narrow down the choices.

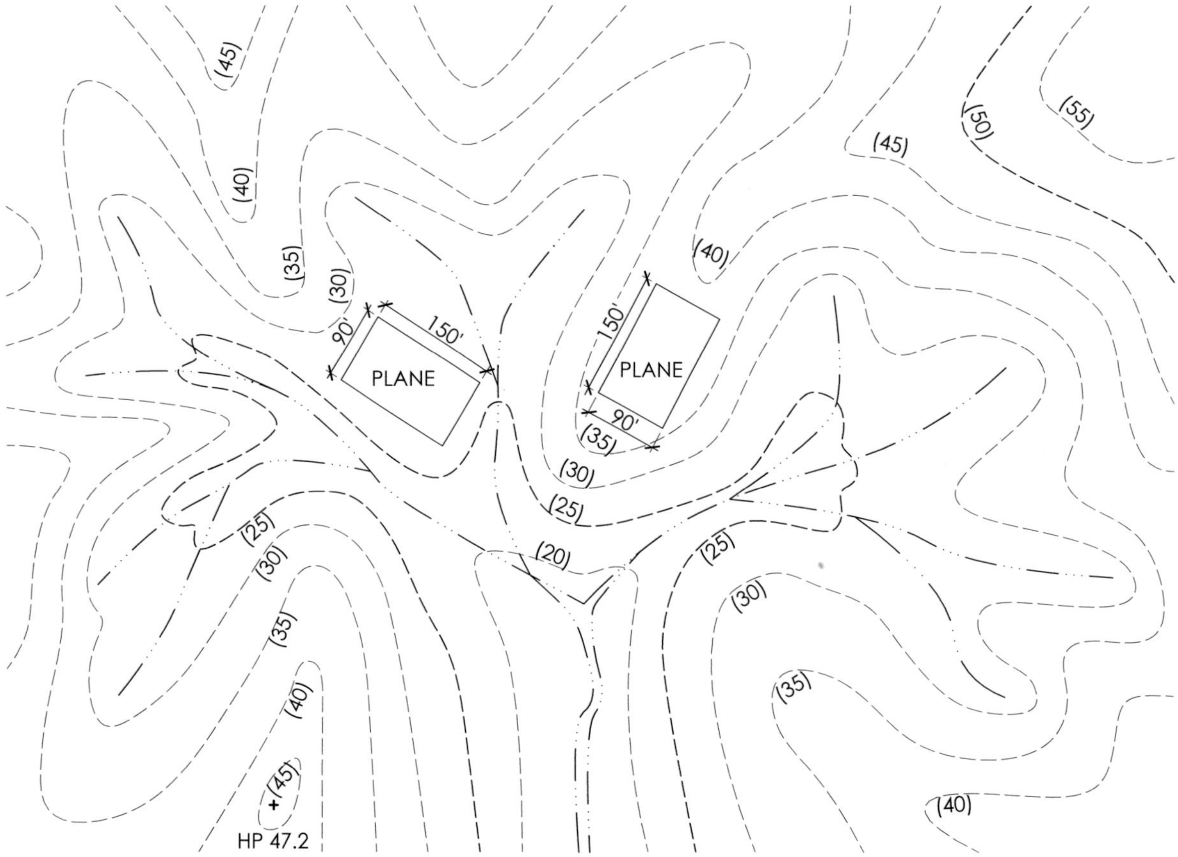

2.5–1
Plane surfaces on the top of a ridge and in a flood plain

2.6 Summary

This section looked at landforms, identifying ridges, valleys, peaks, depressions, plane surfaces, high points and low points, and understanding how they form a larger watershed or drainage basin. These landforms are what we manipulate when we perform landscape grading at a large or regional scale in order to facilitate drainage of a site or watershed. These landforms are what we will mimic when we create site-specific surface drainage.

Ridges and valleys have two slopes, a longitudinal slope that denotes the direction that the ridge or valley is ascending or descending in the landscape, and a cross slope that denotes the flow of water off of a ridge or into a valley. Contour lines show the steepness or shallowness of a ridge or valley. Contours that are closer together are steeper and contours that are further apart are shallower.

2.7 Exercises (answers start on page 182)

2-1) Number and label the existing contour lines at 1' contour intervals to create one peak, two high points (HP) and three valleys.

a) Show center flowlines for the valleys.

b) Darken every fifth contour line; label existing contours on the uphill side.

2-2) Number and label the existing contour lines at 1' contour intervals to create a depression, two ridges and a valley.

a) Show the center flowlines for the valley and ridgelines for the ridges.

b) Darken every fifth contour line; label existing contours on the uphill side.

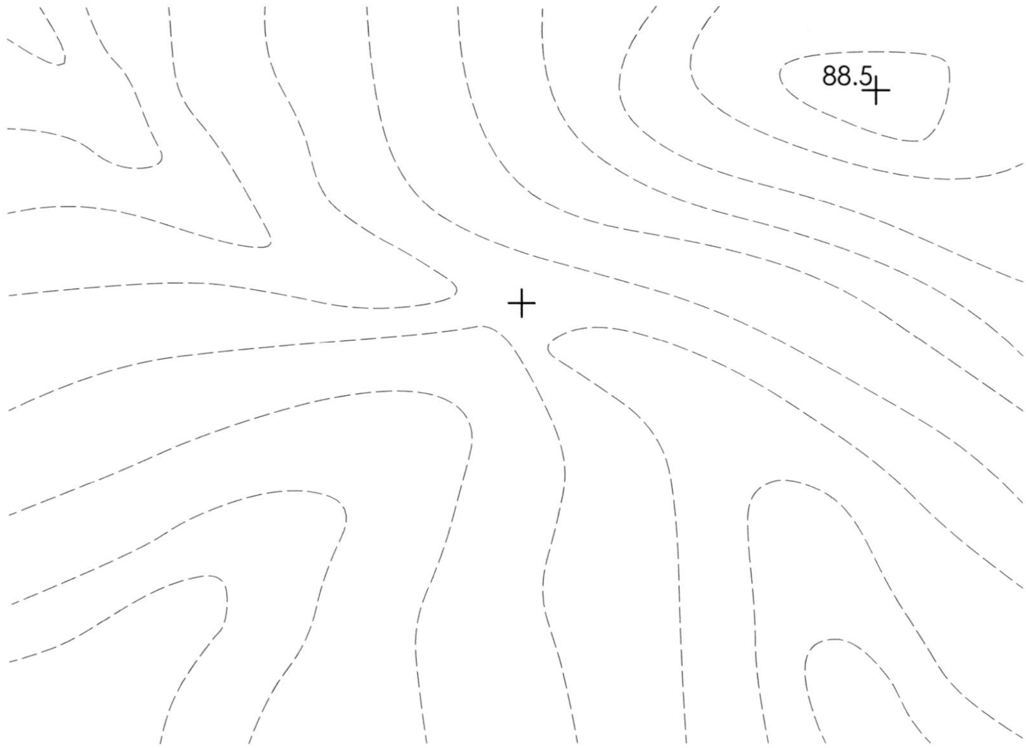

2-3) Draw and label the ridgelines for the three watersheds. Write HP and LP along the ridgelines. Locate four 1000 x 1500 SF planes on ridges with slopes between 2–3%. Draw flowlines for the valleys in the watersheds.

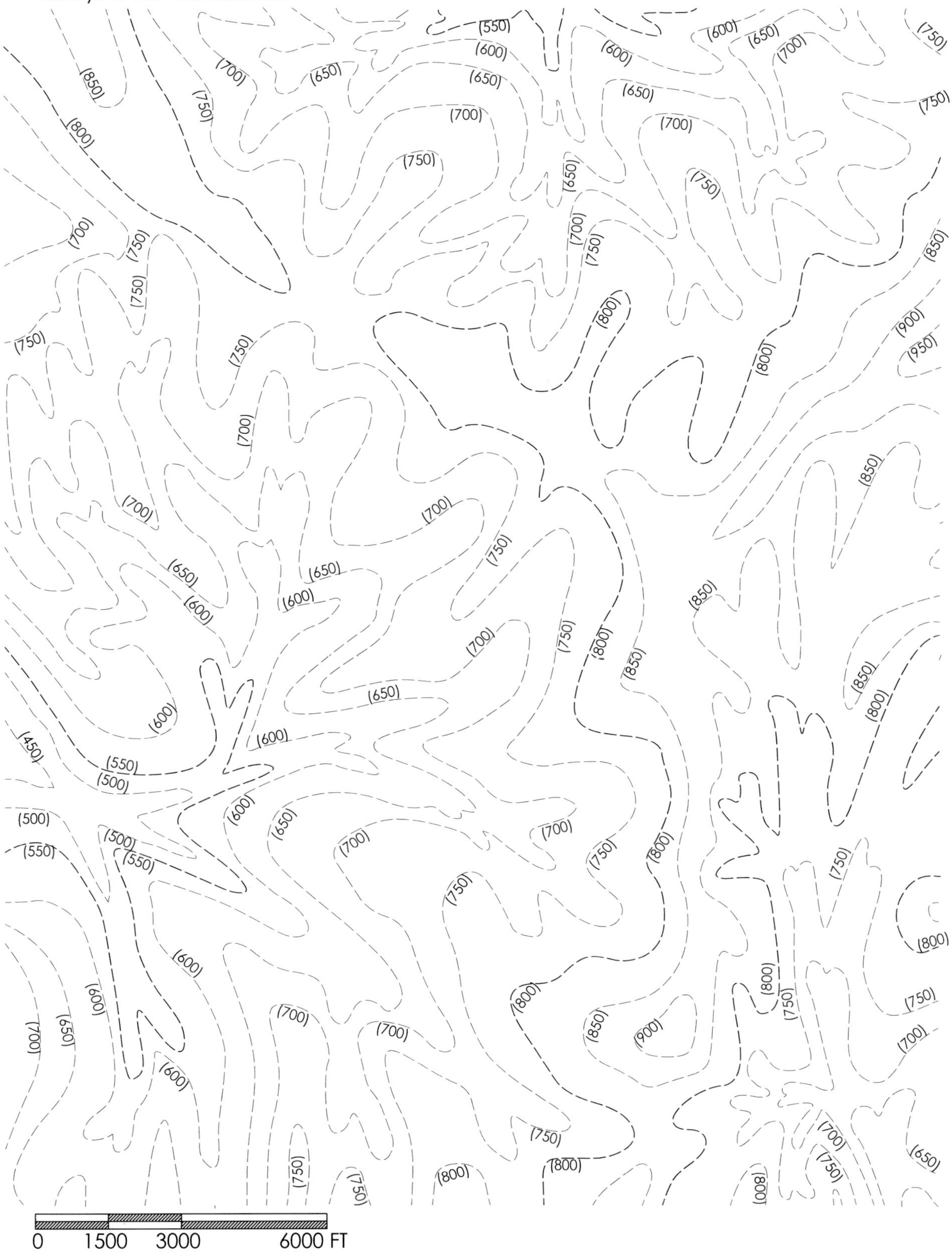

2.8 Literature cited

Harris, Charles W. and Nicholas T. Dines. 1997. *Time-Saver Standards for Landscape Architecture. 2nd Ed.* New York: McGraw-Hill.

Hopper, Edward J. ed. 2006. *Landscape Architectural Graphic Standards.* New York: John Wiley & Sons, Inc.

Strom, Steven and Kurt Nathan. 1998. *Site Engineering for Landscape Architects. 3rd Ed.* New York: John Wiley & Sons, Inc.

Strom, Steven, Kurt Nathan and Jake Woland. 2013. *Site Engineering for Landscape Architects. 6th Ed.* New York: John Wiley & Sons, Inc.

Untermann, Richard K. 1973. *Grade Easy: An Introductory Course in The Principles and Practices of Grading and Drainage.* McLean, VA: Landscape Architecture Foundation.

Section Three
Grading of landscape elements

On a smaller site scale, we mimic landforms and create engineered versions of them in our designs to aid in the drainage of water from the site. Again, landscape grading is about showing how the two-dimensional design fits on a three-dimensional site and understanding how surface water flows through the site. At a smaller scale, we create plane or warped plane surfaces, ridges, swales, berms, and retention or detention ponds. They function similarly to the natural landforms but are man-made landscape elements. To understand how to grade these landscape elements we need to understand first where the water will drain.

3.1 Drainage systems

There are two types of drainage for all sites, surface drainage from rainfall events called runoff, and subsurface drainage which involves percolation and the movement of water through the underlying water table. The speed of surface runoff is directly tied to surface permeability and steepness of slope. With a porous soil, percolation increases, and surface flow is minimized. With an impermeable surface, surface flow increases, and percolation is minimal. For example, a grassy surface has a slower surface runoff than a road simply because some of the runoff will be absorbed into the soil. In landscape grading, we focus on the manipulation of surface runoff through civil engineering and landscape architectural structures that make up two systems, the open and closed drainage systems. Both systems catch surface runoff and move it into lakes, rivers, streams and eventually the ocean.

The open system is characterized by swales, which are grass-covered indentations in the earth, and other structures such as ditches and dry stone beds that enable water to flow across the surface from a higher topography to a lower one tying into streams, lakes, etc. An open system is most often seen in rural areas with minimal infrastructure. Open systems allow for surface water percolation into the underlying water table. The open system is often called green infrastructure because it uses natural elements.

The closed system is characterized by stormwater pipes, intermediary drain inlets, and catch basins creating a stormwater system. A closed system is used in urban environments to minimize surface water ponding and flooding and to quickly remove runoff after precipitation events. This system is called gray infrastructure. Most suburban areas are combinations of open and closed systems. A thorough discussion of these topics and determining the rate and volume of surface drainage can be found in the recommended reading. The important point for the test is recognizing the connection to open and closed drainage systems.

3.2 Plane surfaces and warped planes

A plane surface is the two-dimensional representation of a 'flat' surface. It is seemingly flat but has a uniform slope in at least one direction. Plane surfaces are the underlying landscape element of most built structures, constructed fields, plazas and sidewalks. Topographically, they are characterized by uniform and shallow slopes of 1–5% in one or two directions.

Plane surfaces can be located at higher or lower elevations of a site and may be shown in both areas on a grading problem. Surface water is either allowed to drain across the plane surface as with constructed fields, parking lots, plazas and sidewalks or it must be diverted around the plane surface as with built elements as illustrated in Figure 3.2–1. Surface water that drains across a plane is called sheet drainage. Surface water that drains around a plane is usually diverted into a swale. Sheet drainage and swales can be combined to facilitate surface runoff.

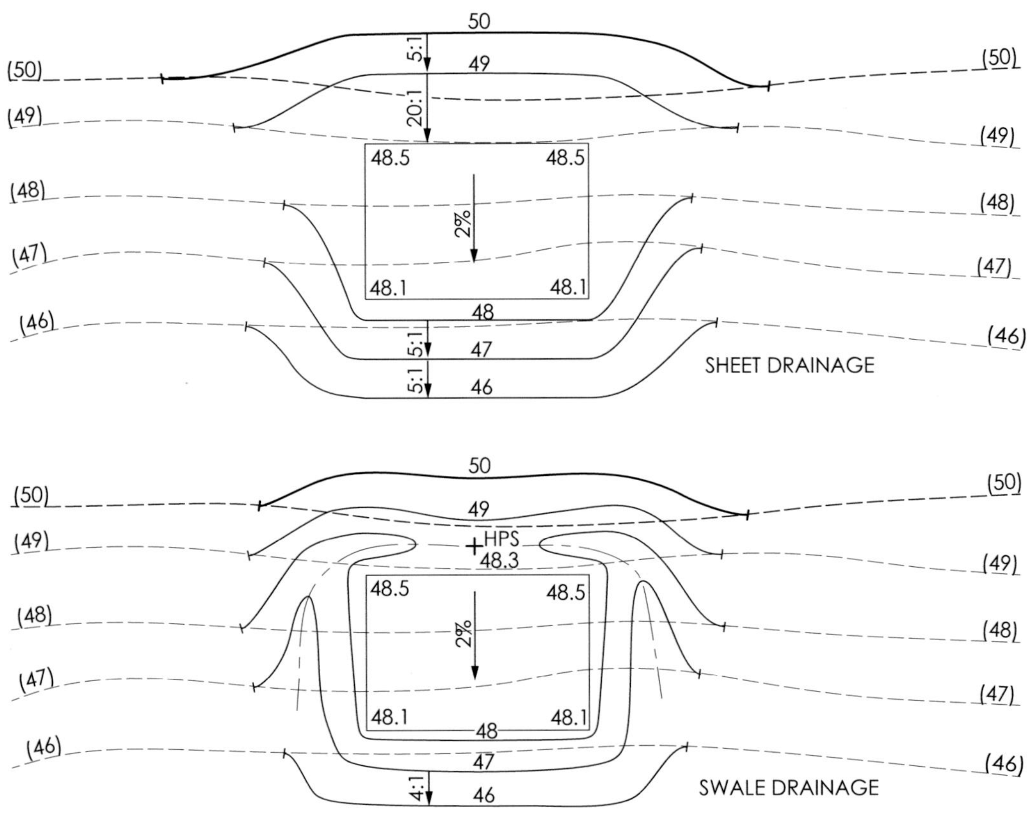

3.2–1
Drainage of a plane surface with sheet drainage and swale drainage

Every plane surface problem has at least two proposed slopes associated with it. The proposed slope(s) for the plane surface and the proposed slope for the land just outside of the plane surface. Remember that when you are grading you are manipulating existing contours. Any existing contours within the plane surface will have to be moved uphill or downhill because you are superimposing the plane surface slopes on those contours. Where plane surface slopes are shallow, between 1 and 5%, slopes outside of and adjacent to the plane surface may be steeper. Most grading problems will provide a maximum slope for proposed contours outside of the plane surface. These are used to tie proposed contour lines back to the existing contour lines outside of the plane surface.

Determining the use for the plane surface will indicate how it should be graded. Athletic fields are usually sited on topography with shallow slopes to minimize grading; sheet drainage is often used. If athletic fields are sited on steep topography a combination of sheet and swale drainage should be used. Plane surfaces that are used as the base for a built structure are flat and swale drainage will be the primary means of drainage around the structure.

To grade a uniformly-sloping plane surface with sheet drainage first determine the parameters that you are given. In some cases, you may be provided with spot elevations and the slope or slopes for a plane, or you may be provided with the slopes and asked to determine the spot elevation from a given point or points.

1 Determine the direction in which the existing topography is sloping. As shown in Figure 3.2–2, note the spot elevation(s) provided for the plane surface and the minimum and maximum slope for the problem. Measure the horizontal distances for the plane using a scale ruler.

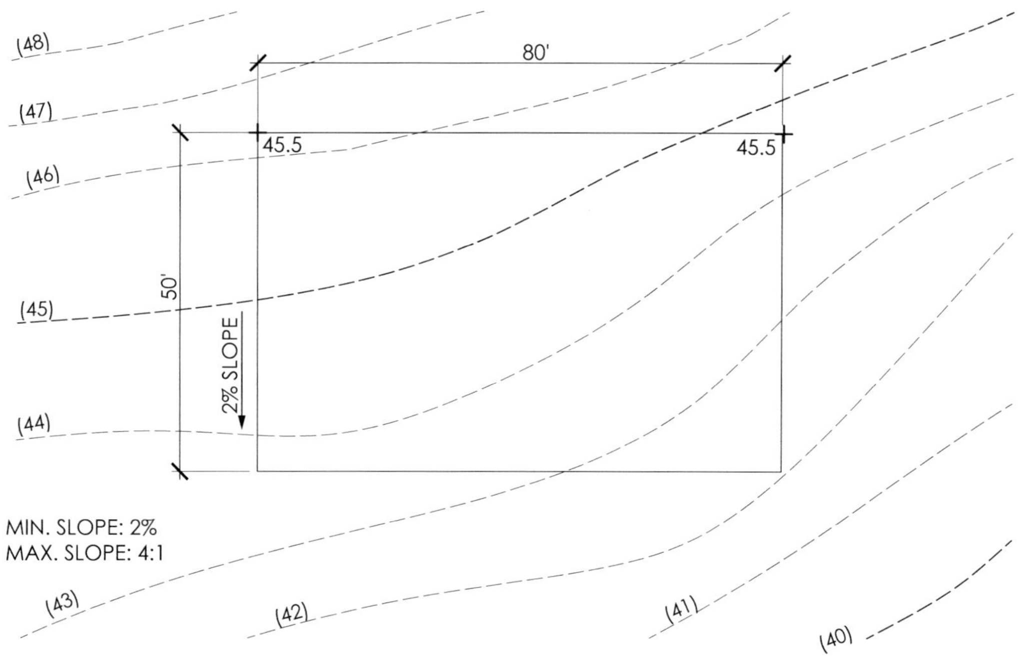

3.2–2
Uniformly graded plane surface with 2% slope

2 Interpolate to find the spot elevations for the missing corners of the plane surface based on the minimum slope or slope provided for the plane surface.
3 Interpolate between the spot elevations at the edge of the plane to determine where the whole number spot elevations are.
4 Connect the whole number spot elevations to form proposed contour lines within the plane surface as indicated in Figure 3.2–3. Only these proposed contours exist within the plane surface; all the existing contours must be adjusted uphill and downhill outside of the plane surface using the maximum slope for the problem.
5 In this example, the existing (46), (44), and (43) contour lines within the plane surface still need to be adjusted. Draw light horizontal and vertical reference lines at the corners of the plane surface. Starting at the top left corner, interpolate to find the proposed 46 contour line above the 45.5' spot elevation using the maximum slope, 46.0'–45.5'=0.5', 0.5'/0.25=2.0'. Place tick marks on the reference lines for both the vertical and horizontal lines at the top left corner. In this instance, the proposed 46 contour line is now uphill of the left 45.5' corner spot elevation.
6 Repeat this procedure for each corner. Remember the spot elevations for all the other corners will be lower than the spot elevations of the plane surface. Use the maximum slope to find the horizontal distance for the proposed 45 contour line adjacent to the spot elevation at the top right corner, 45.5'–45.0'=0.5', 0.5'/0.25=2.0'.
7 Use the slope interval for the maximum slope to evenly space the proposed contours outside of the plane surface once you've found the first whole number spot elevation. Note that although only the (46), (44), and (43) contours have been adjusted, because you have moved them outside of the plane surface, this shift has affected the location of the existing (42) and (41) contours located outside of the

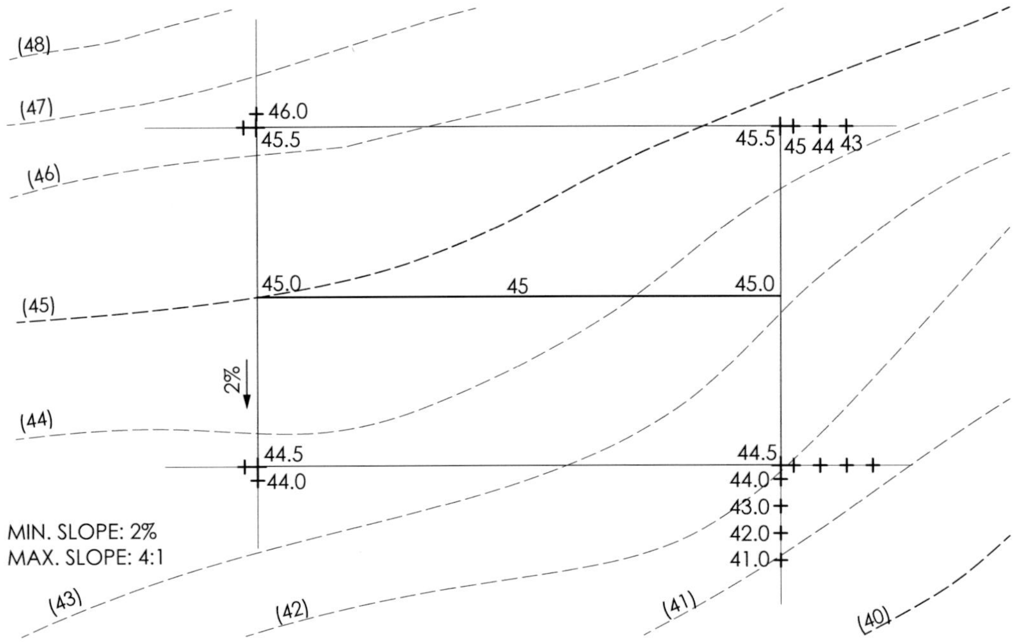

3.2–3
Plane surface with reference lines and hatch marks using the maximum slope

plane surface. These will also need to be adjusted to maintain the maximum slope as shown in the example. Remember to make smooth continuous transitions back to the existing contours and label the proposed contour lines on the uphill side as shown in Figure 3.2–4.

Figure 3.2–5 illustrates a graded plane surface with two slopes. Two slopes will result in a similar outcome to a single sloping plane. The advantage is the plane surface may fit better into the existing landscape with less manipulation of existing contours, minimizing soil cut and fill. Tie back any proposed contours to existing contours with the maximum slope provided on the grading project.

A warped plane has more than two slopes. The additional slope(s) necessitates the use of one or more centerlines. The centerline acts as either a flowline or a ridgeline. As a flowline, it is lower than the surrounding plane surface and water will flow towards it. As a ridgeline it is higher than the surrounding plane surface and water will flow away from it. Either situation is still a representation of sheet drainage over a plane surface. The surface is not uniformly sloping in one or two directions. Figure 3.2–6 adds a third slope to the former plane surface example to create a warped plane.

1 Given a starting spot elevation, Point A, determine the spot elevations for each corner of the plane surface based on the dimensions of the plane surface and the slopes provided.
2 Determine any missing slopes and interpolate along the edge of the plane to find the whole number spot elevations.
3 Draw a centerline from the highest spot elevation to the lowest spot elevation within the plane surface.

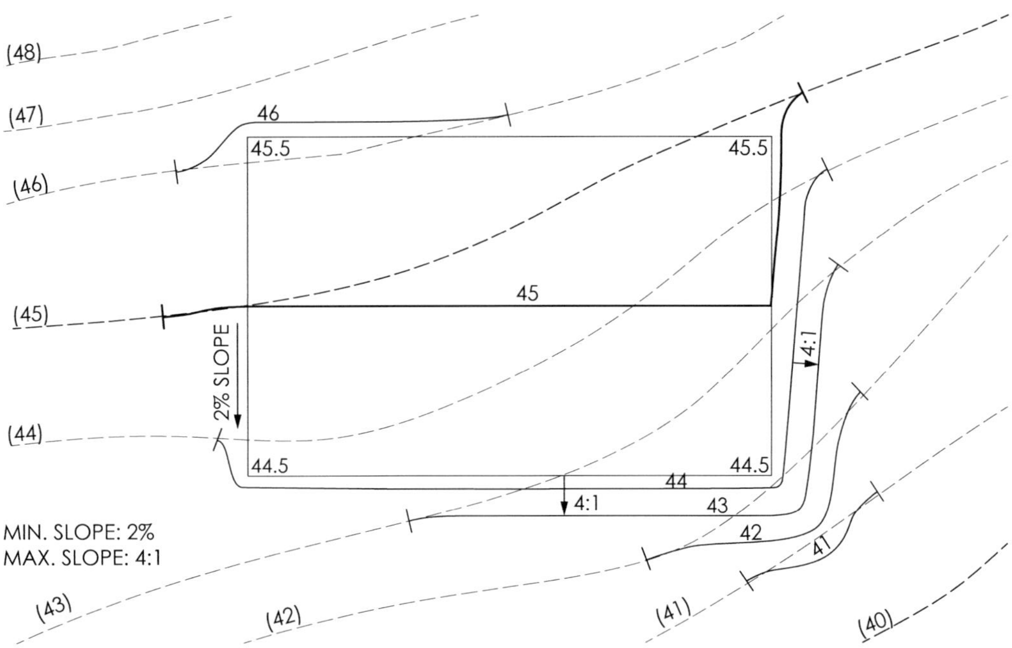

3.2–4
Uniformly sloped plane

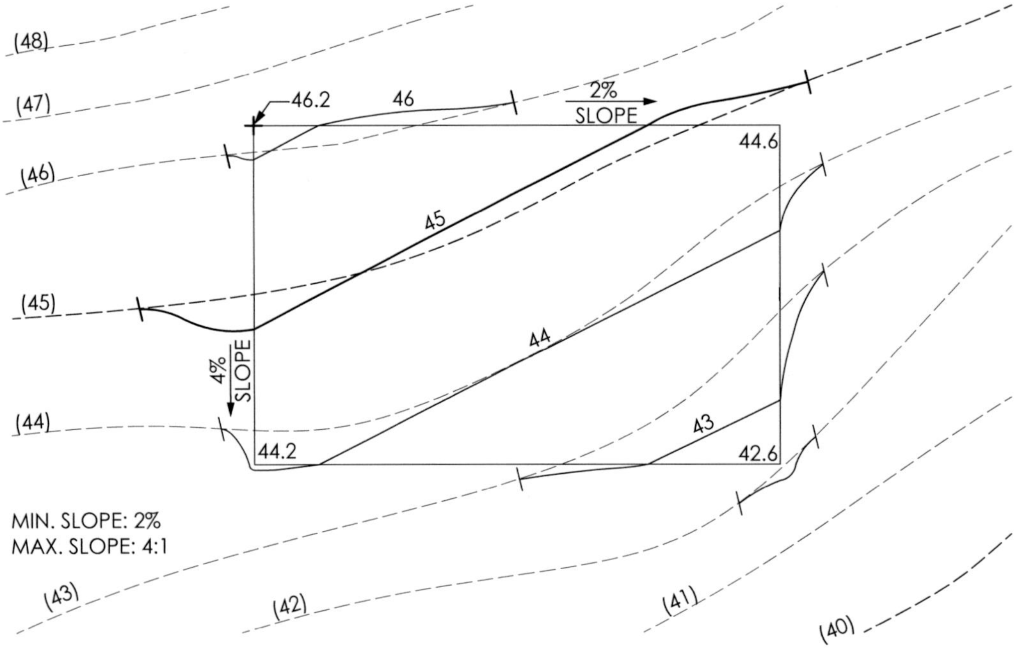

3.2–5
Plane surface with two slopes

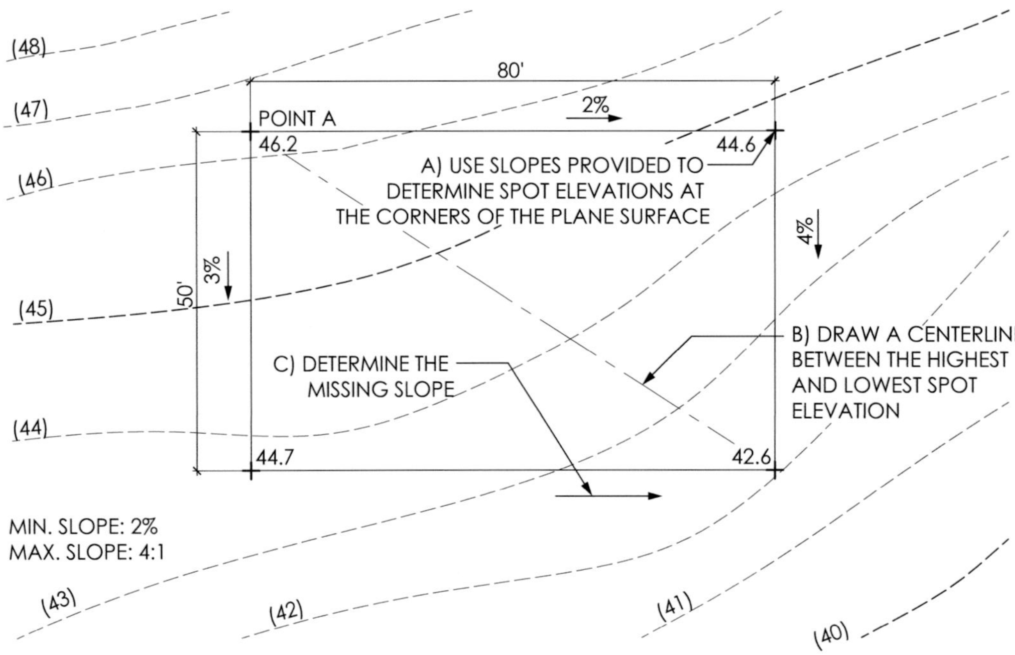

3.2–6
Warped plane surface with three slopes provided

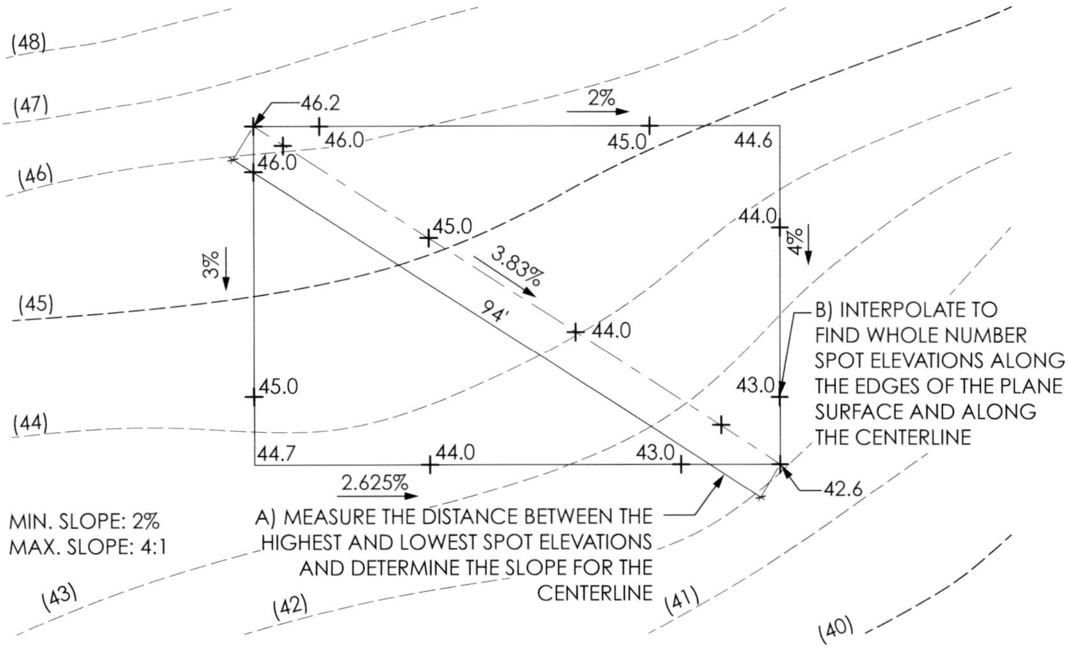

3.2–7
Warped plane showing missing slopes and whole number spot elevations

4 As illustrated in Figure 3.2–7, find the slope for the centerline and interpolate to find the whole number spot elevations along the centerline and around the edges of the plane surface.

5 As in the previous example, connect the whole number spot elevations within the plane surface to create the proposed contour lines. Make sure to connect them to the whole number spot elevations on the centerline, see Figure 3.2–8.

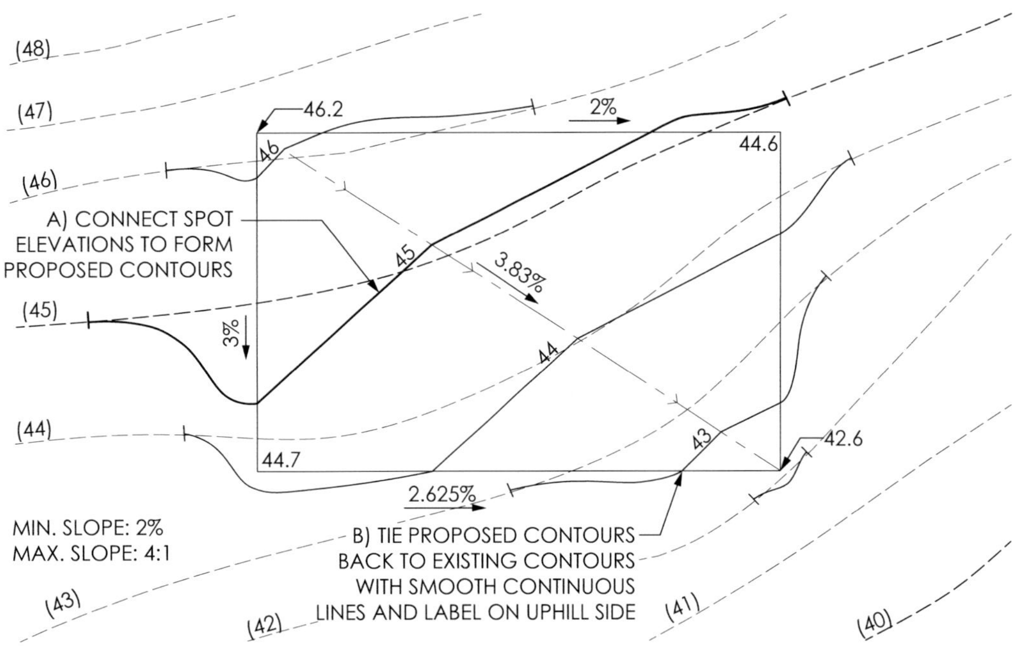

3.2–8
Warped plane with proposed contours and flowline

6 Using the maximum slope for the problem, connect the proposed contours back to the existing contours outside of the plane surface. Label the proposed contours on the uphill side.
7 If necessary, determine the slopes for the warped plane on both sides of the centerline.

Plane surfaces are rarely tested individually on the grading exam and are usually part of a larger more complex problem. They are usually the first landscape element that must be solved on a grading problem whether they are at a higher or lower elevation regardless of whether they are a parking lot or the pad for a built structure. Determine what kind of plane surface the problem is asking you to solve and then apply sheet or swale drainage. Make sure to show any slope for the plane surface and correctly show the direction of surface runoff. It is an easy mistake to make to have water flowing uphill.

3.3 Swales

Swales are a key part of the open drainage system. They are man-made valleys that divert water around built structures. Swales have different topographic signatures as shown in Figure 3.3–1, but they are all designed for water removal and they all point

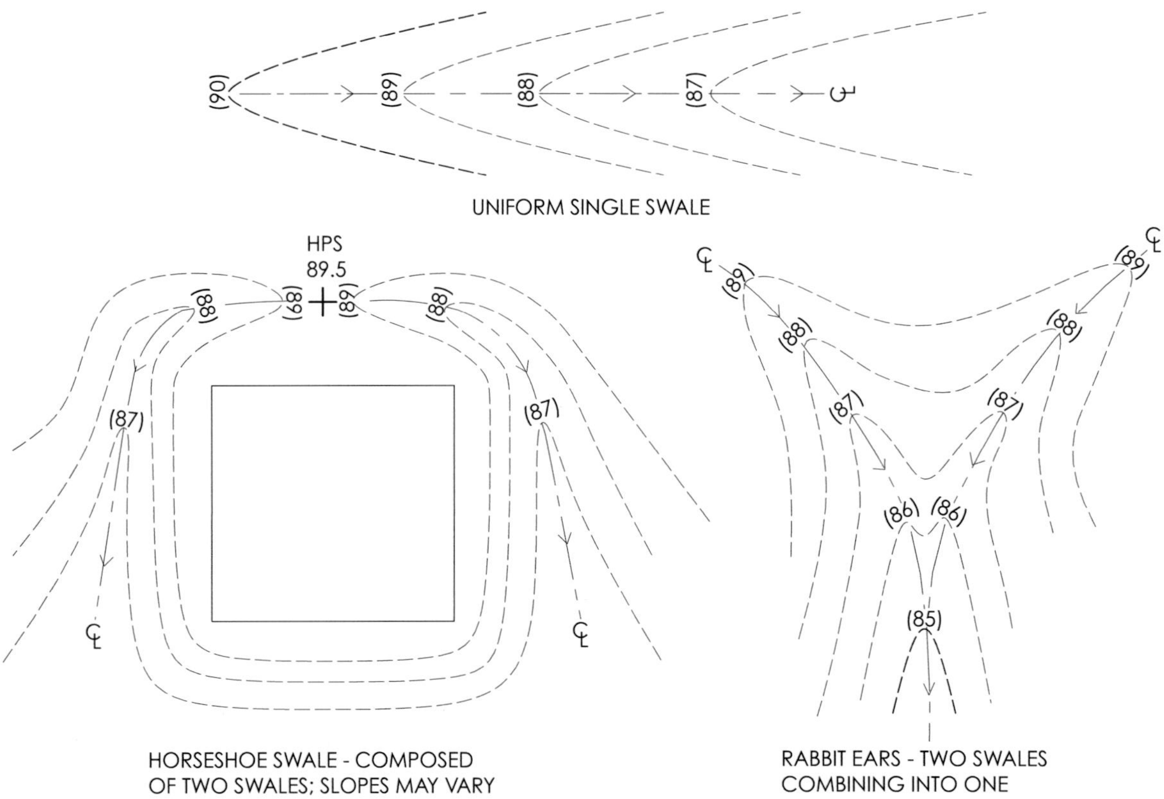

3.3–1
Three types of swale configurations

uphill. On landscape surfaces and asphalt, swales are drawn as parabolas and on hardscape surfaces, such as concrete, they are drawn as chevrons.

Every swale has three parts: a starting spot elevation called the high point of swale or HPS; and two slopes, a longitudinal slope and a cross slope, see Figure 3.3–2. The cross slope, usually the maximum allowable slope, is applied to the sides of the swale. The longitudinal slope, usually the minimum allowable slope for the site, is applied to the length and direction that the swale travels in the landscape. In the exam, you are usually given the minimum and maximum allowable slope for the grading problem, these are the parameters for the longitudinal and cross slope for the swale, respectively.

The cross slope will drain water from the sides of the swale into the center and it will then be carried downstream by the longitudinal slope. Unless specifically stated in the grading problem, the longitudinal slope does not have to be uniformly-graded, it can range slightly to accommodate your solution; try to stay within a range of 2–4% for the longitudinal slope. Remember, the cross slope drains the swale, and the longitudinal slope drains the site.

Once a building has been proposed and the building pad sited, the existing topography that falls within the building no longer exists. The ground must be leveled, and this process removes the existing topography. Earthwork will adjust the soil around the building pad. A swale is needed to divert water around any immovable object such as a building, concrete pad, berm, existing planting, etc. First, a high point of swale (HPS) must be located.

In order to place a HPS correctly, it is necessary to understand what the existing topography is doing in relation to any proposed or existing structures. The HPS should

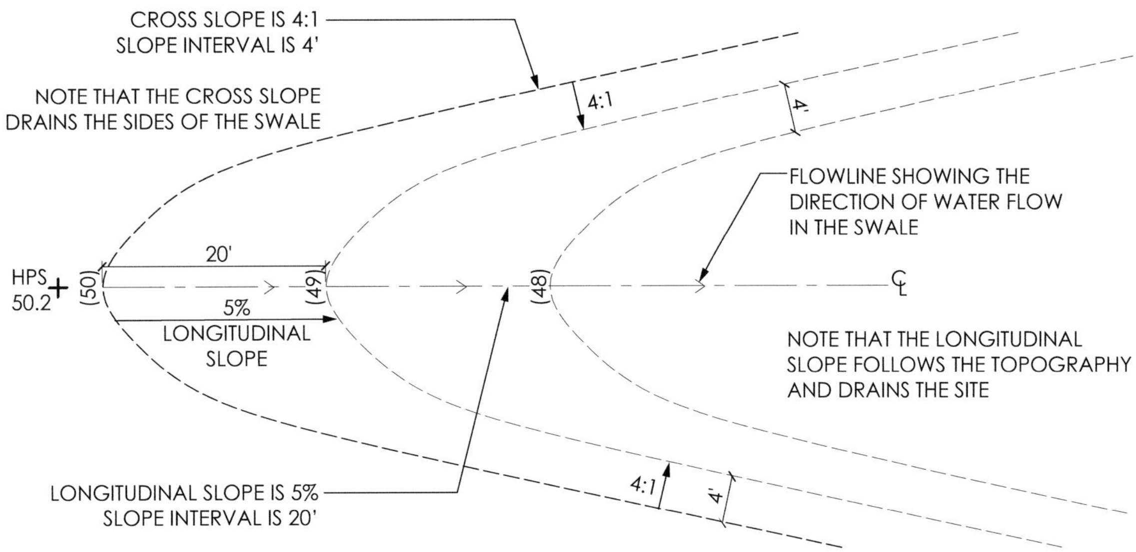

3.3–2
Typical swale showing longitudinal and cross slope

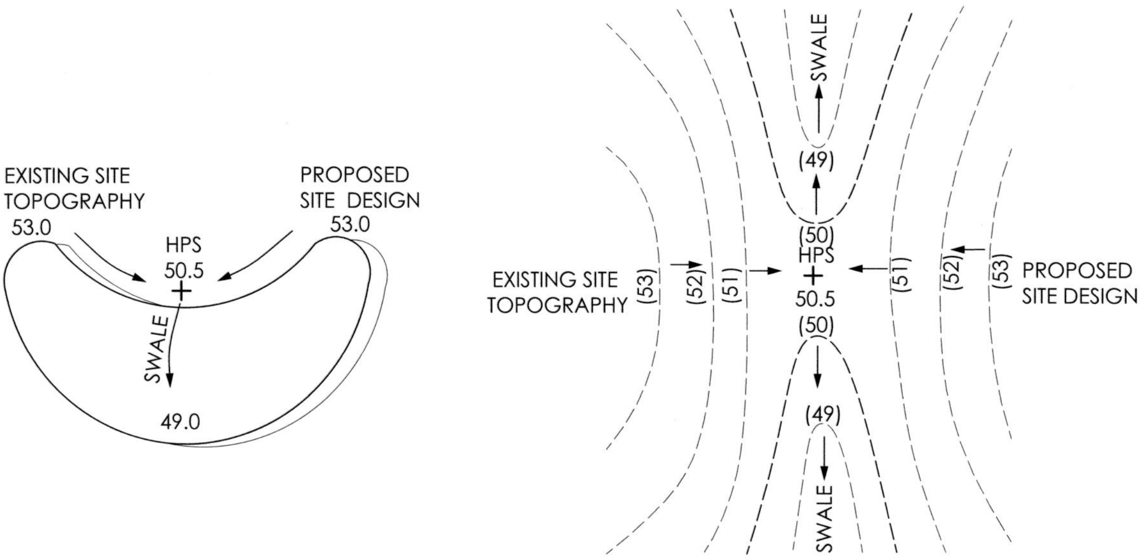

3.3–3
Saddle and a plan view of a saddle point

be located at the saddle point, as illustrated in Figure 3.3–3. It is the point where the descending topography of the site meets the descending topography of the design. This point is the high point for the swale and the low point between the existing site topography and the proposed site design.

When diverting water around a building you will be creating a horseshoe swale. This is two swales not one. Both start at the HPS and they divide the flow of water in both directions around a building. In some examples, one swale may be longer or shallower than the other, but they should both use slopes that fall within the minimum and maximum allowable slopes for the problem. When drawing a horseshoe swale it helps to visualize it with two sides, an open and a closed side.

The open side of the swale establishes the slope intervals for the longitudinal slope of the swale and provides room for the cross slope on the side closest to the proposed design. This is the more flexible side of the swale, because it will change based on adjustments to the proposed grading solution. Start drawing the swale on the open side. The closed side is the opposite side of the swale which completes the parabolic shape of the swale and ties back into the existing topography. Try to maintain smooth, continuous connections to the existing topography when tying proposed contour lines to existing ones. Create swales that are tight in form and take up minimal space in your design solution by using the maximum slope allowable for the cross slope of the swale.

The following example, illustrated in Figure 3.3–8 later, creates a horseshoe swale around a single building pad. Note that the resultant proposed swale is significantly bigger than the building pad. If a problem has more than one proposed building pad, create a large continuous buffer around the pads to minimize the number of swales needed to drain the site.

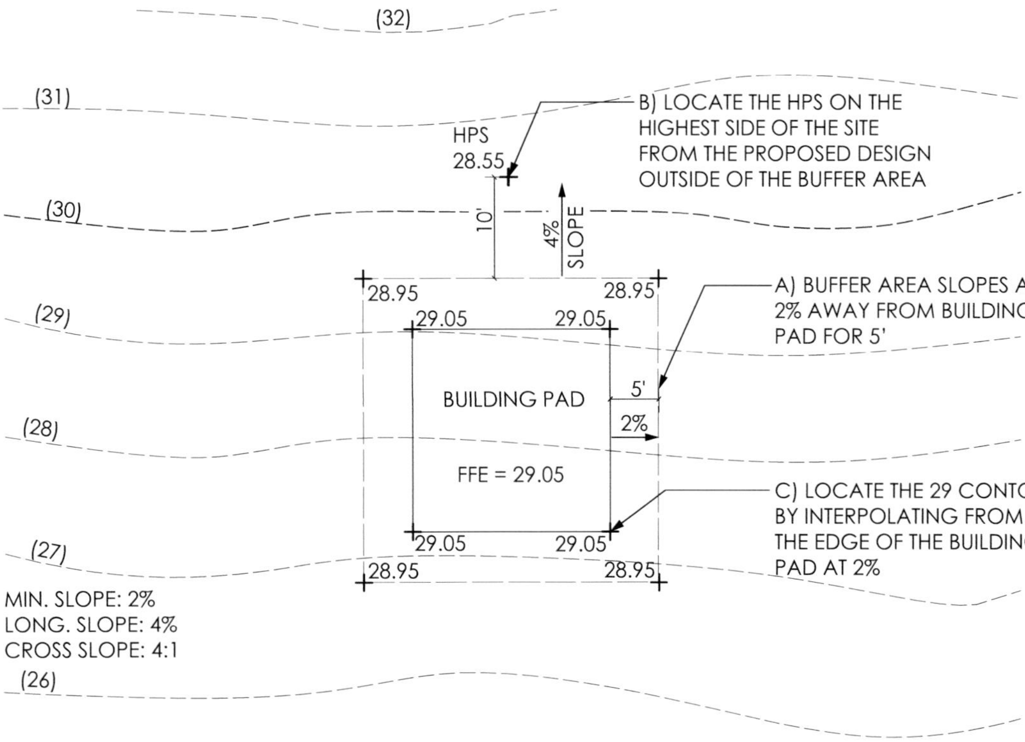

3.3–4
Location of HPS 10' from highest point of the design

1 Place a 5' buffer around the building pad as shown with dashed lines in Figure 3.3–4. Slope the buffer downhill from the building pad at the minimum allowable slope, in this example, a 2% slope. This buffer provides a 'flat' area outside of the building before any site grading changes to the surrounding topography.
2 Locate the HPS at a horizontal distance between 5'–20' from the edge of the buffer. It should slope downhill from the edge of the buffer at a shallow slope between the minimum and the maximum. Make sure to analyze the direction of the contour lines in the problem to determine which way water is flowing and place the HPS on the highest part of the site as it relates to the building pad and the buffer. Note that this may be at a corner of the buffer not in the middle.
3 Next, locate the proposed 29 contour line between the building pad and the buffer by interpolating away from the edge of the building pad at 2%. This will be a closed contour line and will not tie back to the existing topography.
4 After placing the HPS, determine and then locate the whole number spot elevation downhill from it. If, for example, the HPS is 28.55', the first proposed contour line of the swale will be 28. The starting point of the swale's proposed contours will vary depending on the slope used to begin the swale. For example, calculating the first contour line at a 4% slope will bring the first contour line closer to the HPS spot elevation, whilst calculating the first contour line at a 2% slope will start the swale farther away from the HPS. The longitudinal slope you choose will be determined by the parameters of the grading problem. The goal is to be able to tie back into the existing contours with ease.

5 Draw light horizontal and vertical reference lines along the sides of the buffer as shown in Figure 3.3–5. These will help you locate the proposed contour lines as you create the open side of the swale. Mark the location of the whole number spot elevations using the maximum allowable slope, 4:1. These tick marks will be used to create the proposed contours for the sides of the swales. The number of tick marks will depend on how many existing contours fall within the proposed building pad and buffer area. In this example, the tick marks will include the location of the 28 and 27 whole number spot elevations and the location of the 26 whole number spot elevation because it will have to be adjusted to maintain the maximum allowable slope, 4:1.

6 Draw light reference lines from the HPS. Use these as a guide for locating the whole number spot elevations along the longitudinal slope of the swales. Mark off the location of the whole number spot elevations using a uniform slope between the minimum and maximum allowed for the problem, 4% slope. Using the slope interval, 25', you can quickly determine how many proposed contours you will need to create the swale. The width of the swale will increase as it gets bigger so the reference line is only an approximation of location of the centerline for the swale. Adjust it as needed.

7 Connect the proposed contour lines with smooth continuous lines on the open side of the swale using the tick marks on the reference lines as shown in Figure 3.3–6. Make sure to maintain a uniform distance between the contour lines based on the maximum slope including at the corners.

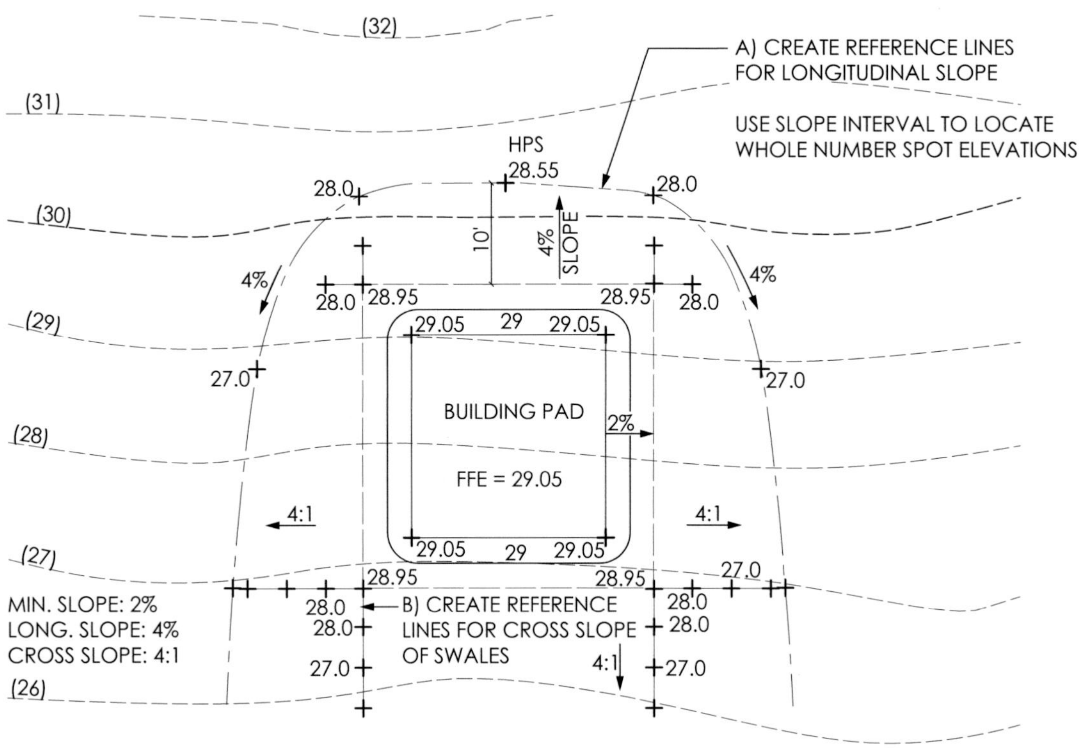

3.3–5
Reference lines for longitudinal and cross slopes

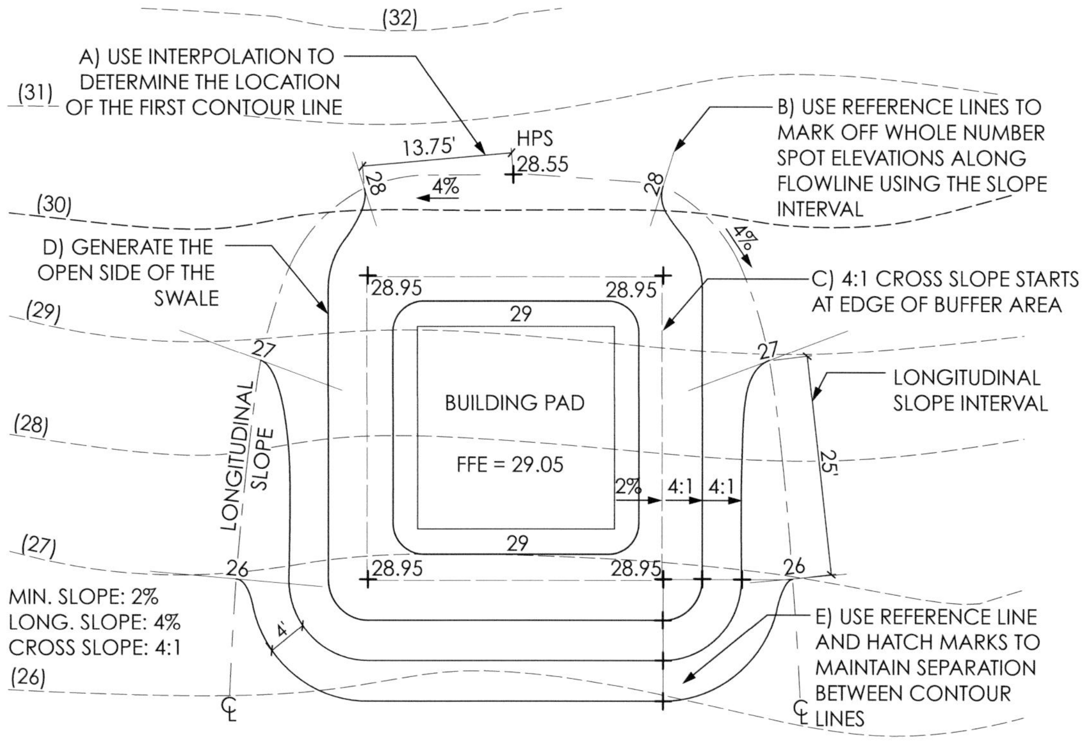

3.3–6
The open side of the swale

8 Once all the proposed contours are complete on the open side, add the closed side for the swale. Make sure to label the proposed contour lines on the uphill side and label the longitudinal and cross slopes on the swales. Make sure you tie the proposed contours back to the correct existing contour lines. Start at the lowest elevation, the existing (26) contour and work uphill. Make sure to maintain at least the maximum slope for the closed side of the swale, see Figure 3.3–7.

9 When the swale is complete, remember to go back and adjust the topography at the top of the HPS on the side of the existing topography as illustrated in Figure 3.3–8. Any existing contour lines (29), (30), and (31) that have been crossed by your proposed swales must be adjusted. These proposed contours should tie back into the existing ones using a slope between the minimum and maximum outlined in the grading problem. Label the proposed contours on the uphill side.

A common mistake when drawing swales is not maintaining at least the minimum longitudinal slope in the swale. Regardless of the scale of the drawing, 1' contours should be no more than 50' apart to maintain at minimum a 2% slope. Continue drawing the swale as needed to maintain the minimum longitudinal slope. This is also a way of checking when to stop drawing proposed contours for your swale.

If a longitudinal slope for the swale is not provided, measure the horizontal distance from the HPS to the lowest existing contour that will be affected by the proposed design – in this example, 68'. Find the vertical change in elevation by subtracting the lowest

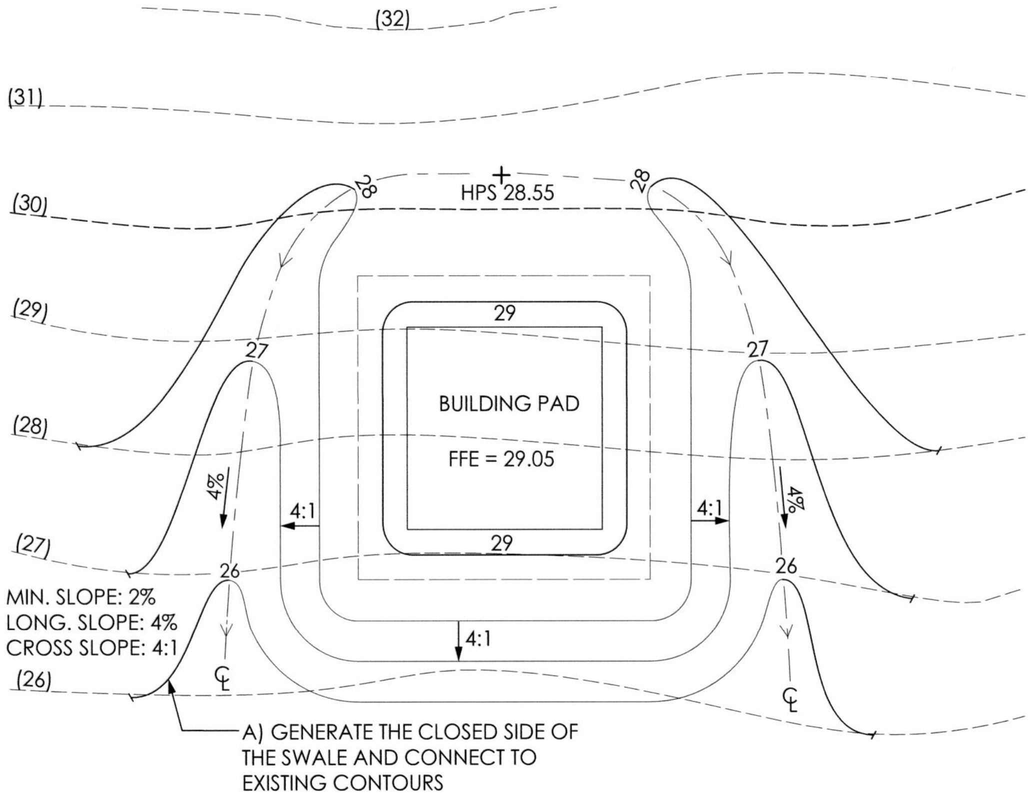

3.3–7
The closed side of the swale

existing contour from the HPS, 28.55'−26' = 2.55'. Find the longitudinal slope for the swale, 2.55'/68' = 0.0375 or 3.75%. Round the resultant longitudinal slope to the nearest whole number to make finding the slope interval easier, 3.75%→4.0%. Remember that the longitudinal slope should be between the minimum and maximum allowable slope.

A swale has a minimum depth of 6". In most cases, you won't need to check the depth of the swale since it will be significantly deeper than 6", however, some problems ask for a minimum required depth of swale. To measure or estimate the depth of a swale, determine where the proposed contour line falls between two existing contours. A 6" deep swale will fall midway between two existing 1' contour lines and a 12" deep swale will measure the full distance between two existing 1' contours lines. Examples are illustrated in Figure 3.3–9.

These types of swales and the uniformity of depth are often seen with road grading and on grading problems with a small drawing scale. Once you understand how a swale functions and how it ties into the design and the existing topography you need to understand how it interacts with other structures in the open and closed drainage systems.

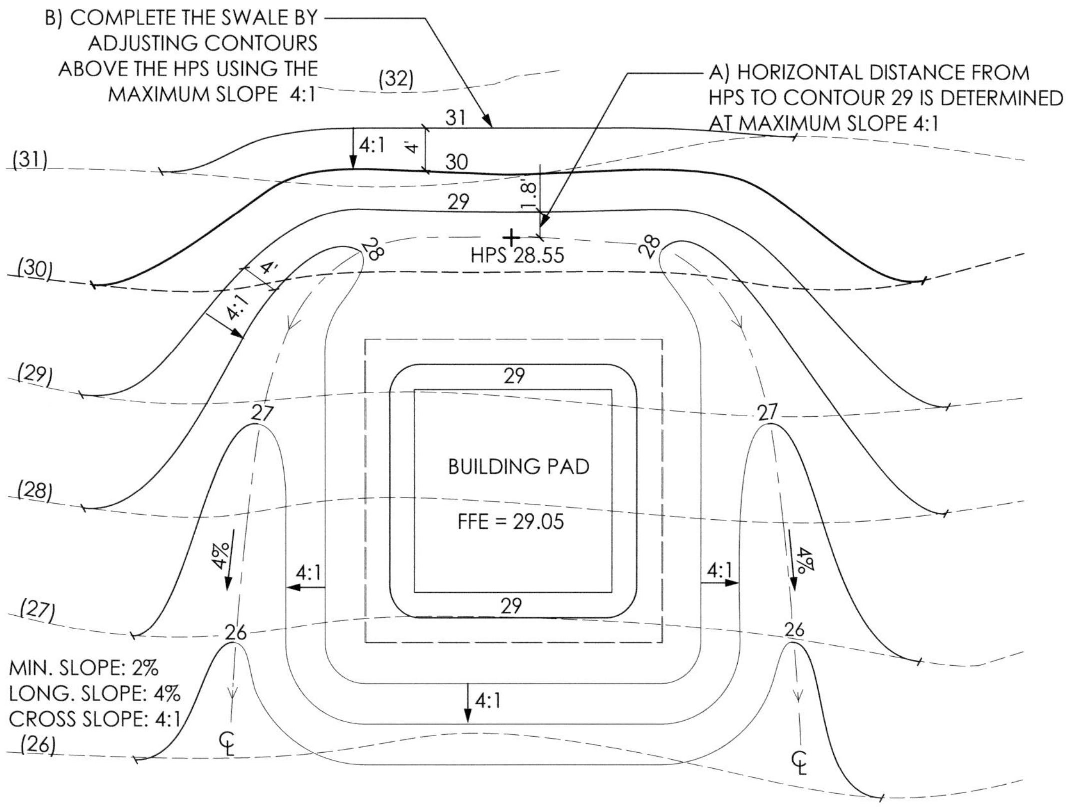

3.3–8
Completed swale with HPS, slopes, flowlines, and proposed contours labelled

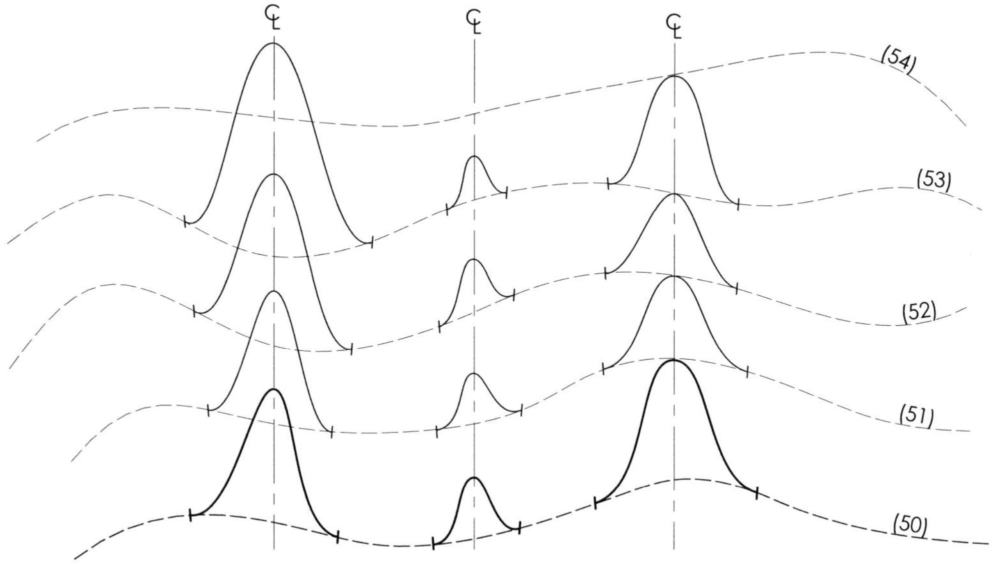

3.3–9
18", 6" and 12" deep swales

3.4 Retention ponds

Most state regulations require that surface runoff be contained on site; because of this, swales often tie into retention and detention ponds. These ponds are man-made collection basins for surface water runoff; they prevent surface flow from leaving a site and aid in the percolation of surface water back into the water table for the existing drainage basin. For the purposes of clarity, they will be referred to as retention ponds throughout this book, but in the examination, they may be called detention ponds, collection basins, etc. A retention pond is a permanently wet pond and a detention pond can be dry and fills depending on rainfall events.

Myriad designs and treatments can be created for retention and detention ponds but the underlying topographic signature remains the same; a closed set of contour lines is nestled between two descending contour lines. Retention ponds are always located topographically on the lower areas of the site on a grading problem. They must meet a specific depth or volume requirement, creating a basin in which water will be held. Position the retention pond in the grading solution to facilitate its attachment to any existing or proposed swales.

A retention pond has three parts: a collection basin made up of closed contour lines of descending topography; a dam or embankment at the lower end; and a connection to a swale at the end opposite the dam. The grading problem will either provide a specific shape for the base of the retention pond or will provide a total area that must be met, as seen in Figure 3.4–1. A problem that gives the shape for the bottom of the pond implies that the basin will be built up at the maximum slope around the bottom of the basin. This edge is the lowest elevation of the pond. A given area implies that the depth of the basin will be encompassed within the area provided. This edge is the surface elevation or highest elevation of the pond.

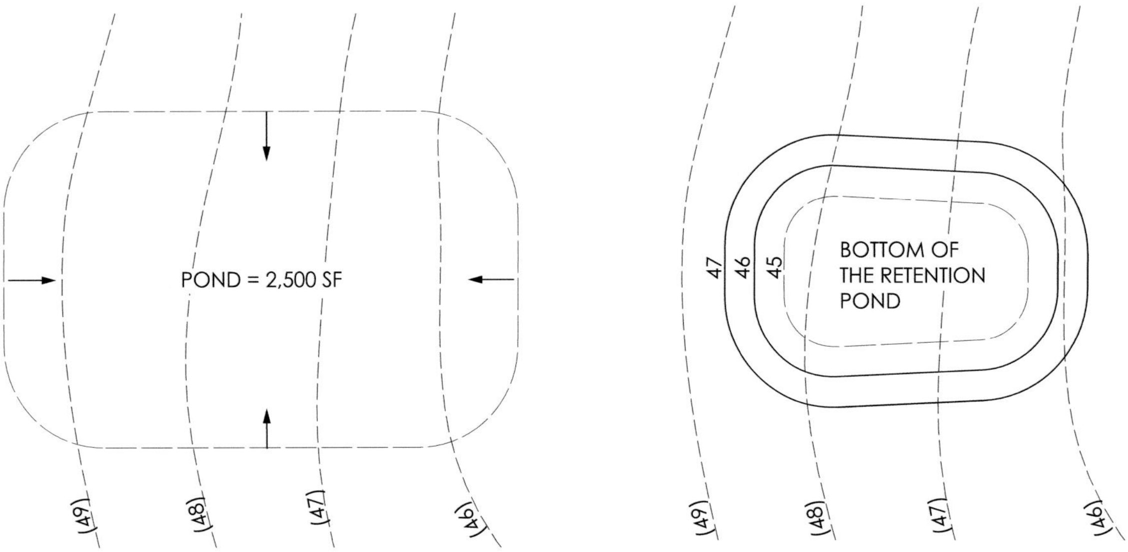

3.4–1
Retention pond's total area for a pond vs. shape of the bottom of a pond

An embankment or dam is a raised edge of the pond at least 6" above the surface level of the proposed pond. It is a ridge located on the lower side of the retention pond as it is sited in the topography. Surface water that hits the top of the embankment will flow into the pond or onto the surrounding landscape. Remember that a retention pond does not collect all of the surface runoff for a site, but is specifically designed to collect the surface runoff from any areas of the site that are affected by your proposed design.

Because rainfall events may exceed the depth of the proposed retention pond, retention ponds may be connected into a series of small wetland features to increase the permeable surface area allowing for surface runoff to percolate slowly into the soil or they may be connected to the stormwater system to capture any overflows. The following example creates a 2' deep basin with an attached swale.

1 Locate the basin in the lower part of the topography in a shallow sloped area, 2–5% as shown in Figure 3.4–2. Draw the basin of the retention pond and label it with an LP and a whole number spot elevation. This will provide space for the basin and the embankment, minimizing the number of proposed contours needed.
2 Give the basin a depth that is 3' lower than the surrounding existing topography. This will provide enough depth to create a 2' deep pond plus an additional 6"–1' for an embankment.

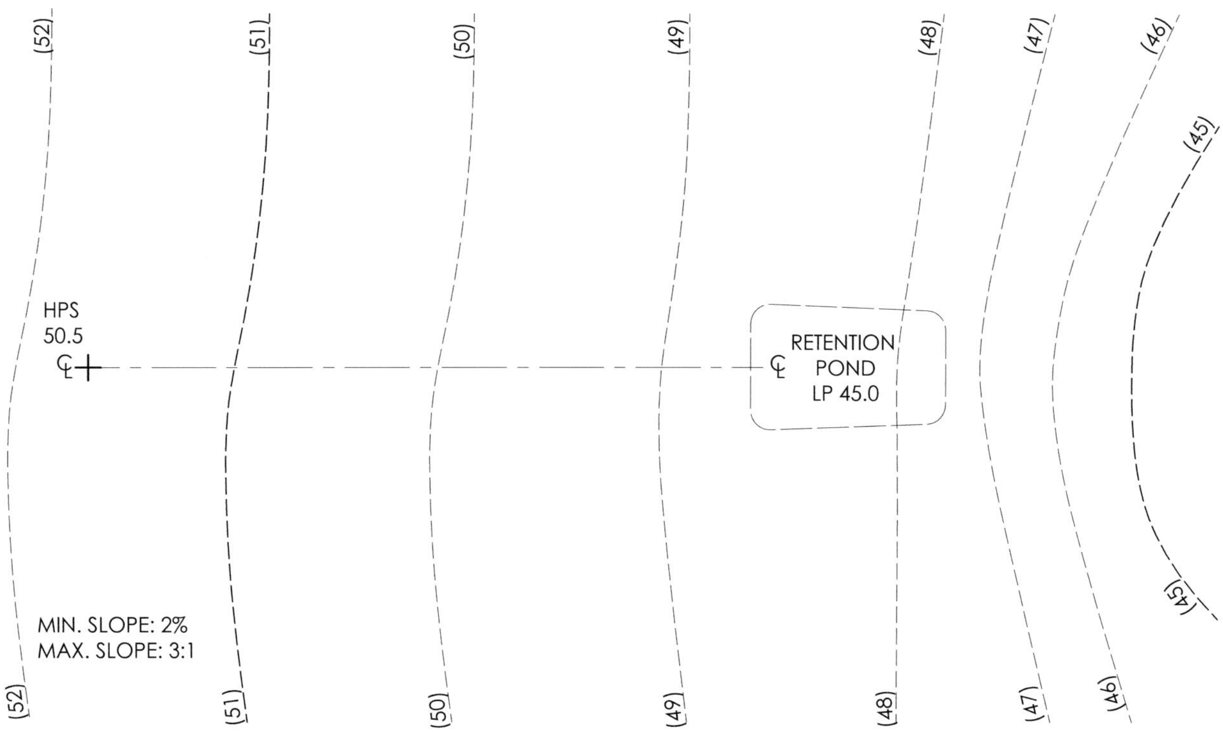

3.4–2
Retention pond basin with centerline and low point

3 Using the maximum allowable slope, draw and label proposed closed contours around the basin until the desired depth is reached. Remember to count the spaces not the lines between the proposed contours to check the depth. On the side of the basin that connects to the swale you can either use the maximum slope or you can vary the slope to create a more gradual transition from swale to pond, see Figure 3.4–3.
4 Measure the horizontal distance from the HPS to the edge of the pond, 65'.
5 Calculate the vertical change in elevation by subtracting the elevation at the edge of the pond from the HPS, 50.5' − 47' = 3.5'.
6 Calculate the longitudinal slope for the swale, 3.5'/65' = 0.0538. Round up the slope if desired, 5.38% → 5.5%.
7 Mark and measure the location of the proposed swale contours based on the slope interval of the longitudinal slope. Label the longitudinal slope along the flowline of the swale.
8 Complete the sides of the swale using the maximum allowable slope as shown in Figure 3.4–4.
9 To create the embankment below the pond, first add a spot elevation that is 2"–6" higher than the last closed contour line for the basin, 47.5'. Note that the spot elevation should be located on the lower side of the existing topography and

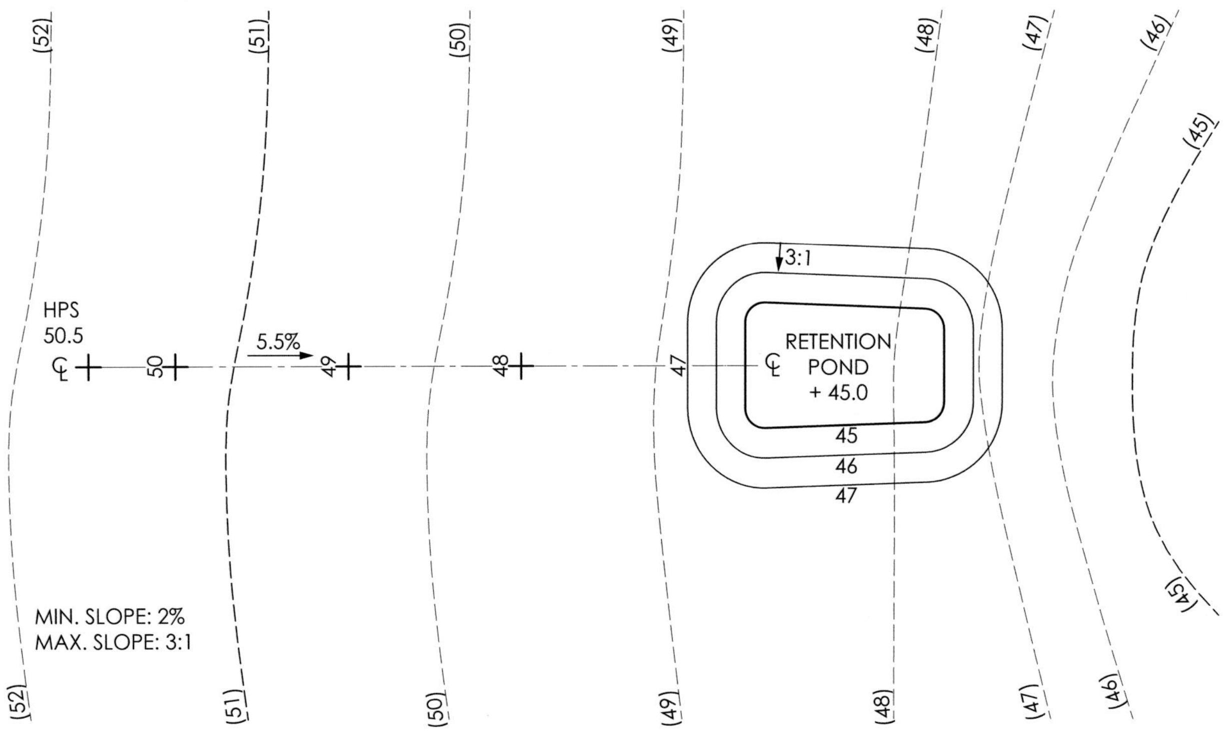

3.4–3
Retention pond with closed contours creating a 2' deep pond

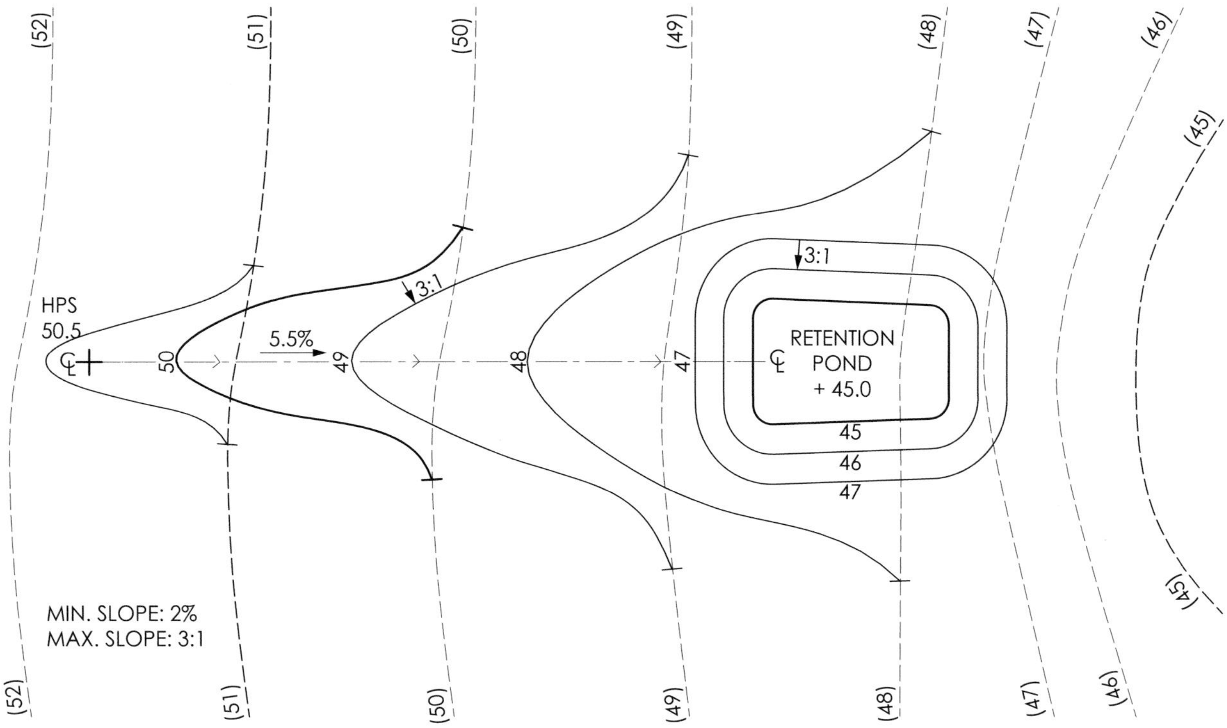

3.4–4
Retention pond with swale

should surround the majority of the pond. You can denote this elevation with a dashed line as shown in Figure 3.4–5. The width of the embankment will be determined by the parameters of the problem, but at minimum it should be twice the width of the slope interval of the maximum allowable slope.

10 Add proposed contour lines below the embankment spot elevation. The first proposed contour line will be at the same elevation as the highest closed contour of the pond. Remember that the pond has closed contours and any existing contours that fall within the bounds of the proposed pond, (47) and (48), must be adjusted using the maximum allowable slope.

11 Label the proposed contours on the uphill side and add the cross slope. A section-elevation of the proposed pond is highlighted in Figure 3.4–6 for clarity.

Remember the swale diverts surface water from the areas around the proposed design and the pond is designed specifically to collect drainage from the swale. There should not be any swales surrounding the retention pond; you are actively trying to get water into the pond not direct it away from the pond. The only area that may require some minimal drainage is the embankment in order to meet the minimum slope requirement between contours, but a full swale is not required.

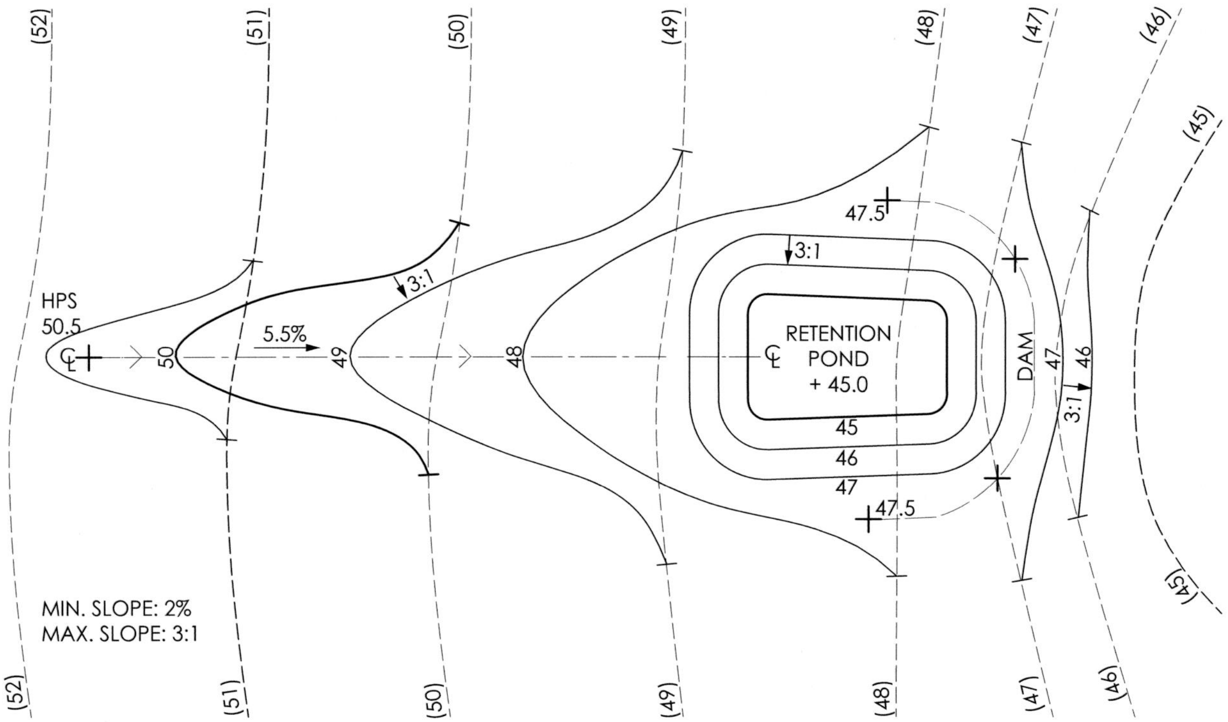

3.4–5
Retention pond with swale and embankment

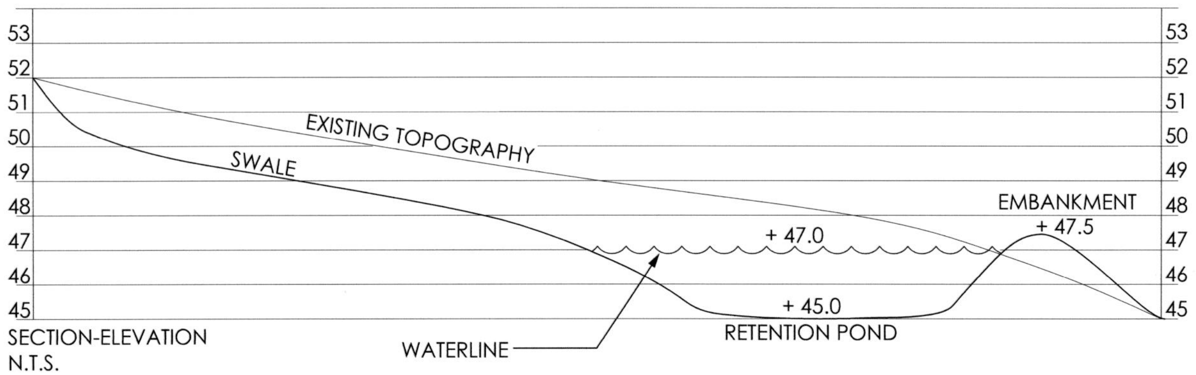

3.4–6
Section-elevation of retention pond

3.5 Drain inlets

Swales and retention ponds are part of the open drainage system, but swales can also tie into the closed drainage system. The easiest connection to the closed drainage system is ending a swale around a drain inlet or catch basin. The following example shows how to connect a swale to a drain inlet and how to set the rim elevation of the drain inlet.

1 Measure the horizontal distance from the HPS to the drain inlet. Using the longitudinal slope for the swale, calculate the vertical change in elevation and subtract it from the HPS, 101' × 0.05 = 5.05'; 5.05' − 5.05' = 45.45' → 45.5'. This will be the rim elevation for the drain inlet in Figure 3.5–1.
2 Using interpolation and the slope interval, mark off the location of the proposed contour lines along the swale's flowline.
3 Create a closed contour around the drain inlet. Remember that the rim elevation is the lowest point along the flowline and the closed contour should be the first whole number contour higher than the rim elevation. This closed contour creates a sump or depression similar to a retention pond. Use the maximum allowable slope to calculate the minimum horizontal distance of the proposed contour on the three sides of the drain inlet opposite the swale's flowline and connect it to the proposed contour along the swale's flowline, 46' − 45.5' = 0.5'; 0.5'/.20 = 2.5'. Refer to the illustration in Figure 3.5–2.

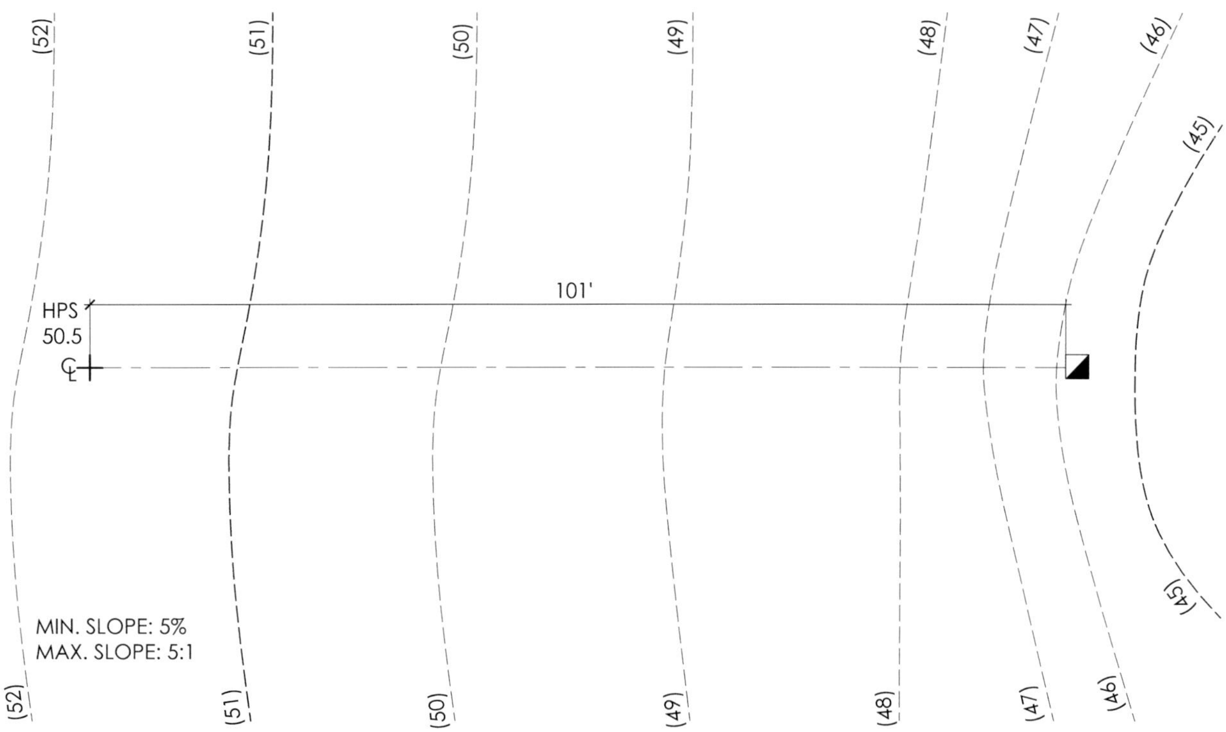

3.5–1
Finding the rim elevation for a drain inlet

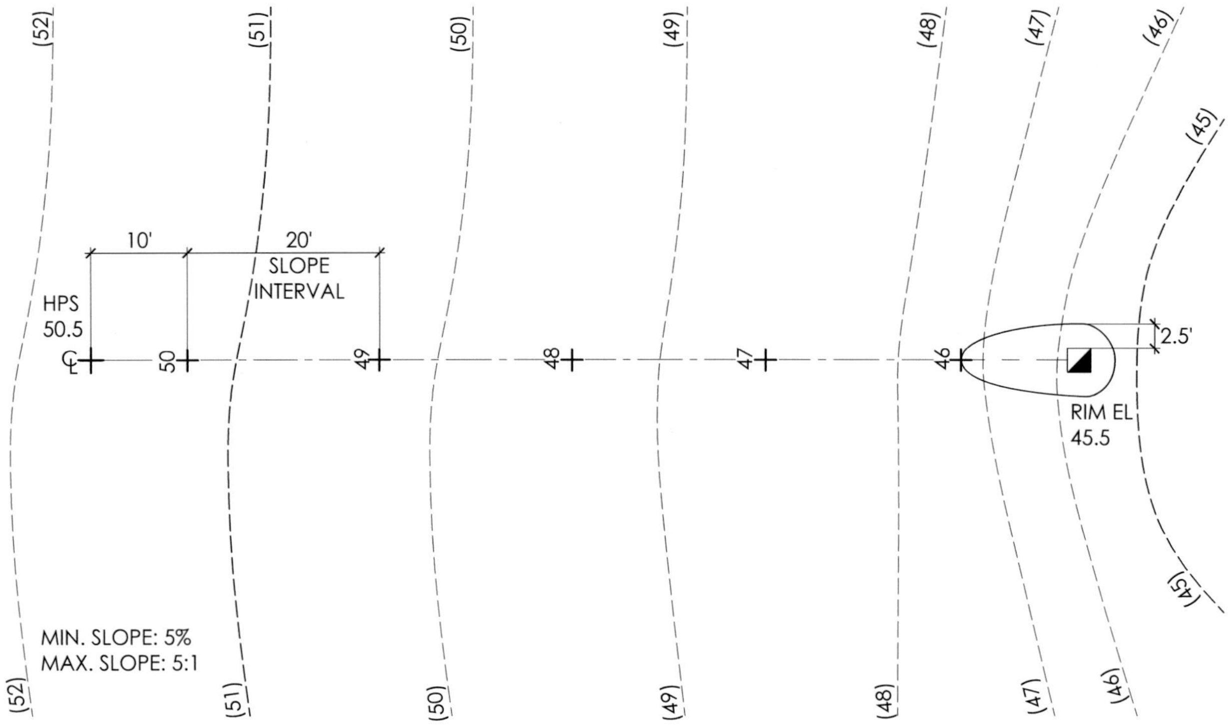

3.5–2
Locating the whole number spot elevations on the swale flowline

4 Once the closed contour is complete, create the remaining proposed contours for the swale using smooth continuous lines. Use the slope interval for the maximum allowable slope for the sides of the swale.

5 Adjust any existing contours that were affected by the setting of the rim elevation for the drain and the creation of the closed contour. Use the maximum allowable slope. Provide a minimum HP spot elevation of 2"–6" between the closed contour and any existing contour if needed. As with the retention pond, this will ensure that water from the swale flows towards the drain, see Figure 3.5–3.

3.6 Berms

A berm is a built structure made of earth, it forms a peak or closed ridge. It may be uniformly sloped or sloped to mimic a natural peak. It should be steeper at the top and shallower at its base to slow down the flow of surface water runoff. It is characterized by one or more high points and its sides are usually sloped according to the maximum allowable slope on a grading problem. It is treated as any immovable object in the landscape with swale drainage around it. On the grading exam, a berm is used to block unwanted views in the landscape. In order to solve a berm problem you need to determine the height and width necessary to create the berm; you're aiming for two objectives, provide enough height and width to hide the objectionable view and provide a way for water that flows off of the berm to get downhill to a drainage structure.

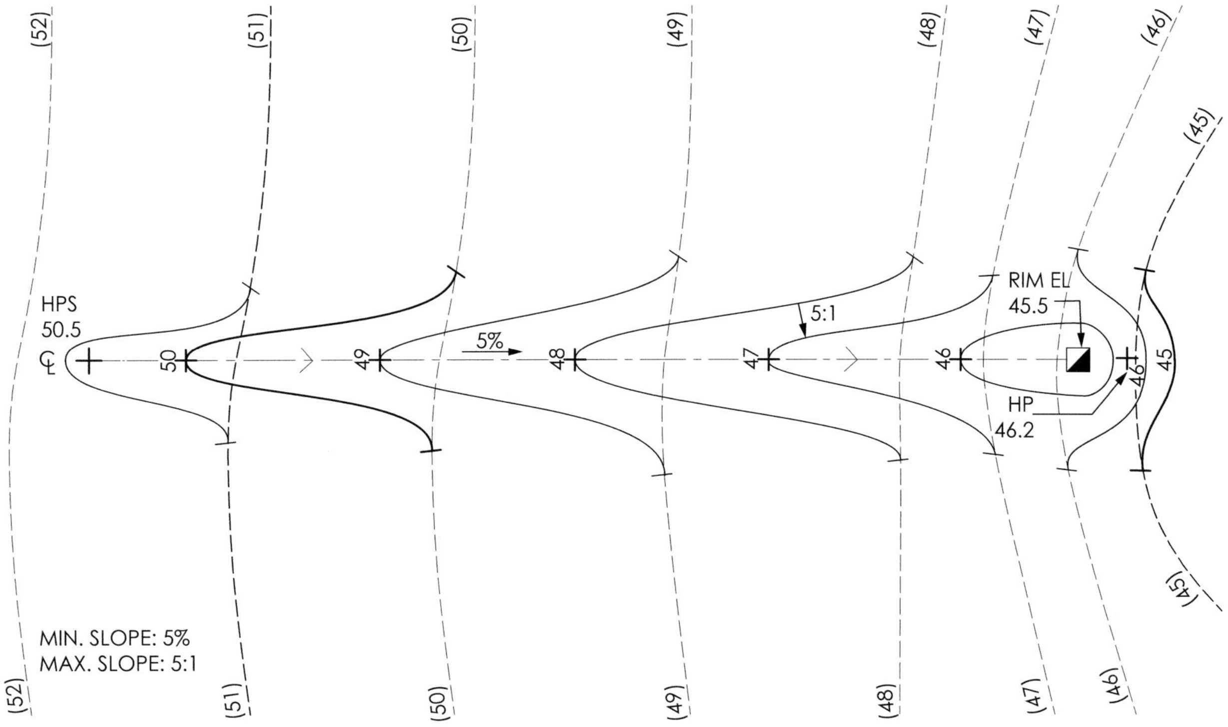

3.5–3
Drain inlet with closed contour at 5% slope

There are three stages to creating the berm: generating the top of the berm with closed contours using the maximum allowable slope; creating the swales to divert water around the berm using a longitudinal slope or the minimum slope; and tying the proposed contours back to the existing contours. The maximum height and width of the berm is tied to the eye height and viewshed of the person standing in front of it, the object that you're trying to hide on the other side of it and the slope of the existing topography. On a grading plan, the viewshed will show up as two dashed lines that encompass the space for the top of the berm. The berm will descend from the edge of the viewshed. In order to solve the berm grading problem, ample space is needed for both the berm and any swales required for the grading solution. The following example creates a berm to hide a new building at a lower elevation.

1 Note or determine the starting elevation at finish grade for the platform, 100.5'. Add to it the eye height of a person, usually 5'; this is the adjusted elevation for the platform, 100.5' + 5' = 105.5'.
2 Note or determine the starting elevation at finish grade for the shed, 94.5'. Add to it the height of the building, 94.5' + 8' = 102.5'. This the adjusted elevation for the shed as illustrated in Figure 3.6–1.
3 Measure the horizontal distance between the two points on the plan.

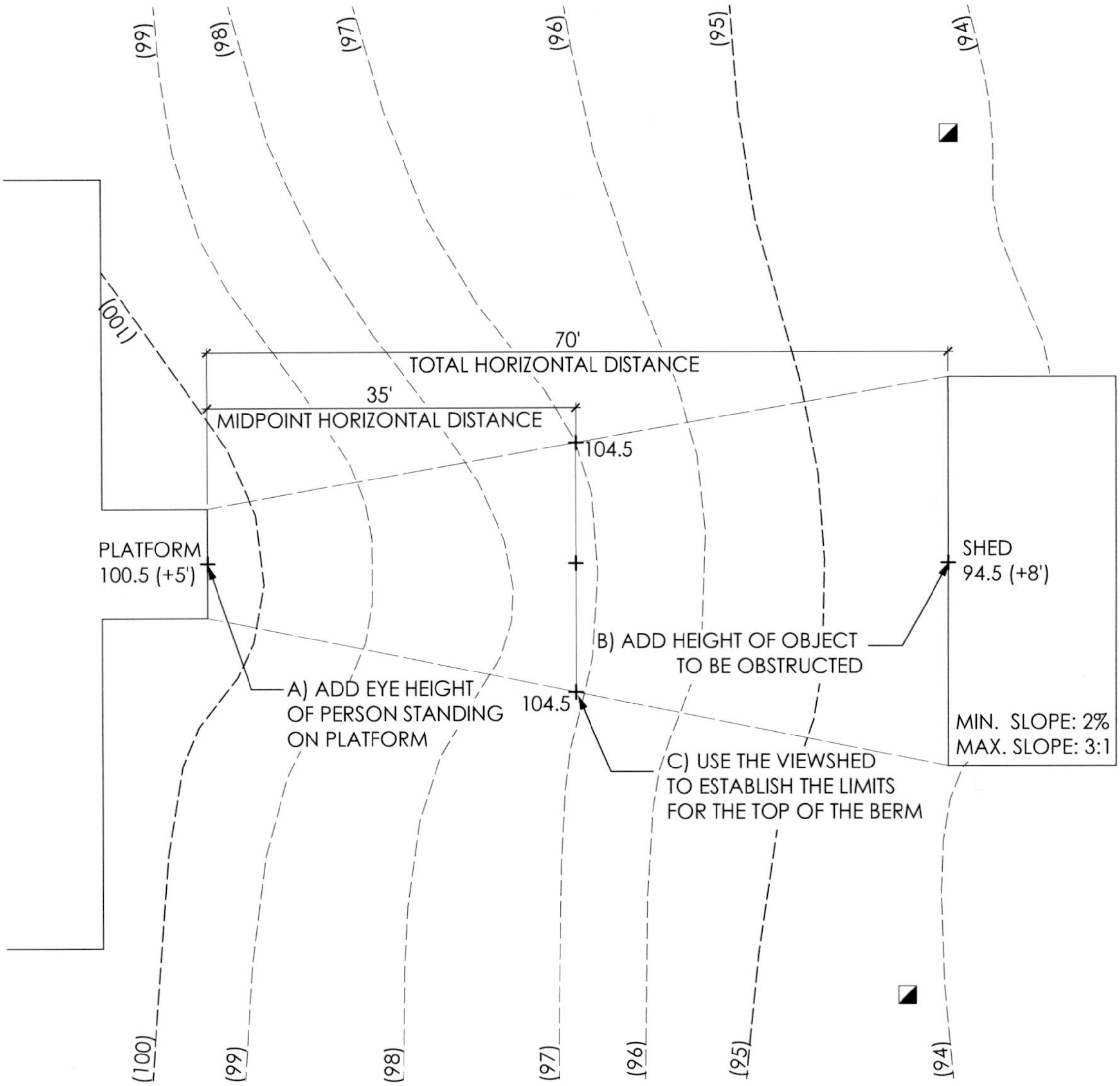

3.6–1
Plan of berm with total length and midpoint distance shown

4 Create a scaled section to determine the height of the berm, see Figure 3.6–2. Use the distance between the platform and the shed as the horizontal distance for the section and use the eye height from the platform as the elevation at one end of the section and the adjusted elevation for the shed at the opposite end of the section.

5 On the section, mark off and label uniform intervals as needed. Using an exaggerated scale makes it easier to visualize the precise spot elevation for the top of the berm. Note the intervals should correspond with the adjusted heights for Point A and Point B respectively, 105.5' and 102.5'.

6 Draw a line from the adjusted elevation for the platform to the adjusted elevation for the shed. This is the line of sight. This line signifies the topmost elevation for a

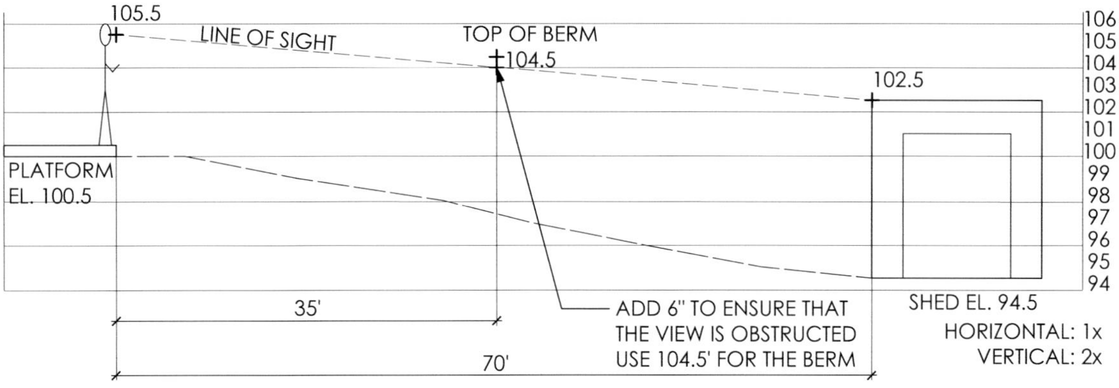

3.6–2
Section of berm to determine the height of the berm

berm. Choosing a point anywhere along this line will yield a berm that hides the objectionable view.

7 Pick a point along this line roughly one third to halfway between the two points and where it intersects with one of the marked elevations. Ensure that there is ample room to provide a swale on both sides of the berm. Sometimes it takes a little trial and error to determine the best spot elevation along the line. It is this spot elevation that is the high point of the proposed berm. For safety, you may want to add 2–6" to the spot elevation.

8 Once you've chosen the spot elevation along the line of sight. Measure the horizontal distance from the platform to the spot elevation on the section and then draw a reference line on the plan using the same distance. This line should intersect the viewshed's dashed lines.

9 Place the spot elevation on the dashed lines at the corners of the horizontal viewshed as provided on the grading problem and label them with 'HP' so it is clear that these are the high points of the berm.

10 Next, determine the first contour line that is lower than the HP for the berm. Encircle the HP spot elevations with an oval or ellipse using the maximum slope to complete the first contour line. Label the contour line with the whole number contour that is downhill from the HP elevation.

11 Create multiple closed contour lines spaced evenly around the ellipse based on the provided maximum slope, 3:1, see Figure 3.6–3. Mark, measure and draw smooth continuous lines. It is important to be precise when making the berm as space can often be an issue when drawing the final problem solution. Stop the closed contours when the proposed contours meet the existing contours at the same elevation, 99 and (99).

12 After drawing the berm, the next step is the swales. Note the direction of water flow from the existing topography. There will be two HPS in this example. The first is to divert water around the berm and the second is to prevent water from the berm flowing back towards the building. Surface water drains all sides of the berm and must be diverted on all sides. The descending existing topography will meet the descending topography of the proposed berm at the saddle points.

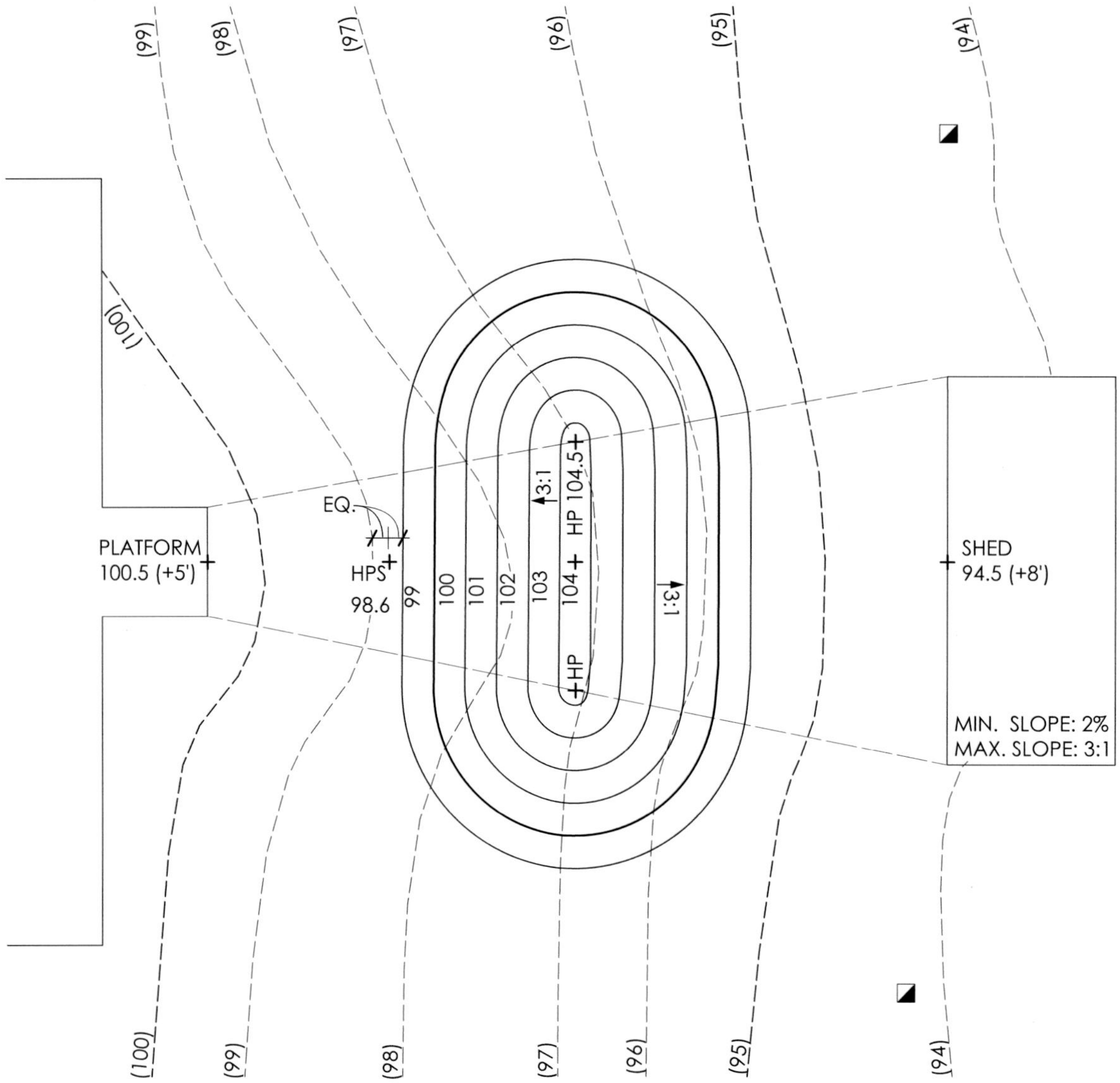

3.6–3
Top of the berm showing closed proposed contours

13 Create the first HPS by measuring the distance from the last proposed closed contour to the existing contour. Divide the distance in half and using the minimum slope determine the elevation at the midpoint. Note that this HPS is higher than the second one, which will be created after the first set of swales are drawn around the berm.

14 From the HPS, draw flowlines to the proposed drain inlets. Although you can determine a workable slope for these swales using the minimum slope for the problem and the overall elevational change, it is better to use the second HPS in front of the building to determine the rim elevations for the drain inlets, to ensure that water in these swales is flowing towards the drains at the minimum slope for the problem.

15 Continue drawing open proposed contour lines at the maximum allowable slope until you have reached the second point where the proposed contours meet the existing elevation. This will be the 95 contour and the existing 94.5' spot elevation for the shed. Measure the distance between the proposed contour and the shed and divide the distance in half. Apply the minimum slope to the lower of the two elevations, 94.5', to determine the second HPS, 94.4', refer to the illustration in Figure 3.6–4.

16 Measure the distance from the second HPS 94.4' to the proposed drain inlets and, using the minimum slope for the problem, calculate the rim elevation for the drain inlets, $37' \times 0.02 = 0.74'$; $94.4' - 0.74' = 93.66'$.

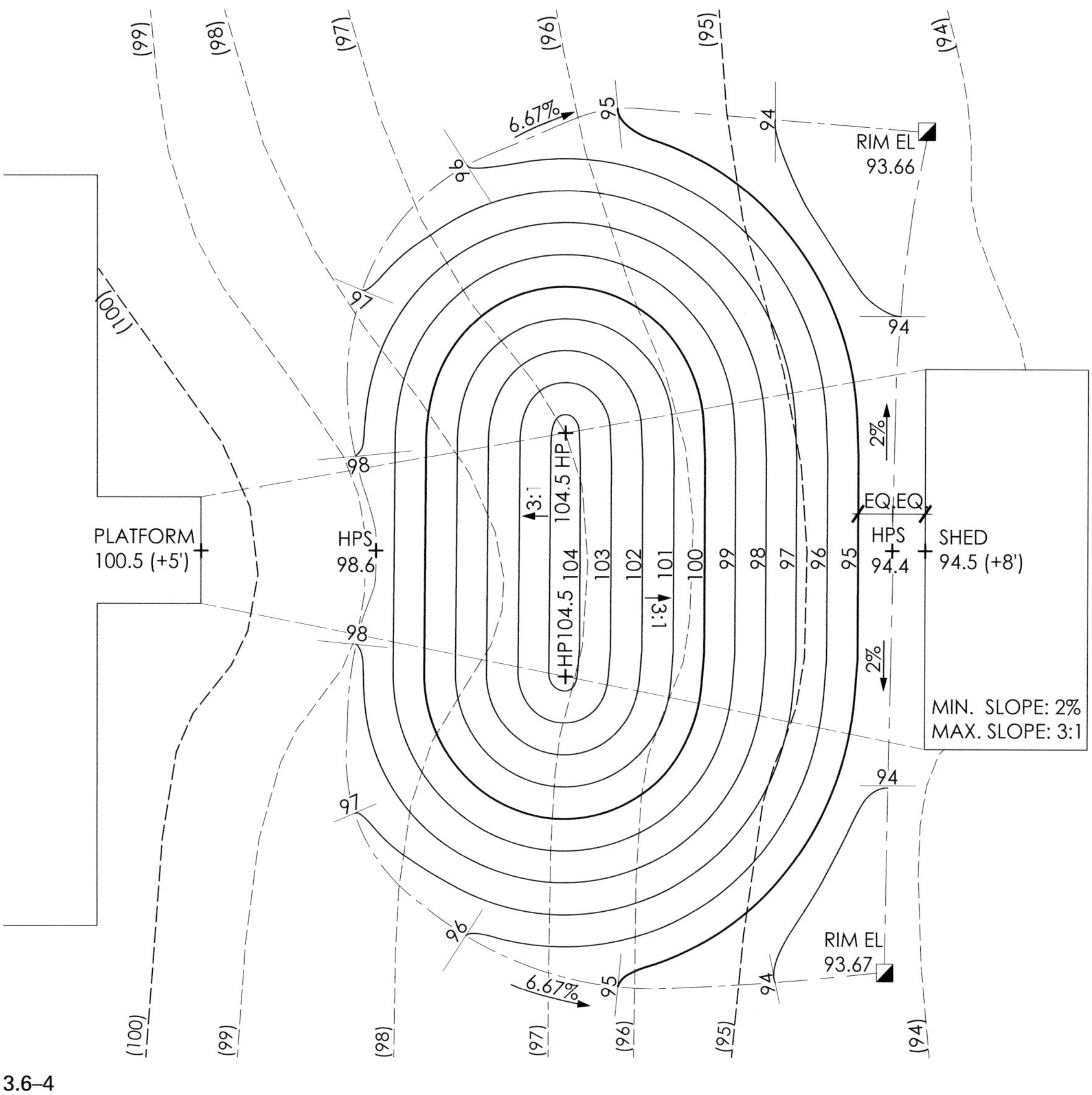

3.6–4
Completed berm with open side of the swales

17 Using the change in elevation between the first HPS 98.6' and the rim elevations, 93.66' and 93.67', determine a workable slope for the two main swales around the berm using the horizontal distance from HPS to RIM EL, 98.6'−93.66'=4.94'; 4.94'/74'=0.0667 or 6.67%.

18 Use tick marks along the flowlines for both sets of swales to show where the proposed contours should meet the flowlines.

19 Create a closed 94 contour line around the drain inlet to ensure that water coming off the berm flows into the drain inlet and adjust the existing (94) contour to allow for a high point elevation between the proposed and existing 94 contour lines. Complete the swales by tying the proposed contours back into the existing contours as shown in Figure 3.6–5. Remember to adjust the existing (99) contour if needed. The completed section of the berm is illustrated in Figure 3.6–6 for clarity.

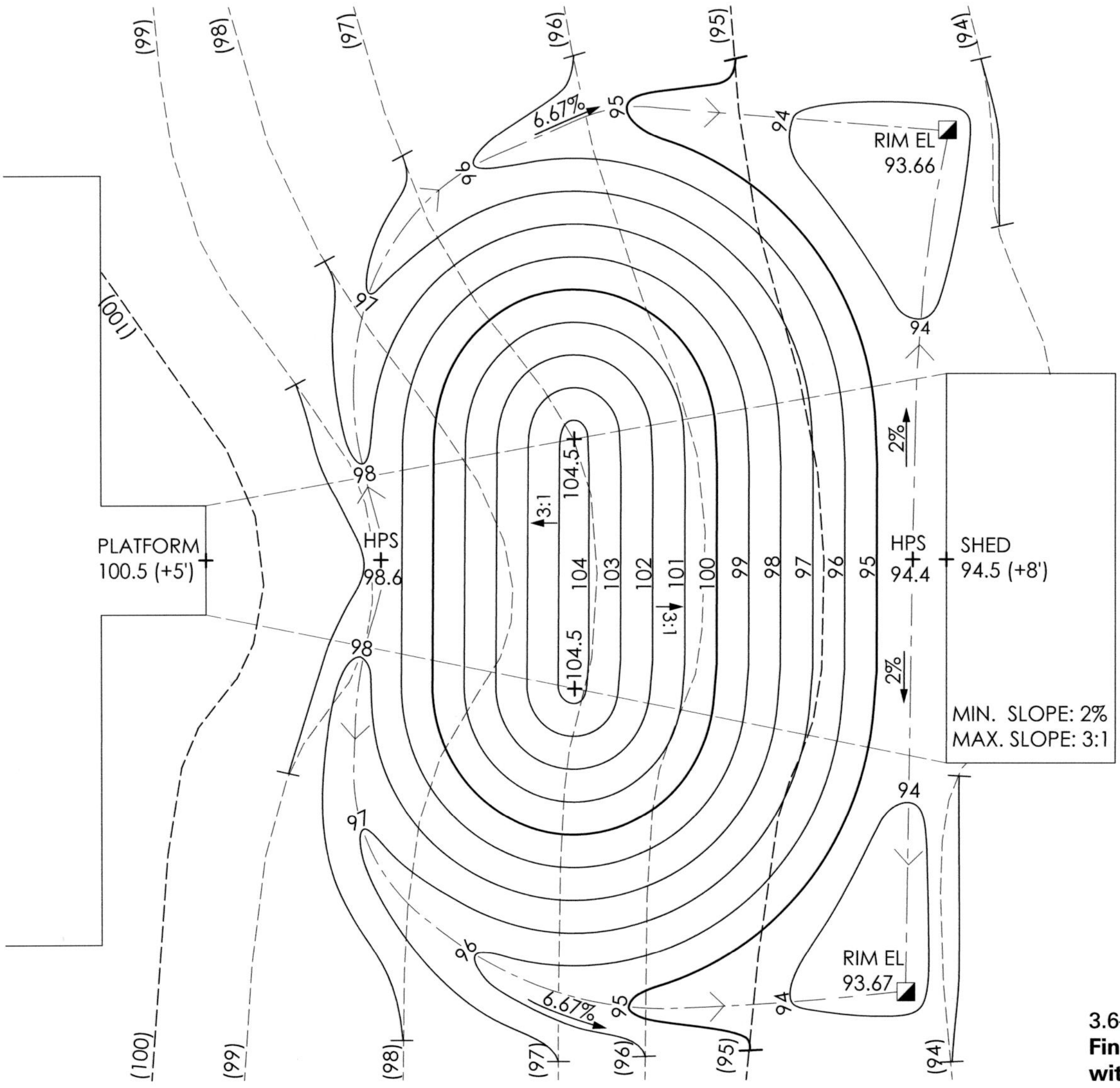

3.6–5
Finished berm with four swales

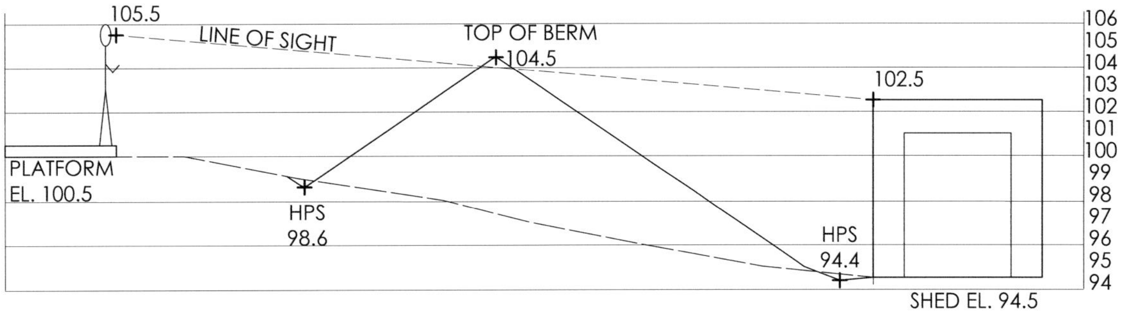

3.6–6
Scaled section of completed berm

3.7 Summary

The open and closed drainage systems direct surface water runoff to streams, rivers, ponds and eventually the ocean. The open drainage system mimics the natural processes found in watersheds. Sheet drainage and swale drainage are key components of the open drainage system. Sheet and swale drainage remove surface runoff from plane surfaces and warped planes. Swale drainage diverts water around built structures and berms, and eventually into retention ponds that are part of the open drainage system or drain inlets that are part of the closed drainage system.

3.8 Exercises (answers start on page 185)

3-1) Grade the two surfaces and show proposed contour lines. Minimum slope is 2%, maximum slope is 5:1; use sheet drainage, no swale is required. Draw two overlapping sections showing the existing and proposed grade.

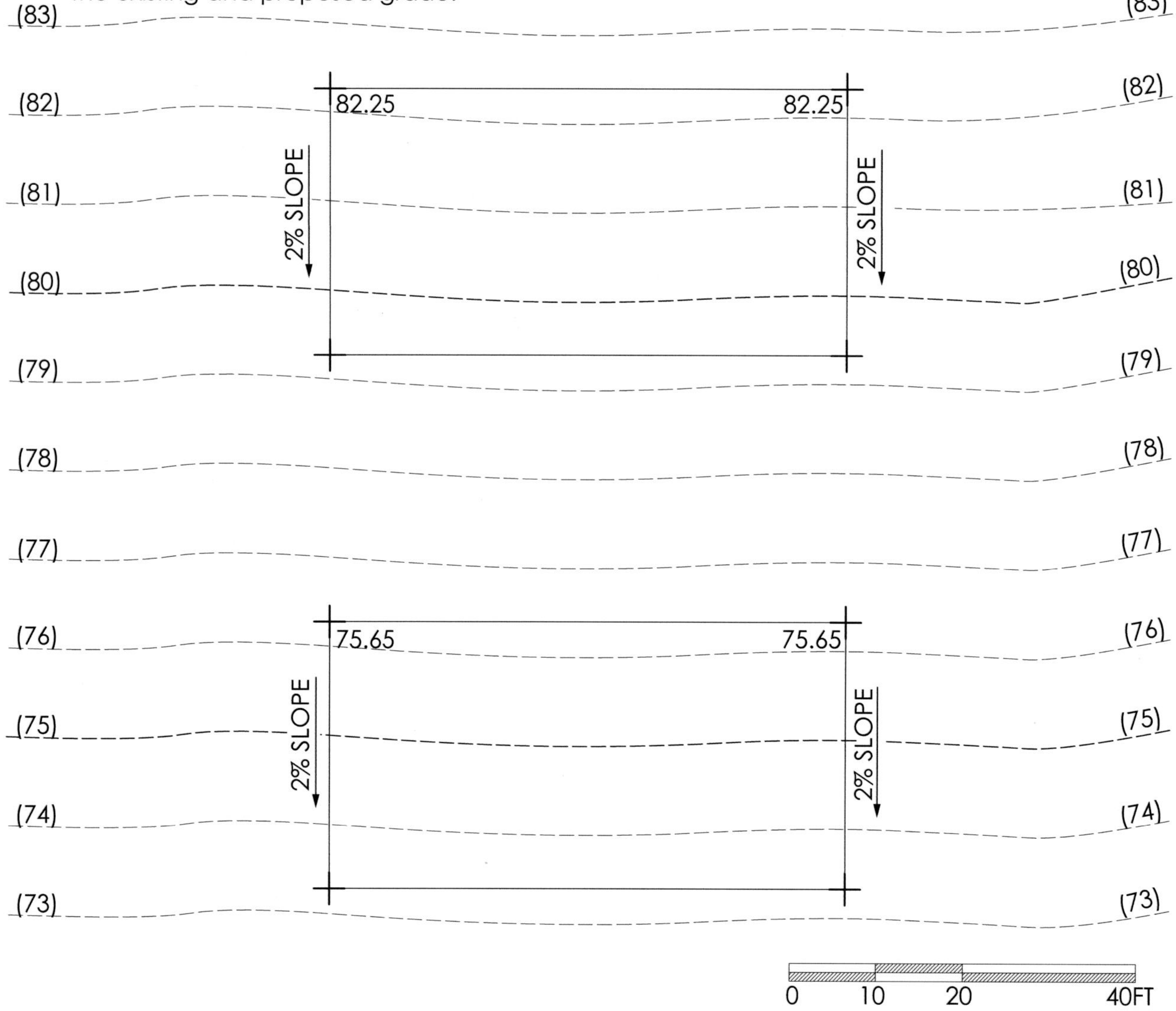

3-2) Draw a 18" swale.

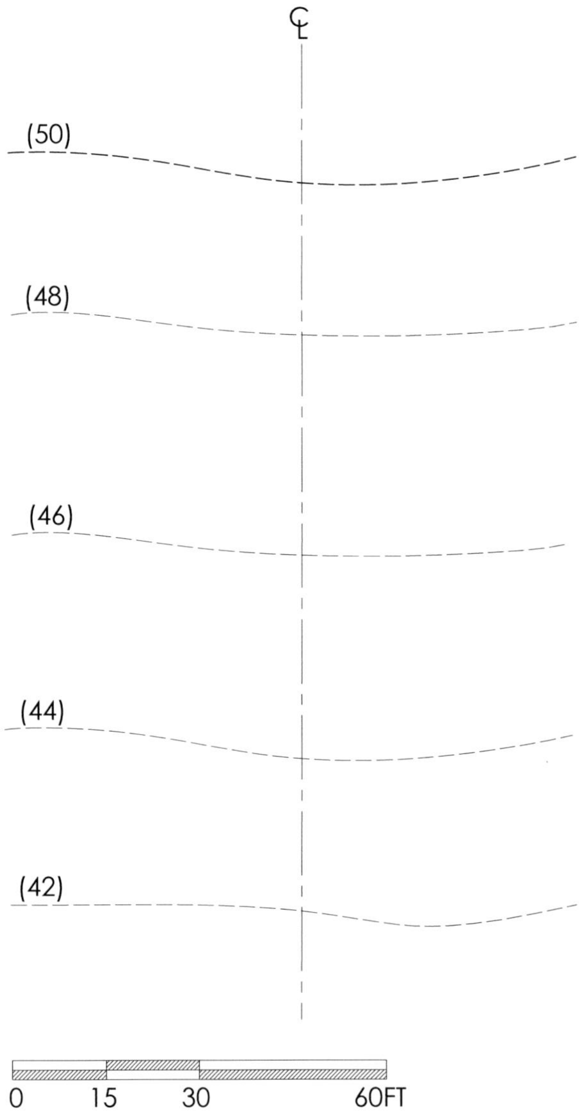

3-3) Draw a swale with an 8% minimum slope and a 3:1 maximum slope.

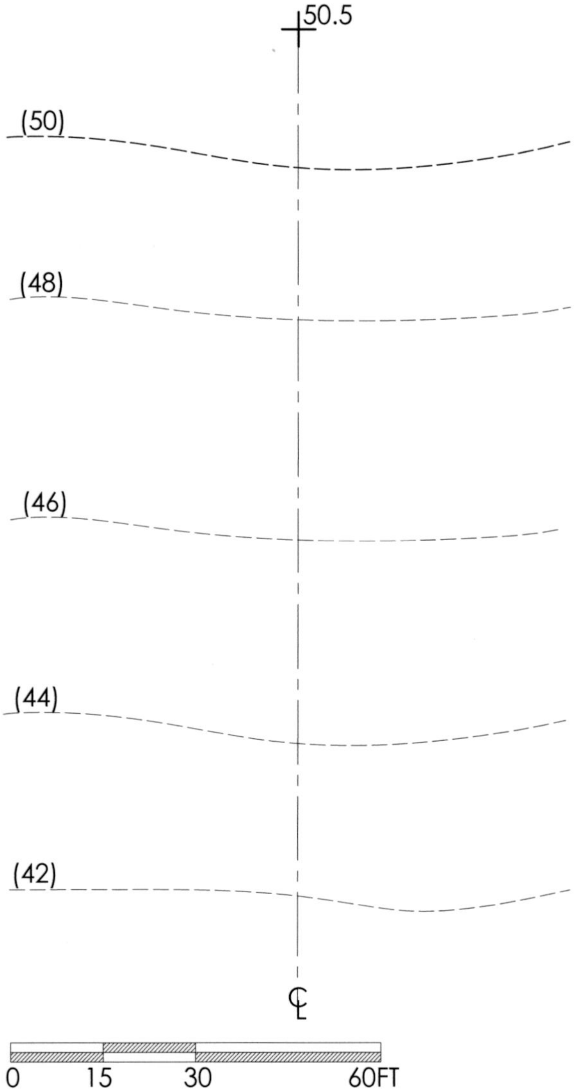

3-4) Draw a 12" ridge.

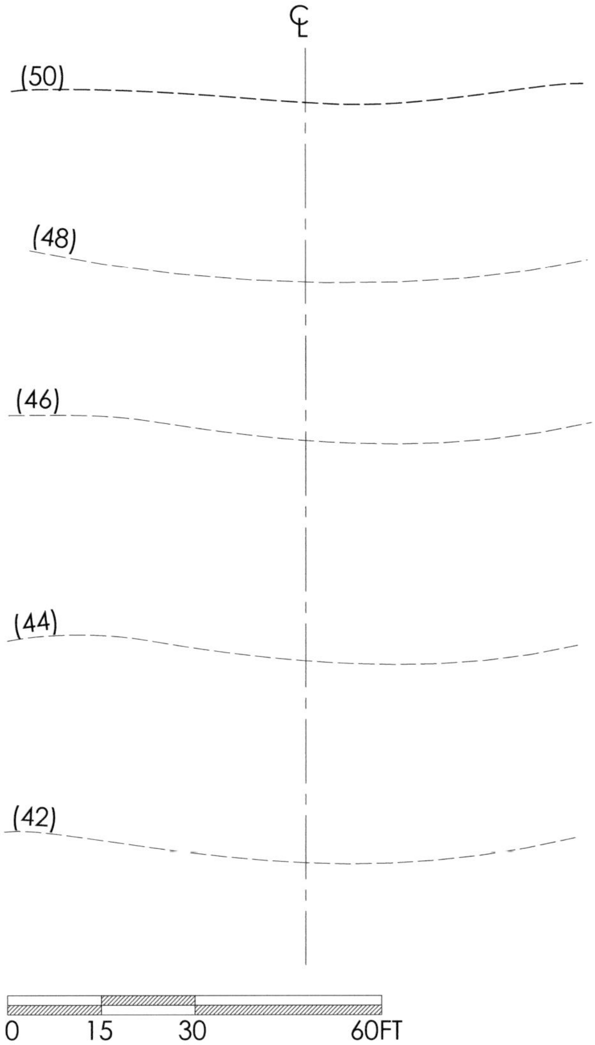

3-5) Draw a ridge with an 8% minimum slope and a 3:1 maximum slope.

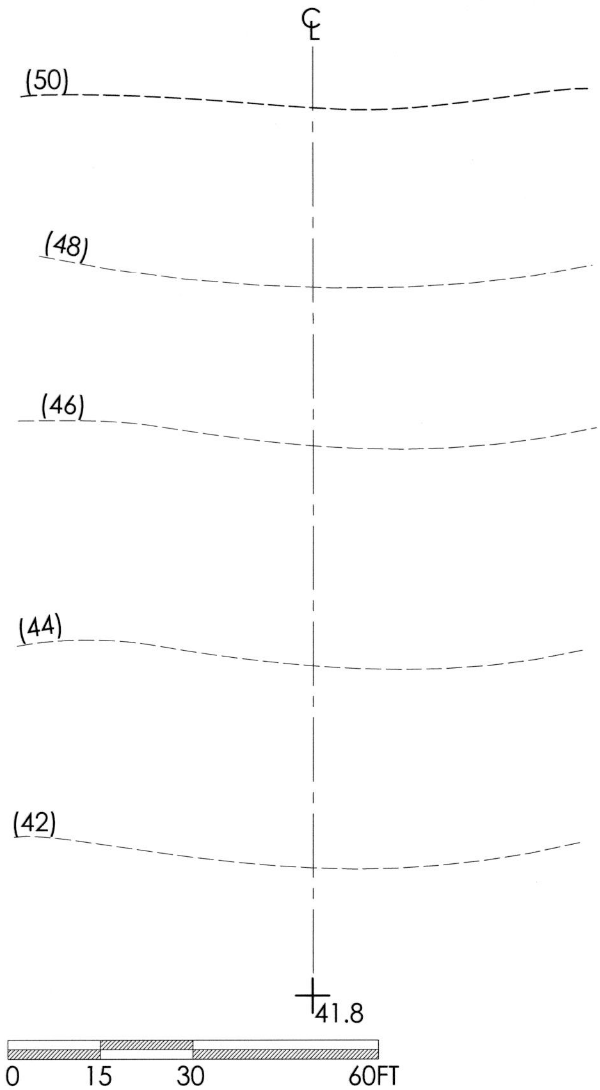

3-6) Draw a swale with a 3' deep retention pond.

a) Swale should have a minimum slope of 4% and a maximum slope of 5:1.
b) Bottom of the retention pond should be 323 and the sides should have a maximum slope of 5:1.
c) Label all parts of the pond and swale.

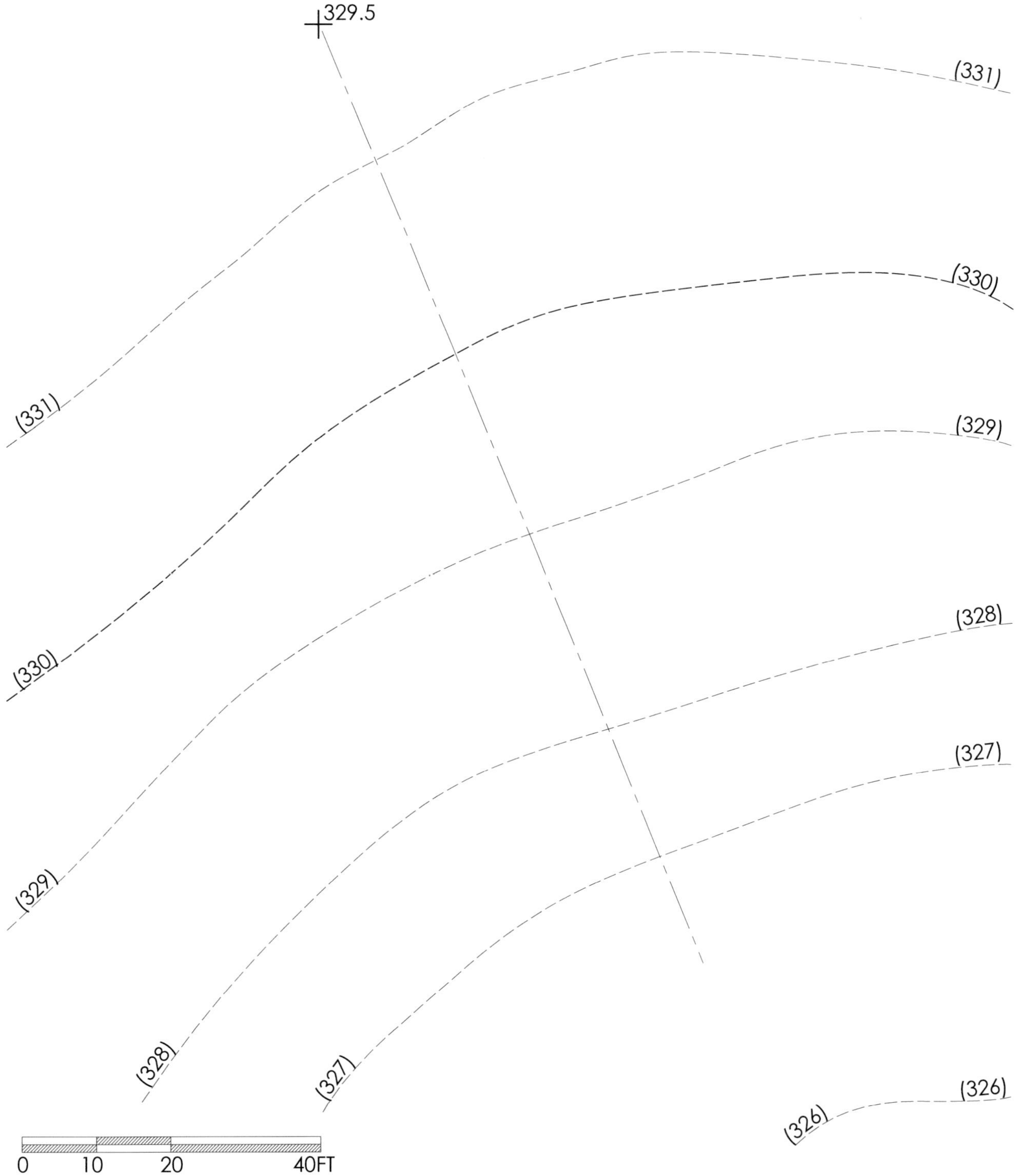

3-7) Draw a horseshoe swale around the concrete pad.

a) Provide a 10' buffer around the pad with a 2% slope.
b) Provide a HPS at 10' from edge of buffer at 2% slope.
c) Minimum slope is 2%; maximum slope is 10:1.

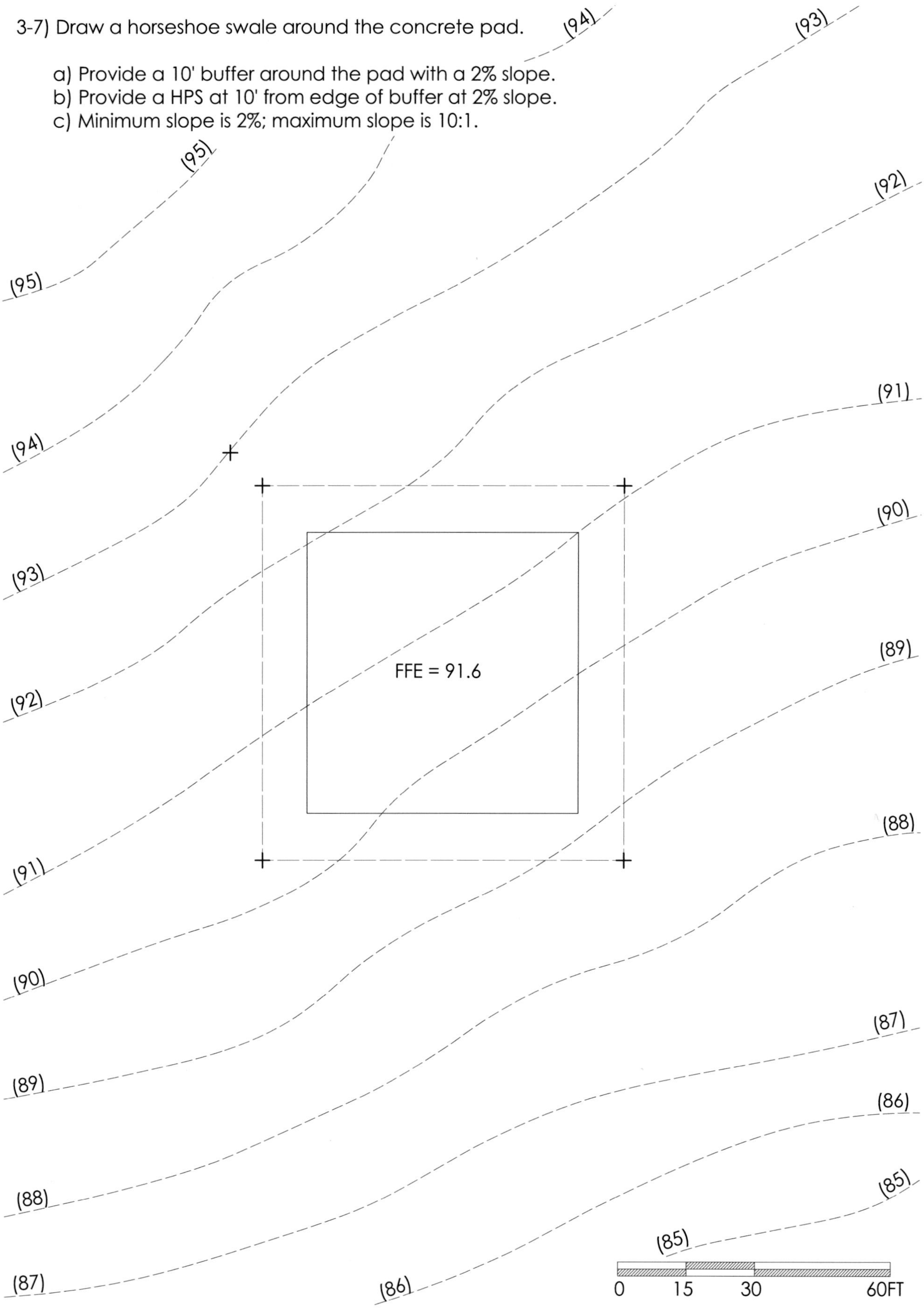

3-8) Draw a berm and swales.

a) Person standing at Point A has an eye height of 5'.
b) Minimum slope is 2%; maximum slope is 4:1.
c) Total building height is 10'.
d) Divert water to two drain inlets.
e) Label RIM EL for both drain inlets.

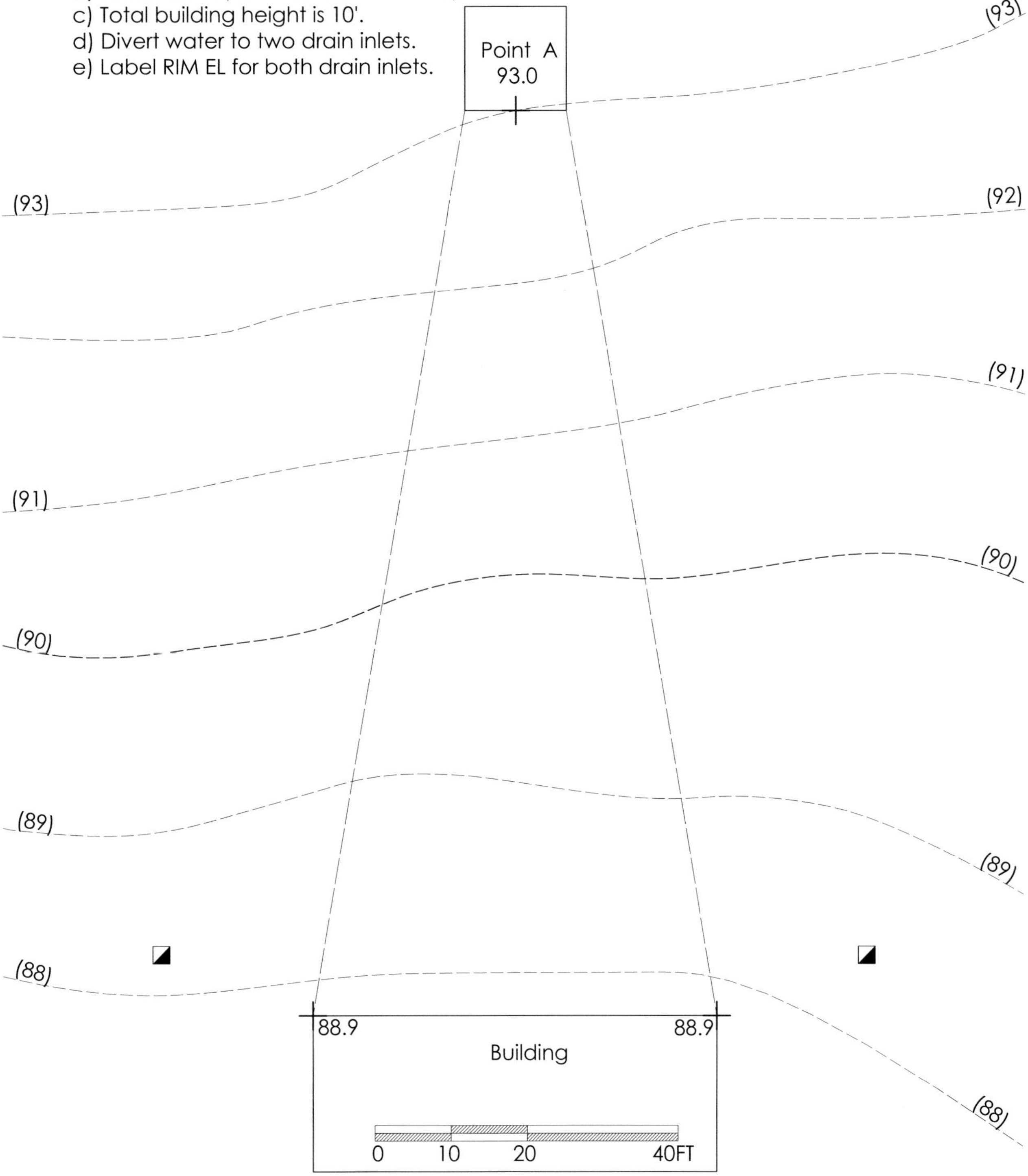

3.9 Literature cited

Harris, Charles W. and Nicholas T. Dines. 1997. *Time-Saver Standards for Landscape Architecture. 2nd Ed.* New York: McGraw-Hill.

Hopper, Edward J. ed. 2006. *Landscape Architectural Graphic Standards.* New York: John Wiley & Sons, Inc.

Strom, Steven and Kurt Nathan. 1998. *Site Engineering for Landscape Architects. 3rd Ed.* New York: John Wiley & Sons, Inc.

Strom, Steven, Kurt Nathan and Jake Woland. 2013. *Site Engineering for Landscape Architects. 6th Ed.* New York: John Wiley & Sons, Inc.

Untermann, Richard K. 1973. *Grade Easy: An Introductory Course in The Principles and Practices of Grading and Drainage*. McLean, VA: Landscape Architecture Foundation.

Section Four
Grading of built elements

The following section builds on the concepts learned in the previous sections, moving from man-made landscape elements to built structures. Regardless of the type of materials used to build these structures the principles of grading are the same. All of these elements have impermeable or semipermeable surfaces which require efficient grading to promptly divert surface water runoff.

4.1 Roads, shoulders, curbs, and sidewalks

In their simplest form, most roads are ridges, shoulders and sidewalks are plane surfaces, and curbs are mini walls. Roads are built as ridges in order to ensure fast surface drainage and to prevent ponding. Roads vary depending on whether they are located in rural or urban areas, see Figure 4.1–1. Many rural roads have a shoulder and a swale and all urban roads have a gutter and connect to a closed drainage system. Some urban roads have shoulders and curbs, suburban roads are a combination of both types.

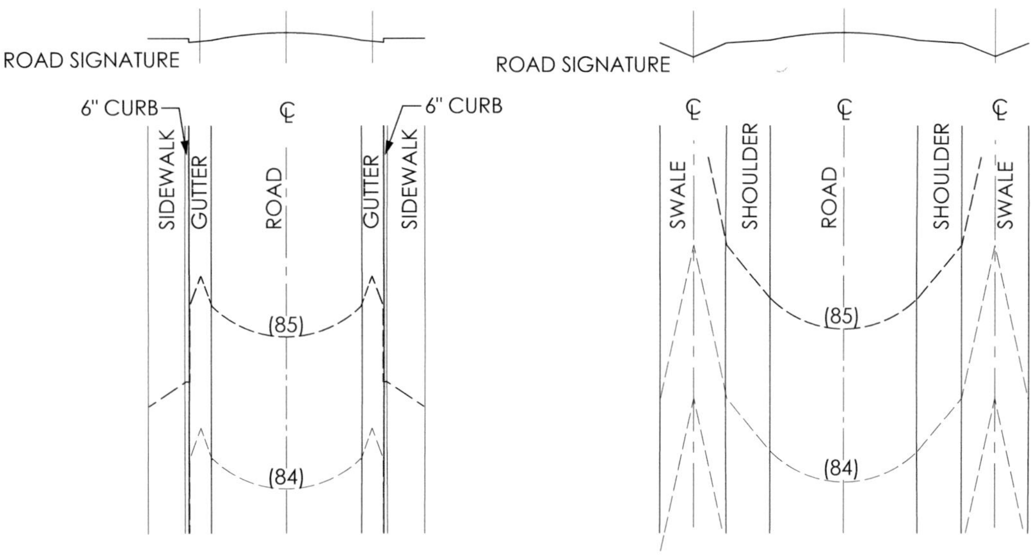

4.1–1
Urban road and rural road

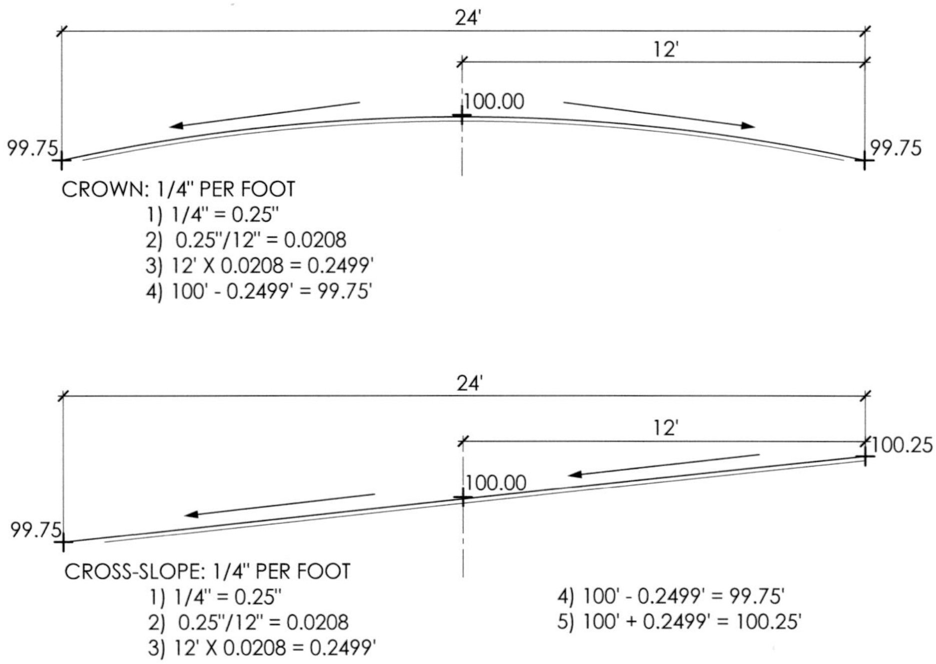

4.1–2
Crowned road and road with a uniform cross slope

Roads have three distinct parts: a longitudinal slope called a gradient, which determines whether a road is ascending or descending in relation to the surrounding topography; the centerline which is often a ridgeline called the crown of the road, and the cross slope which makes the road slope away from the crown on both sides or creates a uniform cross slope across the road.

Roads will have a crown, a reverse crown or a cross slope. From a crown, the width of the road will usually slope evenly on both sides. In Figure 4.1–2, for example, if a 24' wide road has a crown of 1/4" per foot then the road slopes evenly on both sides from the centerline at 1/4" per foot. A reverse crown is a swale with contours that point uphill. Although these are not as common in practice, they are sometimes shown on the LARE so make sure to confirm what is required by the problem. Most road problems have crowned roads that are ridges. Not all roads have crowns, some are sloped to one side. Roads that are pitched to one side have a cross slope. Grading a road with a cross slope is similar to grading a uniformly-sloped plane surface. The crown of a road may be called out in a variety of ways; as inches per foot, as a percentage, or in inches. Figure 4.1–3 illustrates the different calculations for the crown of a road.

It is often the case when you see a road in plan view that the contours lines are spaced evenly; this suggests a road that is rising or falling at a consistent gradient. If you see the contour lines of the road grow closer together or farther apart then you are looking at the effects of two different gradients combining to create a vertical curve. If you see the ridges of the road elongating on one side and contracting on the other then you are looking at the effects of a horizontal curve. Section 6 has additional information about horizontal and vertical curves.

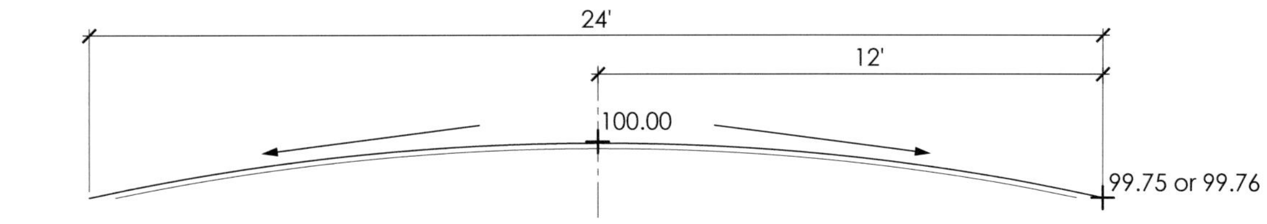

CROWN: 1/4" PER FOOT
1) 1/4" = 0.25"
2) 0.25"/12" = 0.0208
2) 12' X 0.0208 = 0.2499'
3) 100' - 0.2499' = 99.75'

CROWN: 3"
1) 3" = 0.25'
2) 100' - 0.25' = 99.75'

CROWN: 2%
1) 12' x 0.02 = 0.24'
2) 100' - 0.24' = 99.76'

4.1–3
Calculations for the slope of a crowned road

The one thing that is consistent with all roads is their uniformity. They remove the randomness of natural ridge topography and apply a very definite recognizable shape. It takes multiple steps to determine the contour lines for a road, but because of the road's uniformity the calculations only have to be done once and then the resultant contour line is copied for each subsequent contour line of the road.

Some grading problems will provide a road cross-section as illustrated in Figure 4.1–4. The cross section outlines the cross slopes of the different elements of the road and some will provide written text of the different longitudinal and cross slopes for the road, others still will provide only spot elevations from which you must determine the appropriate slopes.

Regardless of the way it is presented, the solution is essentially the same. Solve for each of the necessary parameters, draw one contour line, and use the slope interval for the road gradient to complete the subsequent contour lines along the road. The road gradient is often provided; however, sometimes two or more spot elevations are provided along the road and the longitudinal slope has to be calculated. The following example shows the steps for creating proposed contours for a road with a 4% gradient and its supporting structures, swales, curbs, gutters and sidewalks.

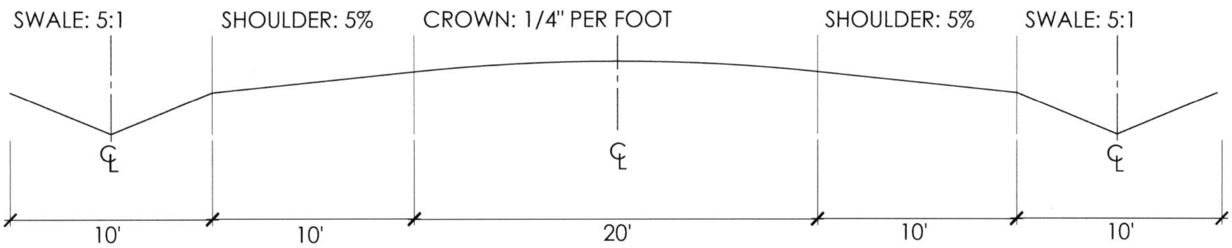

HORIZONTAL SCALE: 1x
VERTICAL SCALE: 2x

4.1–4
Typical road cross section showing crown, shoulder, and swale

Grading of built elements

1 Create a centerline down the middle of the road and using the slope formula and interpolation mark off whole number spot elevations.
2 Label these whole number spot elevations on the uphill side of the line so that the downhill direction is readily apparent, see Figure 4.1–5.
3 Draw a very light perpendicular line through the starting spot elevation across the width of the road include any curb, shoulder, swale or gutter areas on the grading problem. This guideline is used for the cross slopes of the different elements and will make copying the answer for each contour line easier. The road gradient will apply to all elements of the road.
4 Next, determine how much the topography will descend or ascend on the reference line from the crown of the road. Interpolate using the crown or cross slope for the road to determine the spot elevation at the edge of the road on both sides, see Figure 4.1–6.

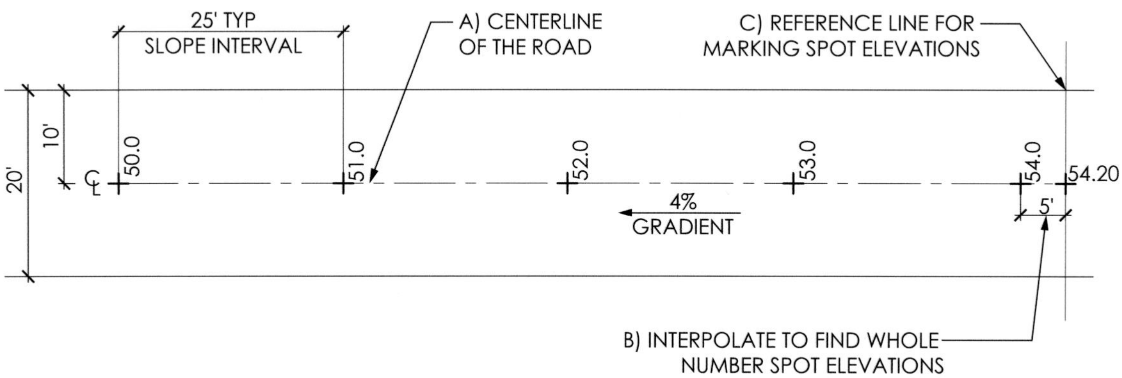

4.1–5
Road with centerline, light reference line, and whole number spot elevations

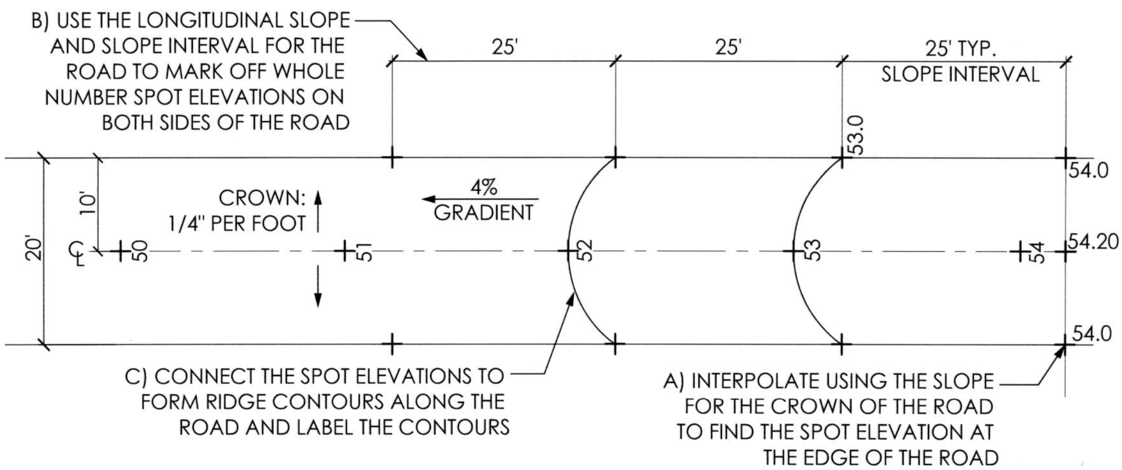

4.1–6
Road with spot elevations marked off along reference line

5 Mark the spot elevation at the edge of the road where it intersects with the reference line.
6 Determine where the whole number spot elevations occur along the edge of the road using the slope interval of the road gradient.
7 Mark off the whole number spot elevations for all whole number spot elevations on both sides of the road. Before completing the road contour line, determine if the road is asphalt or concrete. A concrete road will create a chevron topographically and an asphalt road will create a parabola, see Figure 4.1–7. Remember that a ridge points downhill. Use arrows to help visualize the direction of water flow and the correct curve for the crown of the road.
8 Connect the whole number spot elevations from the crown of the road to the edges of the road, creating contour lines for the road; use smooth, continuous lines.
9 The procedure is the same for the other elements along the road, the shoulder, the gutter and the swale. Using the reference line, calculate the vertical change for the shoulder or swale based on their cross slopes and add the spot elevation to the reference line, see Figure 4.1–8.
10 Use interpolation to find the nearest whole number spot elevations uphill or downhill from the reference line. Use the road gradient slope interval to quickly mark off the other whole number spot elevations along the shoulder or swale.
11 Connect the whole number spot elevations to the road contour line. Repeat for each contour line.

If a deep swale is required, make sure to mark off the whole number spot elevations along the reference line as well as along the centerline for the swale. In Figure 4.1–9 for example, if a slope of 3:1 is provided for a 2' deep swale at the edge of the road, make sure that between the spot elevations on the reference line you interpolate to determine the whole number contours. These will be used to help draw the swale. If a swale is not present a gutter may be provided to direct water into a closed drainage system; the calculation is the same as for a swale.

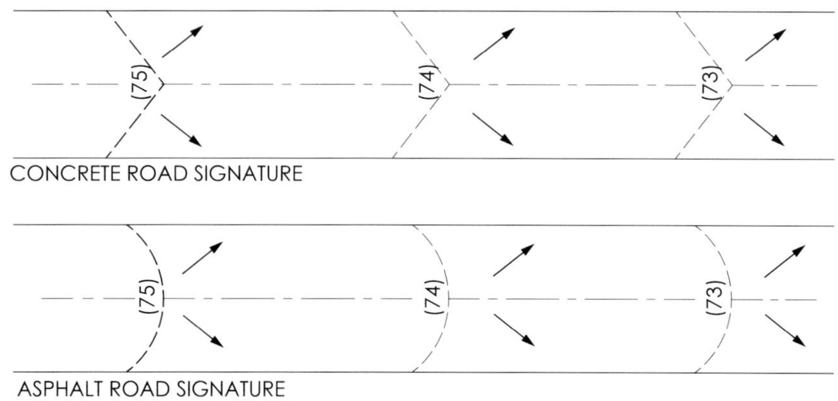

4.1–7
Concrete and asphalt road signatures

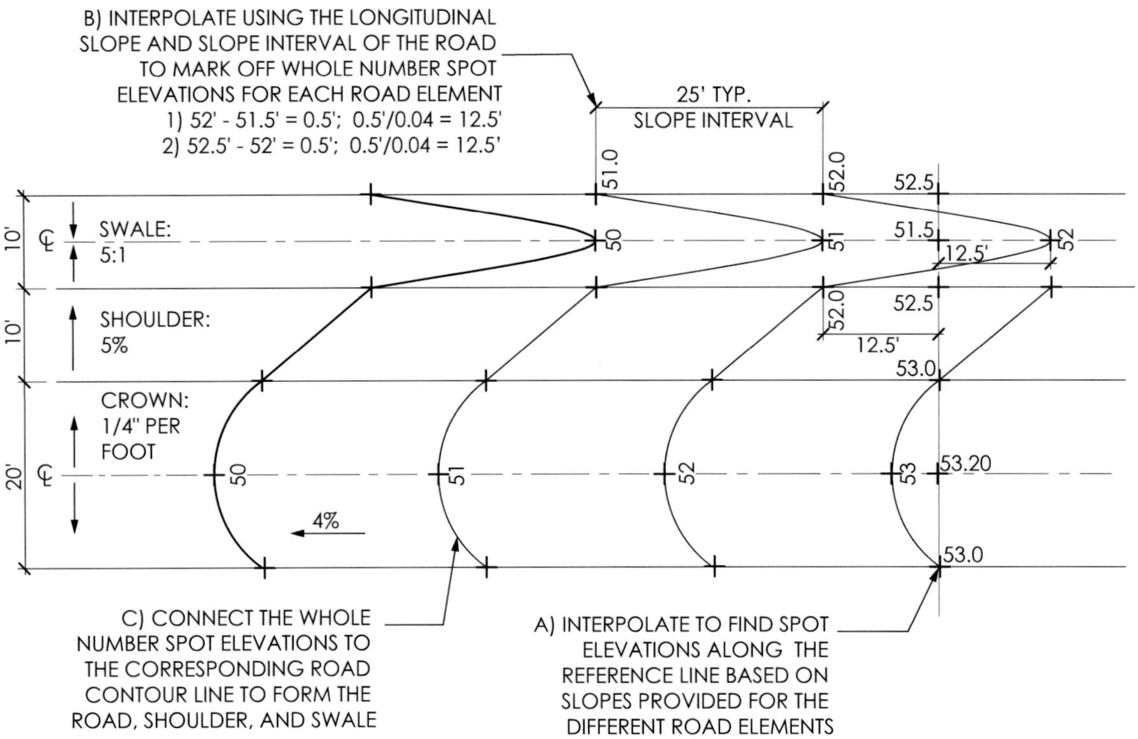

4.1–8
Road with shoulder and swale

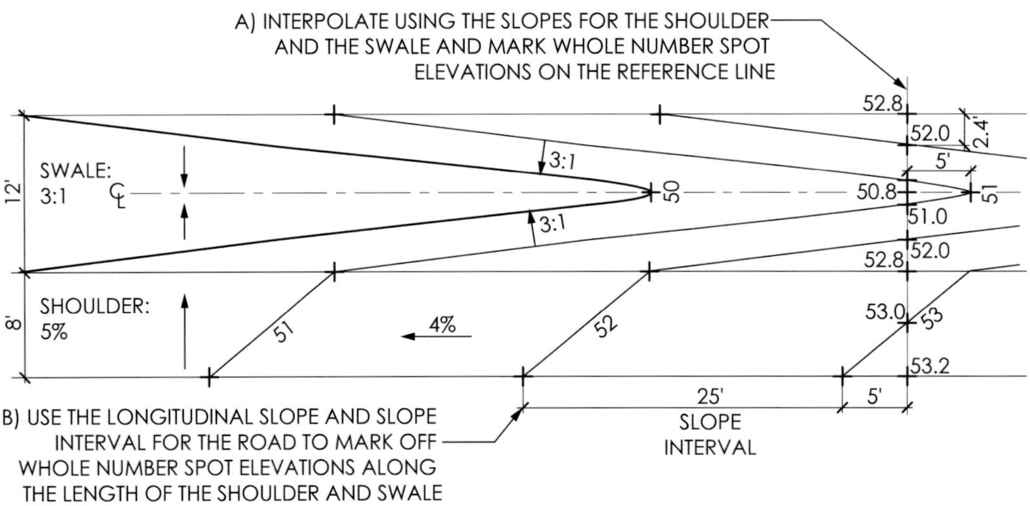

4.1–9
Two-foot-deep swale with a 3:1 slope

For a road problem with a curb, remember that the curb will have a specified height of 6"–8" and may or may not be connected to a sidewalk.

1 Convert the height of the curb into decimal feet and add the height to the spot elevation along the reference line at the bottom of the curb. Label these as TC and BC for clarity, see Figure 4.1–10.
2 Use interpolation and the road gradient slope interval to find the whole number spot elevations at the top of the curb. The proposed contour line will ride along the front face of the curb until it meets the elevation at the top of the curb, see Figure 4.1–11. Show the connection of the contour line along the length of the curb.

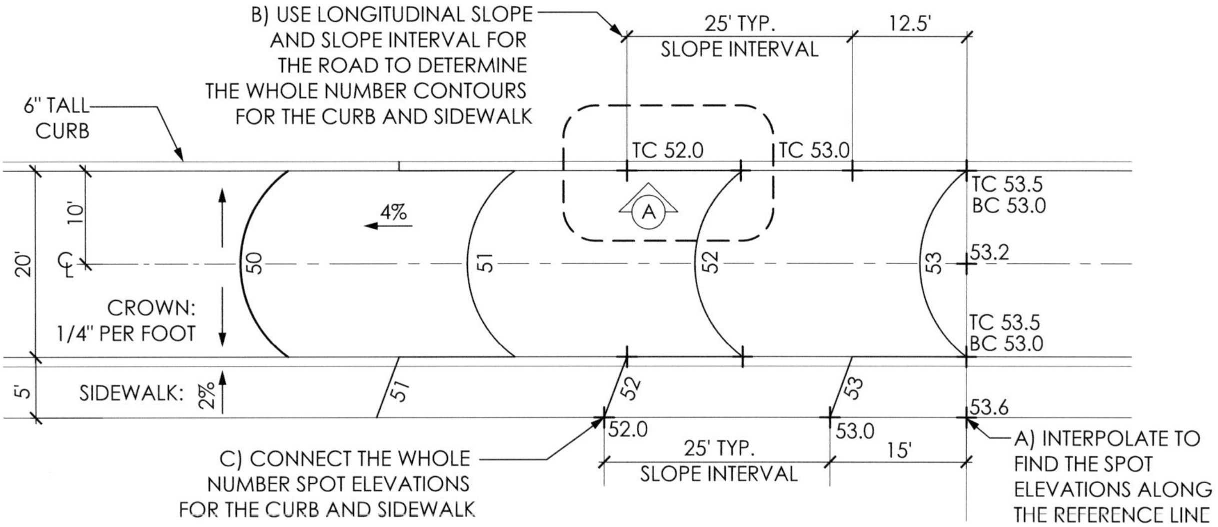

4.1–10
Road with curb and sidewalk

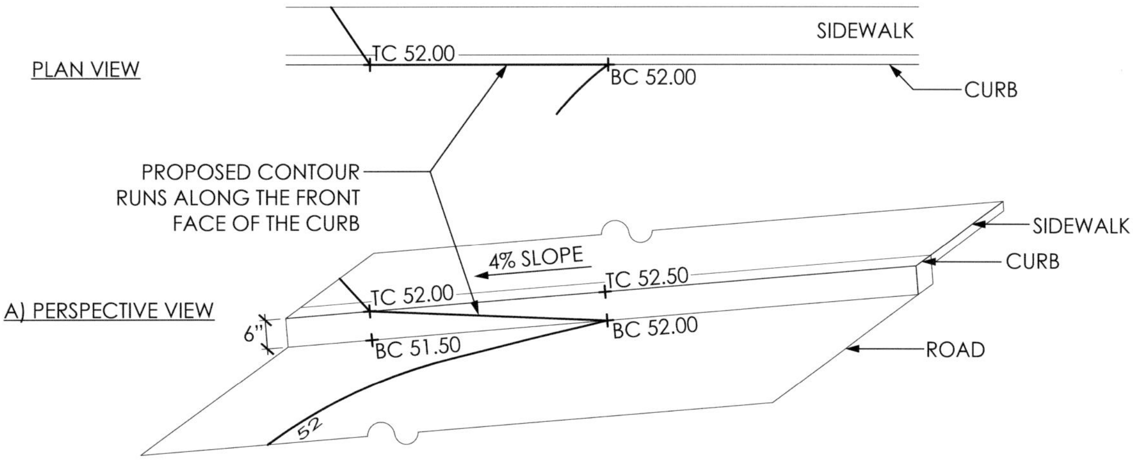

4.1–11
Proposed contour rising along the face of a curb

3 If the curb is attached to a sidewalk, use the sidewalk's cross slope and the road gradient (or the sidewalk slope if it differs from the road) to determine the contour lines on the sidewalk. Sidewalks usually pitch towards the curb and gutter.

When you have completed several of these problems you will begin to recognize the signature and develop a rhythm for solving them. It is always a good idea, if a road section is not given, to draw one so that you know what you're aiming for as a final grading solution. Always double-check your math and specifically the slope and slope interval calculations because one mistake will be compounded in all the subsequent contour lines. When presented with these types of problems, the starting point is always the centreline of the road working out from the crown of the road. The road problem is very rarely shown as a separate problem, warped planes such as parking lots and driveways are often shown together when solving a road problem. These will be covered in Section 5.

4.2 Paths

In addition to sidewalks, there are multiuse paths and walkways that are not tied to road structures. The standard widths for paths are 4' wide for one-way travel, 5' wide minimum for two-way travel with a preferred width of 6' and 8'–12' wide for a multiuse path. Paths make it easier for people to traverse a landscape. In addition, guidelines set by the Americans with Disabilities Act of 1990 provide standards for minimum and maximum slopes for accessible paths. Refer to the current ADA requirements at www.ada.gov.

As with roads, uniformity of slope on a path makes solving a path problem easier. A path has a longitudinal slope that follows the surrounding ascending or descending topography and a cross slope or a crown that enables sheet drainage of surface water. A concrete path has a cross slope and an asphalt path has a crown. The following example shows the process of finding the slope along a concrete path.

1 Paths and roads are started with a centerline using horizontal alignment for layout. Determine the path width and note the longitudinal slope. Draw a centerline for the path.
2 Starting at the given spot elevation, determine if you are going downhill or uphill. Use interpolation to locate the nearest whole number spot elevation using the path's longitudinal slope.
3 Use the slope interval for the longitudinal slope to mark off each subsequent whole number spot elevation along the centerline as shown in Figure 4.2–1. Label each spot elevation on the uphill side.
4 To ensure sufficient drainage of water off of the path use 0.75–2% slope for the path cross slope. The standard is 2%, this will ensure sheet drainage of surface water. Determine to which side of the path you want to drain surface water. On a grading problem this will be determined by the existing topography, landscape structures, and plants that can't be moved or changed. Note that it is best to drain water to the lower side of the existing topography.

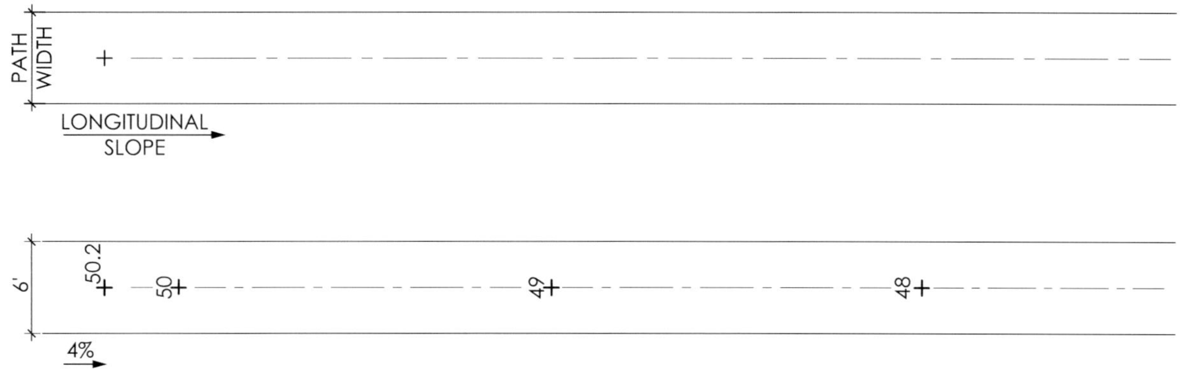

4.2–1
Path centerline with the location of whole number spot elevations

5 With spot elevations along the centerline, draw a light perpendicular reference line through the first whole number spot elevation across the width of the path as illustrated in Figure 4.2–2.
6 Using the path cross slope determine the spot elevation on both edges of the path along the reference line. One side will be higher than the other.

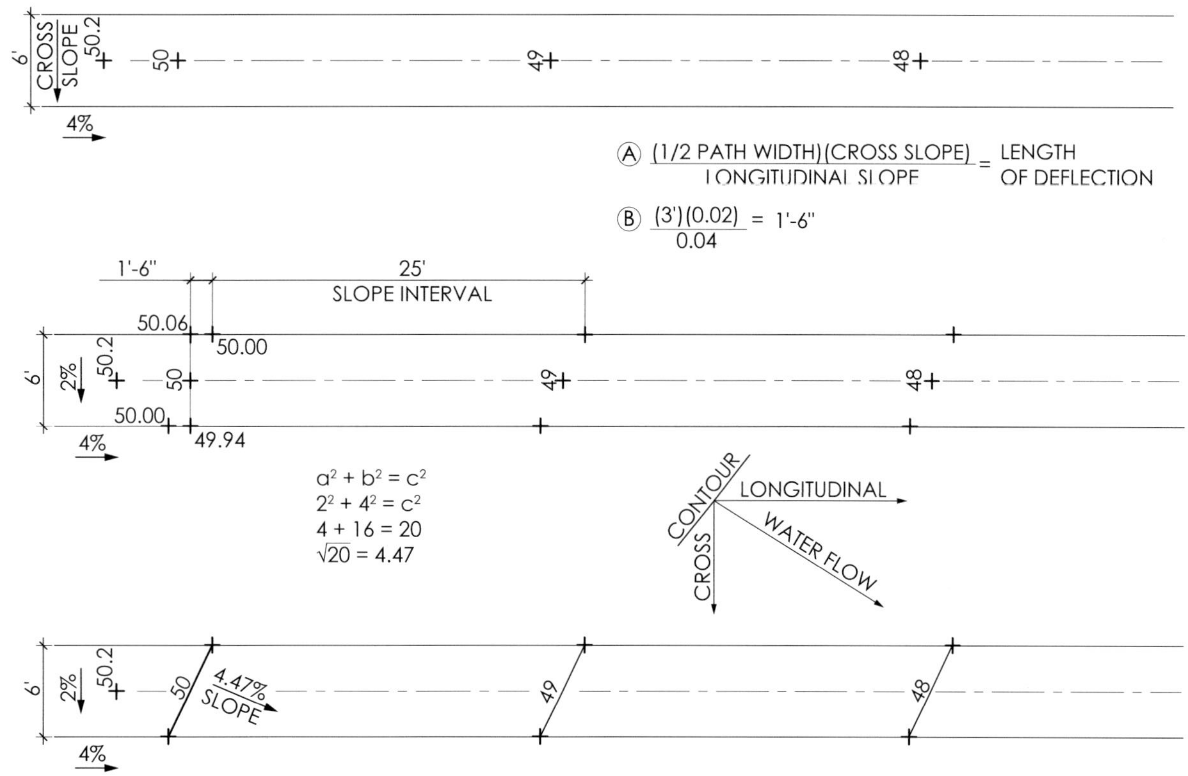

4.2–2
Using cross slope and longitudinal slope to create proposed contours on a path

7 Using the longitudinal slope for the path, interpolate to find the whole number spot elevation at the edge of the path that corresponds to the whole number spot elevation on the centerline.
8 Connect the whole number spot elevations to form the proposed path contour. Double-check that the proposed contour is angled in the correct direction to ensure surface water will drain off the path in the direction that you want. Remember that water flows perpendicular to the contour. Use arrows to clearly show the direction of water flow across the path.
9 Using the slope interval for the longitudinal slope of the path, mark off the location of all the other whole number spot elevations along the edges of the path.
10 Connect the proposed contours across the path.

You can simplify Steps 6 and 7 with the following equation: 1/2 path width × path cross slope divided by the path longitudinal slope. This will give you the length of deflection or horizontal distance uphill or downhill from the reference line to the nearest whole number spot elevation on the edges of the path. This equation is most useful when drawing a curving path. It eliminates the need to use the longitudinal slope interval to find whole number spot elevations.

If necessary, you can determine the slope across the path using the Pythagorean theorem, $a^2+b^2=c^2$. The resultant c value will be the slope that the surface water will take across the path. Note that this value should be within the limits of the minimum and maximum values set in the grading problem.

There are two ways to add a proposed accessible path to a landscape. The following example uses the slope interval of a given longitudinal slope to locate a path. The resultant path allows for minimal disturbance of the surrounding landscape to tie the proposed path back to the existing topography.

1 Calculate the total vertical change that the path will take through the topography from Point A to Point B, 79.3' − 74' = 5.3'.
2 Use the maximum allowable slope for the path to determine the minimum required length of path needed, 5.3'/0.04 = 132.5'.
3 Calculate the slope interval for the maximum allowable slope, 1'/0.04 = 25'.
4 Measure and mark the slope interval as the horizontal distance between the existing contours as shown in Figure 4.2–3. This distance may be perpendicular to the existing contours or it may be angled. This line is the centerline for the proposed path.
5 Draw the width of the path and determine the direction of the cross slope for the path based on the surrounding topography and vegetation.
6 Draw a light perpendicular reference line across the width of the path at each contour line.
7 Use the length of deflection equation to determine the location of the whole number spot elevations along the edges of the path as shown in Figure 4.2–4. Remember to check the direction of the cross slope for the path and orient the proposed contours to drain water off of the path.

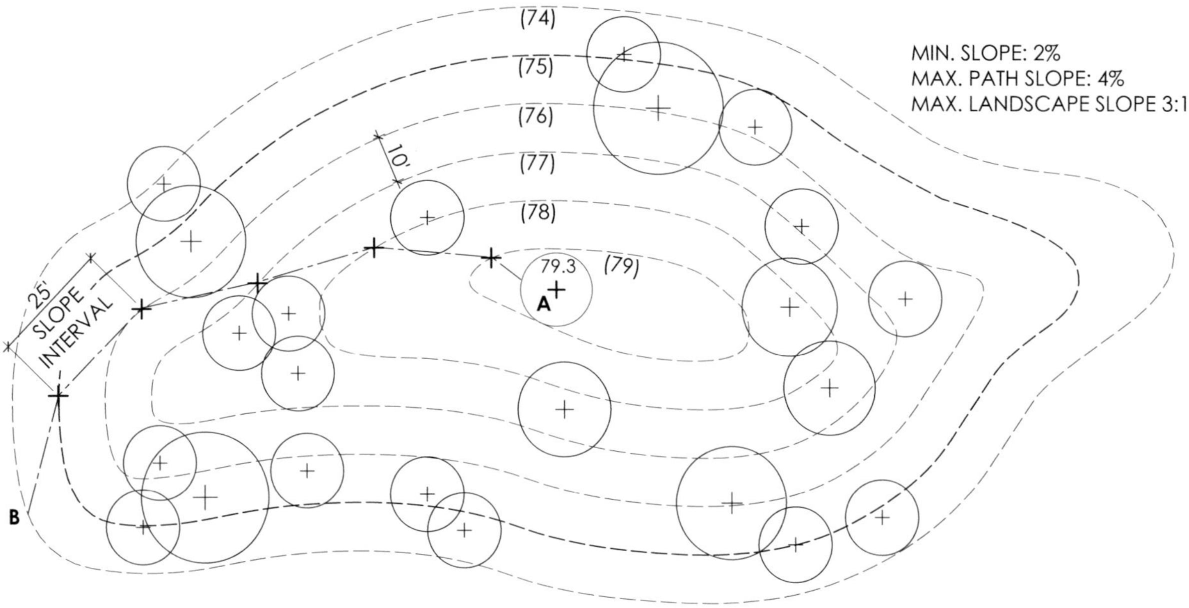

4.2–3
Path centerline drawn with longitudinal slope interval

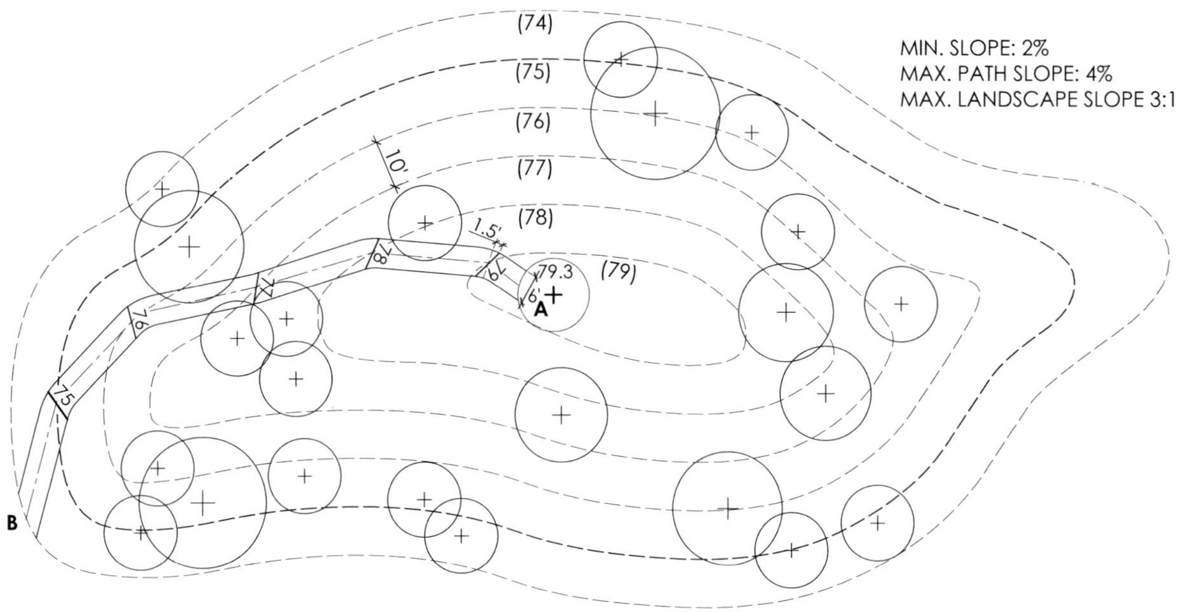

4.2–4
Path width and calculation of proposed contours

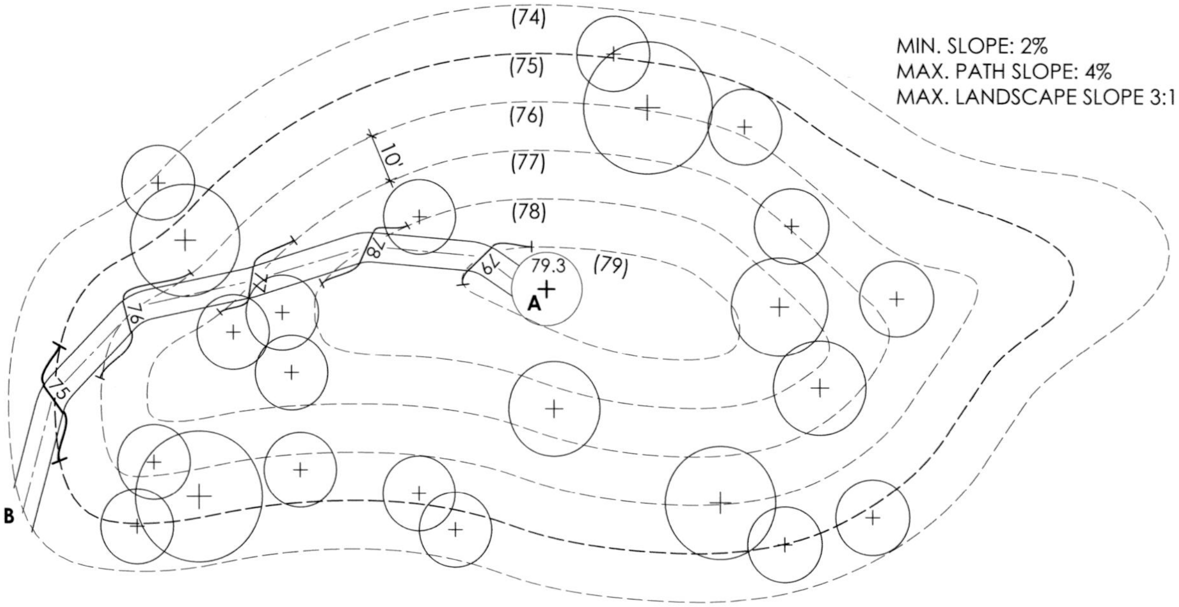

4.2–5
Completed path with proposed contours

8 Connect the spot elevations to show the proposed contours for the path.
9 Connect the proposed path contours back to the existing contours using the maximum slope for the topography, see Figure 4.2–5.

The second method of adding a proposed accessible path to the landscape is to create a path length that falls within the range of maximum and minimum allowable slope for the path. This allows you to create a path that fits into the landscape more efficiently. In practice, this method may generate greater disturbance of existing topography and it will require calculations for horizontal and vertical curve alignment. Ample room needs to be available to adjust the topography at the maximum allowable slope without disturbing existing landscape features. Refer to Section 6 for information about curve alignment.

1 Calculate the total vertical change that the path will take through the topography from Point A to Point C, 79.3'−74' = 5.3'.
2 Using minimum and maximum allowable slopes, determine the maximum and minimum required path lengths, max: 5.3'/0.02 = 265'; min: 5.3'/0.04 = 132.5'.
3 Measure and mark a path length and location based on the existing topography as shown in Figure 4.2–6, 139.14'. This is the centerline of the proposed path. The path length falls within the range of the minimum and maximum path lengths.
4 Calculate the slope for the desired path using the length of the proposed path and the vertical change between Points A and C, 5.3'/139.14' = 3.81%.
5 Calculate the slope interval for the path and mark off the location of the proposed contours along the centerline as shown in Figure 4.2–7, 1'/0.0381 = 26.25' → 26'.

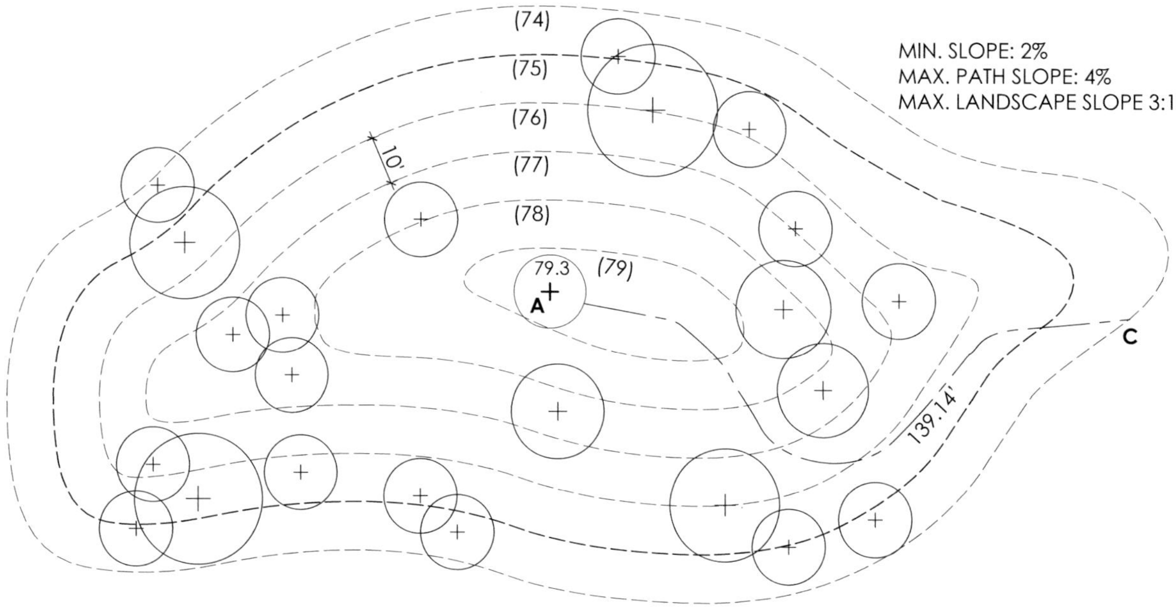

4.2–6
Path length that falls within the range of minimum and maximum allowable path length

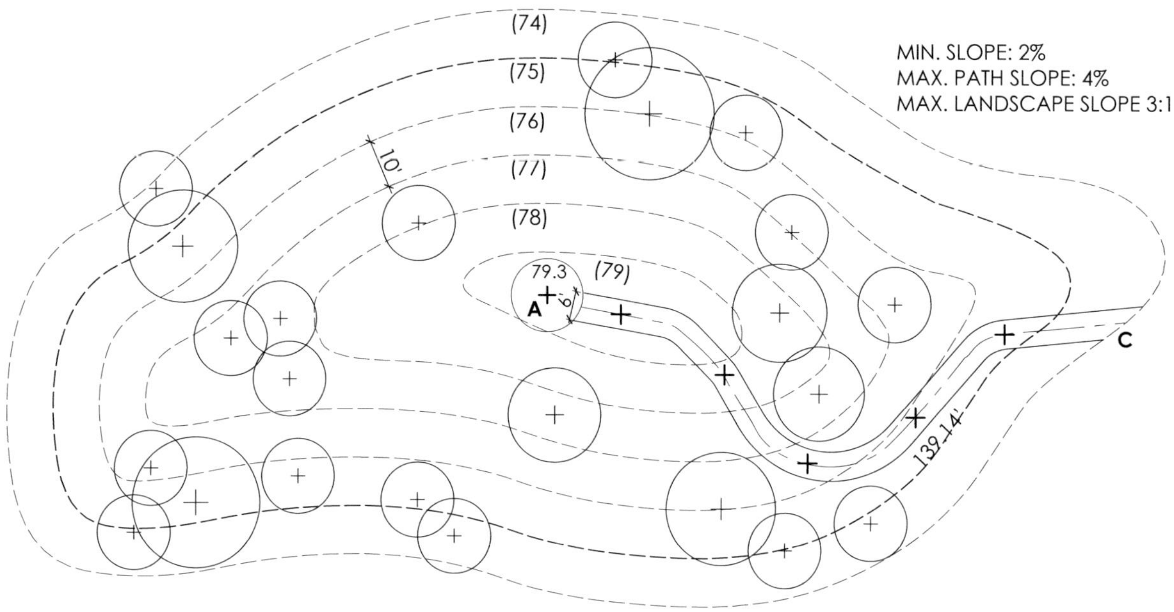

4.2–7
Path showing location of proposed whole number spot elevations

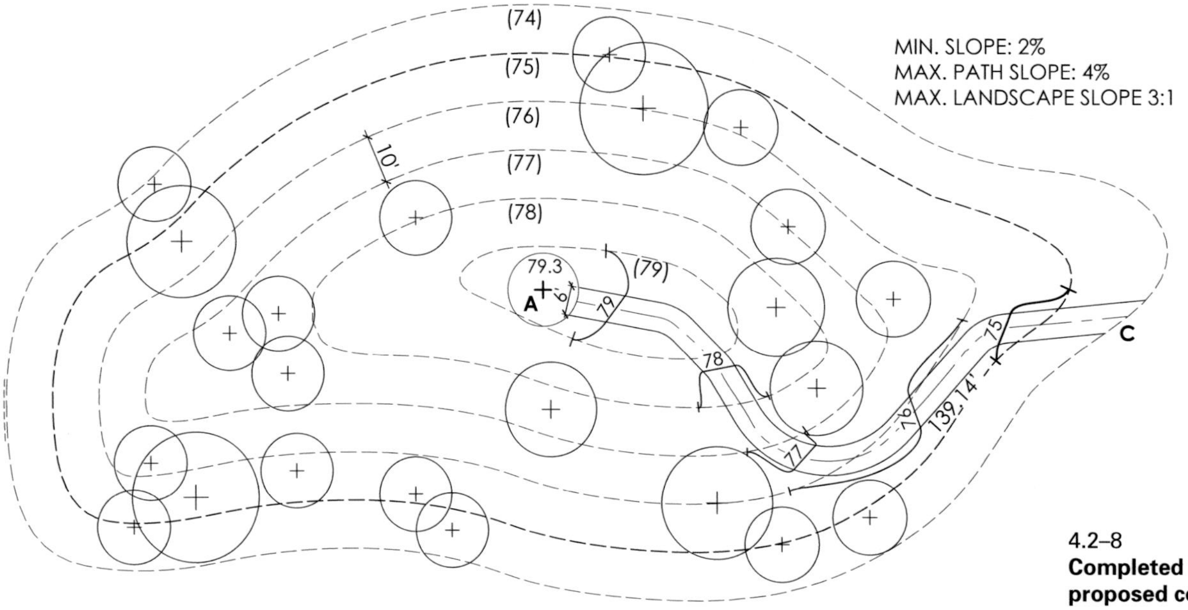

4.2–8
Completed path with proposed contours

6 Draw the width of the path and determine the direction of the cross slope for the path based on the existing topography and vegetation.
7 Draw a light perpendicular reference line across the path and using the length of deflection equation determine the location of the whole number spot elevations on both sides of the path.
8 Connect the spot elevations to show the proposed contours for the path and connect the proposed contours back to the existing contours using the maximum slope for the topography as shown in Figure 4.2–8.

Concrete paths in the landscape are rarely flush with the surrounding topography. They are usually elevated by 1"–3" above the finished grade to prevent surface runoff from the landscape flowing over the path, see Figure 4.2–9. When drawing the proposed contours for a path there is often a stagger between the path and the surrounding landscape. This is miniscule in most grading problems, however it is useful to understand that the edge of the path acts as a curb. It is not necessary to add a swale at the edge of the paths, the edge of the path is a sufficient barrier.

4.3 Stairs

Stairs and ramps provide the same function within the landscape. They move people from higher to lower elevations when the slope of the land exceeds 5%. Stairs are generally more efficient than ramps, taking up less room in the landscape. Stairs and ramps in the landscape are governed by two sets of guidelines. The 2010 ADA Standards for Accessible Design (ADA) as set forth in the Americans with Disabilities Act of 1990 covers the basics of stair design and the 2015 International Building Code (IBC)

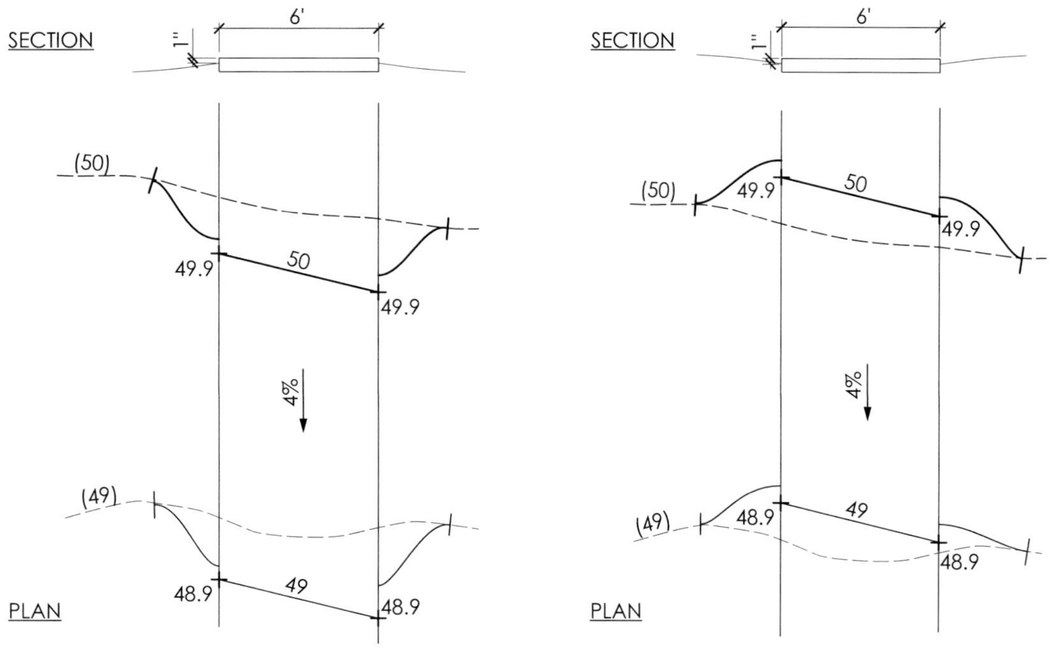

4.2–9
Plan and section of proposed paths with surrounding existing topography

looks specifically at stairs for emergency egress from buildings. For a full understanding of the rules for building stairs and ramps please see the recommended reading.

Stairs and ramps may be designed as separate structures, or they may be combined in a design at specific intervals. The former is easier to do, and the latter is based largely on aesthetics rather than functionality. For the purposes of this book, we will use the rules that most generally apply to all stairs and ramps.

Stairs consist of four parts: risers, which provide the change in vertical distance, and treads, which provide the change in horizontal distance, together these create a step; landings, which occur between flights of stairs and at the top and bottom of a staircase; and handrails along both sides of a stair. Stairs have a longitudinal slope but not a cross slope. They will always be level at the top and bottom of the stairs.

A flight of stairs in the landscape should have a minimum of 3 steps and ideally no more than 10 steps or 5' of vertical change before a landing to allow pedestrians places to stop and rest. Steps vary in height from a minimum of 4" to a maximum of 7". All steps in a flight of stairs must be of equivalent height; in other words, the riser height must be uniform for the flight of stairs. For example, if you have two flights of stairs, one may have steps that are all 6" in height and the other may have steps that are all 5" in height. Most landscape steps in the US are 6" tall, largely because it is easier to calculate vertical distance based on 0.5' increments.

Treads vary in length from a minimum of 11" to a maximum of 18" for standard stairs. The most common tread length is 12", again because it is easier to calculate the horizontal distance based on 1' increments. All treads have a wash of 2% (1:48). A wash is simply a slope from the back corner of the tread to the front toe of the riser. The wash prevents tripping, and ponding of water on the stairs. When calculating the height of a flight of stairs, this 2% wash does not need to be figured into the calculations since the height of the step includes the 2% wash. A common formula of 2R+T=24"–26" is used to give a workable riser height to tread length. Note that tread lengths can exceed 18" and be several feet apart for decorative stairs.

To determine the total number of steps in plan view, count the number of solid lines in the flight of stairs, each line corresponds to a step. Multiply the total number of lines by the height of one riser and you will know the vertical distance covered by the stairs. Multiply the total number of lines by the tread length and you will know the horizontal distance covered by the stairs. Note that the dashed line shown in Figure 4.3–1 and subsequent stair images throughout this book is not an additional step, it represents the expansion joint between the top back edge of the stairs and the adjacent landing. It is shown to add clarity.

Stair landings are evenly spaced between flights of stairs and at the top and bottom of stairs. They are a minimum of 4' long on a straight run of stairs and should be the same width as the stairs. Landings have a longitudinal slope allowing surface runoff to flow down the stairs. In practice, a landing may be a warped plane and slope towards a drain in the center of the landing. Current ADA requirements recommend a maximum slope on a landing of 1:48 or 2%. A slope between 1.5% and 2% should drain surface water efficiently from a landing.

Handrails are required for all flights of stairs with more than three steps. The number of handrails needed for a flight of stairs varies based on the width of the stairs. One

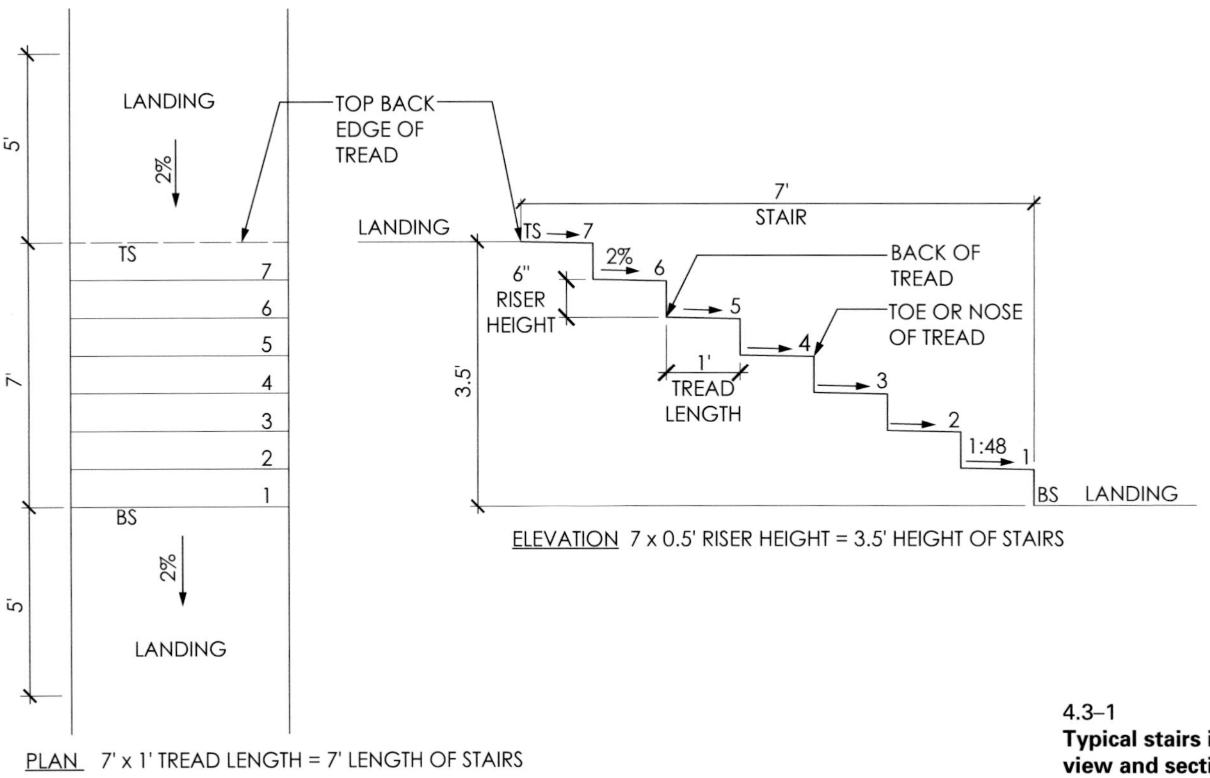

4.3–1
Typical stairs in plan view and section

handrail is required for every 2.5' (30") of stair width. Therefore stairs that are greater than 6' wide will require an intermediate handrail, see Figure 4.3–2. The stair handrail is attached to the stair or cheek-wall and extends one tread length past the bottom of the stairs and 1' at the top of the stairs. The height of the handrail should fall between 34"–38" above the risers. See the recommended reading for additional guidelines on width and design of handrails. All the elements of a stair: risers, treads, landings, handrails and cheek-walls (if permitted) must be shown on the grading solution.

In addition to the riser, tread, landing and handrail components that make up all stairs, there are other elements for stairs that must be shown on the grading solution. A cheek-wall may be permitted on both sides of the stair. If the grading problem allows a cheek-wall, draw one. It is likely that the proposed contour lines that should be drawn for the problem will need to be tied into the cheek-wall. A cheek-wall ties into the stair and generally extends above and below it.

Always label the top and bottom of the stair spot elevations with top of stair, TS, and bottom of stair, BS. These spot elevations plus the number of steps will enable you to calculate spot elevations for each stair if needed. Don't show contour lines along the face of the stairs in plan view; this can be misconstrued as an additional step. Contour lines converge with the sides of the stair or cheek-wall and are picked up on the other side of the stair or cheek-wall.

To solve for a flight of stairs, first determine if stairs are needed. In the area that has been demarcated for the stairs, you will either have to determine spot elevations or you will be provided with the top elevation and bottom elevation for the distance you need to cover. In the example in Figure 4.3–3, the change in elevation is 7.4' and the horizontal distance for that change is 34'. The riser height and tread length for each step are 0.5' and 1.0' respectively.

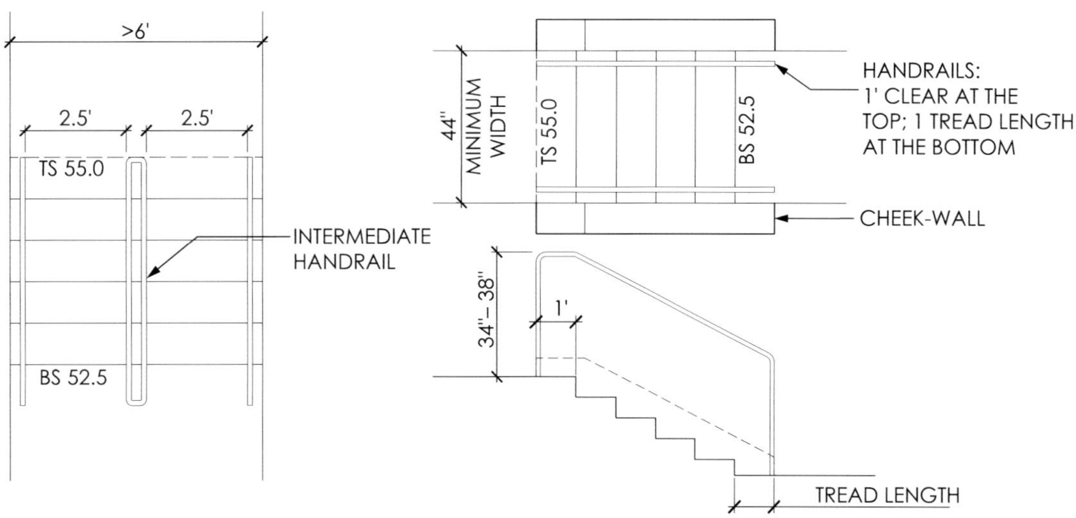

4.3–2
Stairs with handrails and cheek-wall

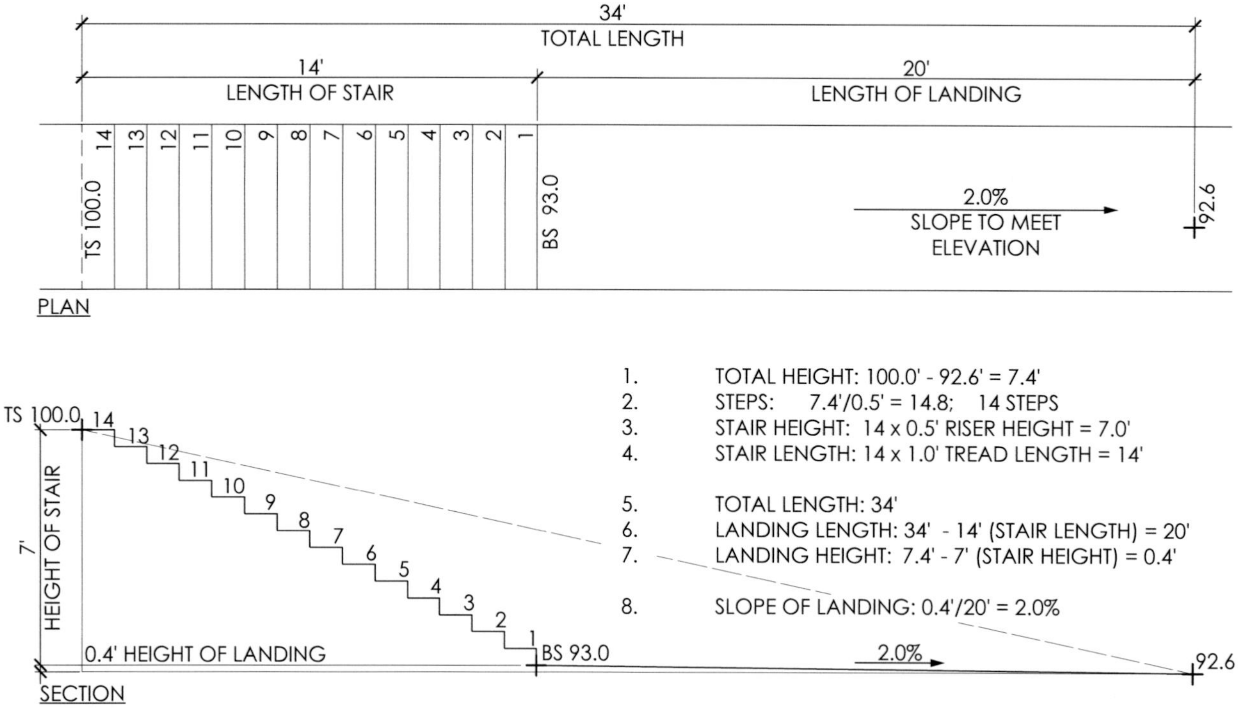

4.3–3
Steps required to meet a vertical change of 7.4'

1 Calculate the overall slope for the area and make sure it is greater than 5%. If it is less than 5%, you won't need either stairs or a ramp, anything greater will require both, 7.4'/34' = 0.217647 or 21.8% slope.
2 To determine the number of steps needed for the stairs divide the change in elevation by the height for one riser, 7.4'/0.5' = 14.8 or 14 steps.
3 Multiply the number of steps by the riser height to determine the total vertical change that will be encompassed by the stairs, 14 steps × 0.5' = 7'. The remaining 0.4' of vertical change will be used to determine the slope for the landings.
4 Multiply the number of steps by the tread length to determine the total horizontal distance that will be encompassed by the stairs, 14 steps × 1.0' = 14'.
5 Subtract the total length of the stairs from the total horizontal distance provided to determine the landing length, 34' − 14' = 20'.
6 Calculate the slope for the landing length. Divide the remaining height by the remaining length to determine the slope of the landing, 0.4'/20' = 0.02 or 2.0%.
7 The total step count is 14 steps with a 20' long landing sloped at 2.0%.

The general rule is to provide no more than 10 steps per flight of stairs; therefore, you will need to divide the stairs into two or more flights. The decision to divide the stairs will be determined by the amount of room provided for a grading solution. Avoid dividing the stairs based on aesthetic reasons; choose the simplest solution for the problem and divide them as evenly as possible. For example, it is better to use two flights of 7 steps; or two flights of 5 steps and one flight of 4 steps rather than to have one flight of 4 steps and one flight of 10 steps.

Each landing must be a minimum of 4' long. For two flights of stairs, we will use two 5' long landings at the top and bottom of the stairs and a 10' long landing between the stairs. For three flights of stairs we will use four 5' long landings, see Figure 4.3–4.

For the two flights of stairs consisting of 7 steps, the vertical change is 3.5' for each flight of stairs. The top and bottom landing will be 5' long and have a vertical change of .1' each. The 10' long landing will have a vertical change of 0.2'. For the three flights of stairs, one flight of stairs will have 2' of vertical change and the other two will have 2.5' of vertical change. Each of the four landings will be 5' long with a vertical change of 0.1'. It is helpful to calculate the vertical distance and the horizontal distance for each flight of stairs prior to drawing a solution. Quick sections or a table as shown in Figure 4.3–4 will enable you to visualize the calculations.

A second stair problem, illustrated in Figure 4.3–5, uses the same principles as the first example but with added complexity. It is unlikely that a grading problem on the exam will use a complex stair problem since the goal of the examination is to test

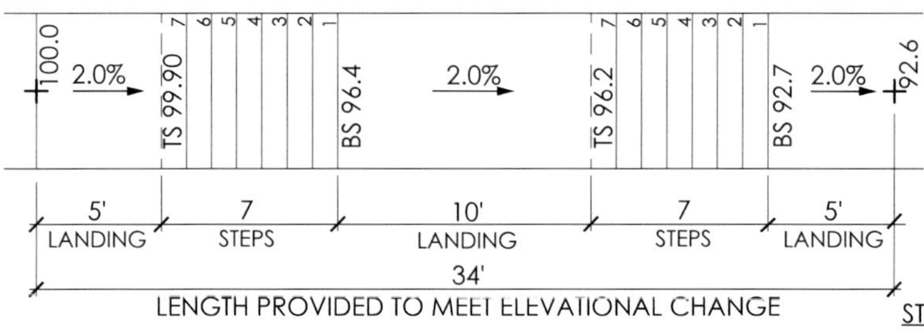

	STAIRS (7 STEPS)	STAIRS (7 STEPS)	LANDINGS 1	LANDINGS 2	LANDINGS 3	TOTAL
VERTICAL	3.5'	3.5'	0.1'	0.2'	0.1'	7.4'
HORIZONTAL	7'	7'	5'	10'	5'	34'

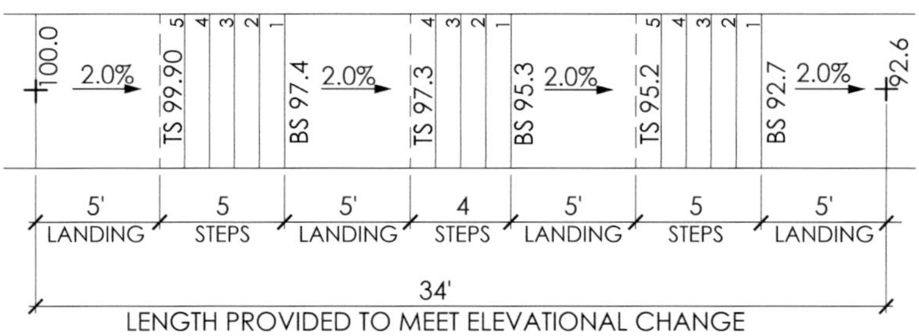

	STAIRS (5 STEPS)	STAIRS (4 STEPS)	STAIRS (5 STEPS)	LANDINGS 1	LANDINGS 2	LANDINGS 3	LANDINGS 4	TOTAL
VERTICAL	2.5'	2'	2.5'	0.1'	0.1'	0.1'	0.1'	7.4'
HORIZONTAL	5'	4'	5'	5'	5'	5'	5'	34'

4.3–4

Evenly divided steps: two and three flights of stairs

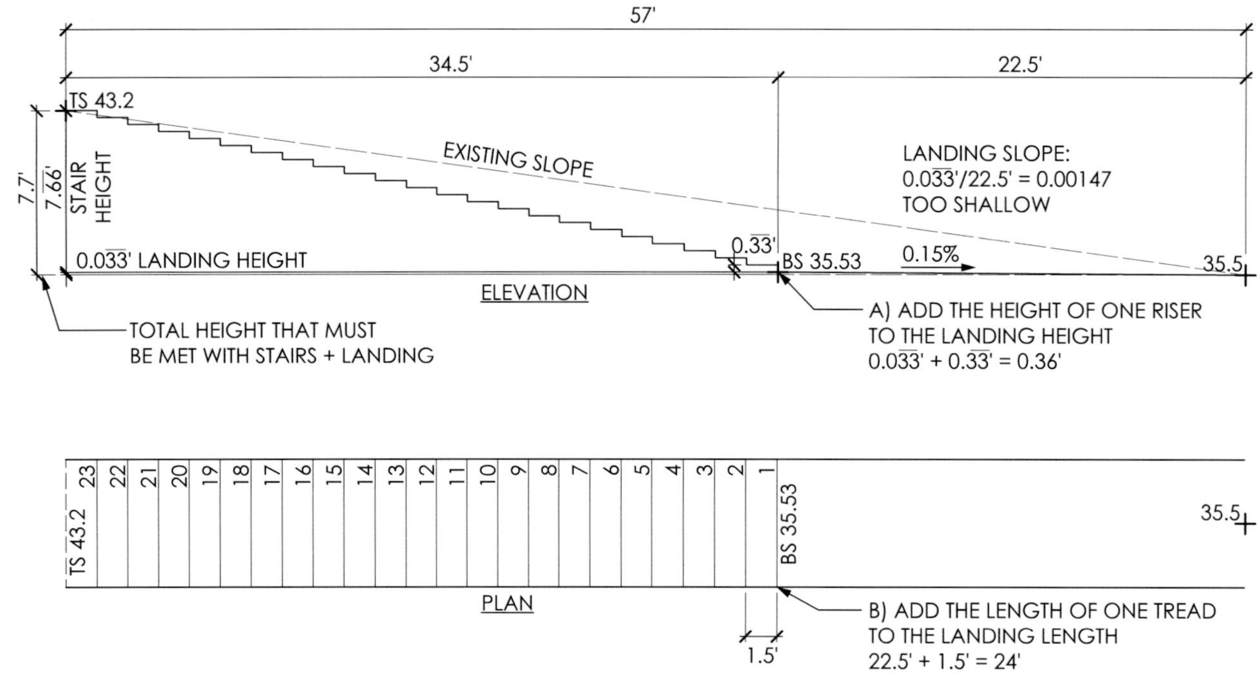

4.3–5
Twenty-three steps and a landing with a shallow slope

minimum competency and there are time constraints, but these sets of equations will always produce evenly-spaced stairs which is useful in practice. In this example, the change in elevation is 7.7' and the horizontal distance for that change is 57'. The riser height and tread length for each step are 4" (0.33') and 18" (1.5') respectively.

1 Calculate the number of steps needed for the stairs by dividing the change in elevation by the height for one riser, 7.7'/0.33' = 23.10 or 23 steps.
2 Multiply the number of steps by the height of one riser to determine the total vertical change that will be encompassed by the stairs, 23 steps × 0.33' = 7.66'.
3 Multiply the number of steps by the tread length to determine the total horizontal distance that will be encompassed by the stairs, 23 steps × 1.5' = 34.5'.
4 Subtract the total length of the stairs from the total horizontal distance provided to determine the landing length, 57' − 34.5' = 22.5'.
5 Subtract the total height of the stairs from the total elevation provided to determine the landing vertical change, 7.7' − 7.66' = 0.033'.
6 Divide the height of the landing by the landing length to find the slope of the landing, 0.033'/22.5' = 0.00148 or 0.15%. This yields a landing slope which is too shallow and will not drain water efficiently. These calculations are illustrated in Figure 4.3–5. We must subtract one or two steps from the total height of the stairs and add the tread length and riser height to the landing to increase the slope to a minimum of 1% and a maximum of 2% (1:48).

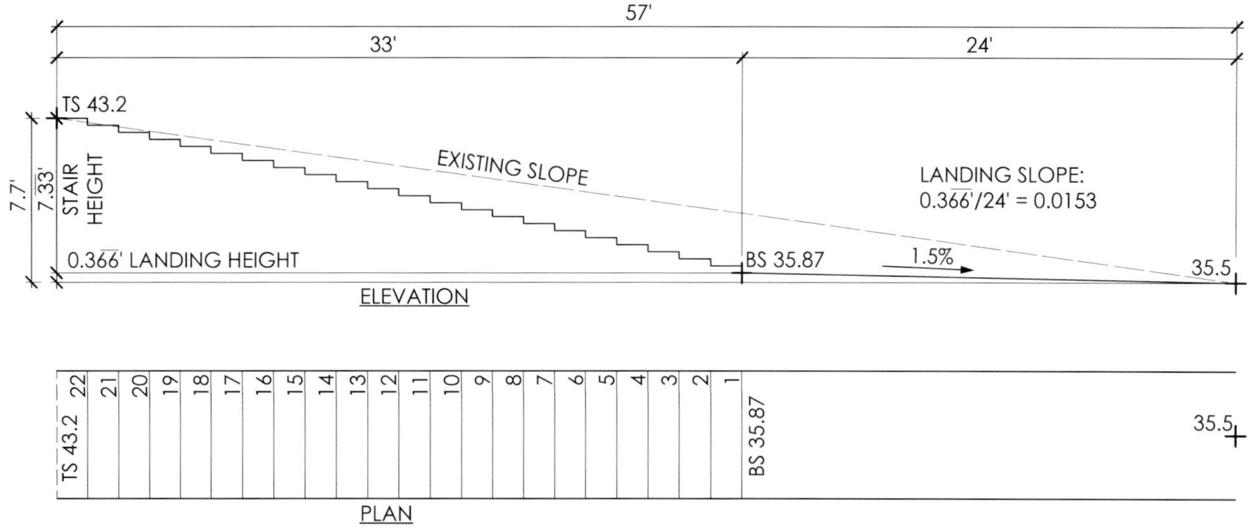

4.3–6
Twenty-two steps and a landing with a 1.5% slope

7 Subtract the height of one riser from the total stair height and add it to the landing height, 7.66' − 0.33' = 7.33'; 0.033' + 0.33' = 0.36'.
8 Subtract one tread length from the total stair length and add it to the landing length, 34.5' − 1.5' = 33'; 22.5' + 1.5' = 24'.
9 Calculate the slope for the new landing by dividing the new height of the landing, by the new landing length, 0.36'/24' = 0.015 or 1.5%. This slope falls within the range of 1% minimum and 2% maximum slope for a landing.
10 The total step count is 22 steps with a 24' long landing sloped at 1.5% as shown in Figure 4.3–6.

To finish the problem, divide the number of steps so that no more than 10 steps are in each flight of stairs and add handrails. You can create three or four flights of stairs. Remember that as you divide the steps into groups you are also dividing the landing into groups; the minimum length for any individual landing is 4'. Note that you also want to ensure the best fit of stairs into the existing landscape. This will make tying proposed contours back to existing contours easier. Use the existing slope for the surrounding topography to help determine how best to divide the stairs. Figure 4.3–7 shows the layout and slope for three flights of steps and the calculations needed to confirm that the vertical and horizontal distances of the grading problem have been achieved. It may be useful to calculate the spot elevations for the stairs and the landings without rounding up or down until all the numbers have been calculated. When all the calculations are complete, round the changes in elevation to the nearest hundredth for greater accuracy.

Stairs will be shown in two ways on the exam. The first is the typical view of stairs descending from a building FFE or a higher elevation to a lower elevation. The other shows stairs in the landscape descending towards the building FFE. To determine which

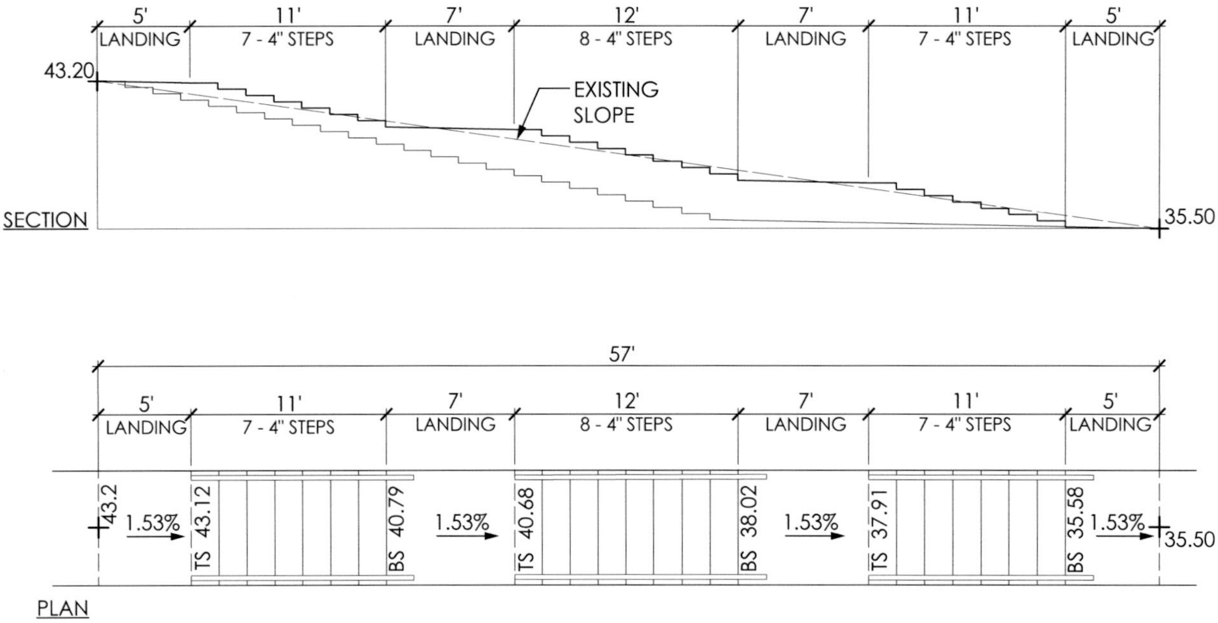

	STAIRS (7 STEPS)	STAIRS (8 STEPS)	STAIRS (7 STEPS)	LANDINGS 1	LANDINGS 2	LANDINGS 3	LANDINGS 4	TOTAL
VERTICAL	2.34'	2.66'	2.34'	0.075'	0.105'	0.105'	0.075'	7.7'
HORIZONTAL	10.5'	12'	10.5'	5'	7'	7'	5'	57'

4.3–7
Three flights of stairs with two 5' and two 7' long landings

one to use, note which way the topography is sloping. If the grading problem asks for stairs and it seems that stairs are not needed because the topography slopes towards the building, this is usually an indication that stairs are needed in the landscape to lower the topography so water will flow away from the building, thus creating a lowered courtyard. Stairs are rarely shown separately on the exam and are often combined with ramps, courtyards or building entrances. Remember to tie the existing contours into the side of the stairs or the cheek-wall based on the maximum slope provided in the grading problem. Although contour lines may not be shown on stairs, they are shown on the landings between stairs so make sure to tie in proposed contours on the landings to any existing or proposed contours in the landscape.

4.4 Ramps

As illustrated in Figure 4.4–1, ramps consist of three parts: a horizontal distance called a ramp run that corresponds to a minimum and maximum allowable longitudinal slope; 5'×5' square landings; and continuous handrails. The minimum and maximum allowable longitudinal slopes for ADA-compliant ramps are 5% and 8.33% respectively. Each ramp run has a maximum horizontal length of 30' before a landing and a maximum height of 30" before a landing. Therefore, the maximum rise: run is 1":12" or 1:12 or

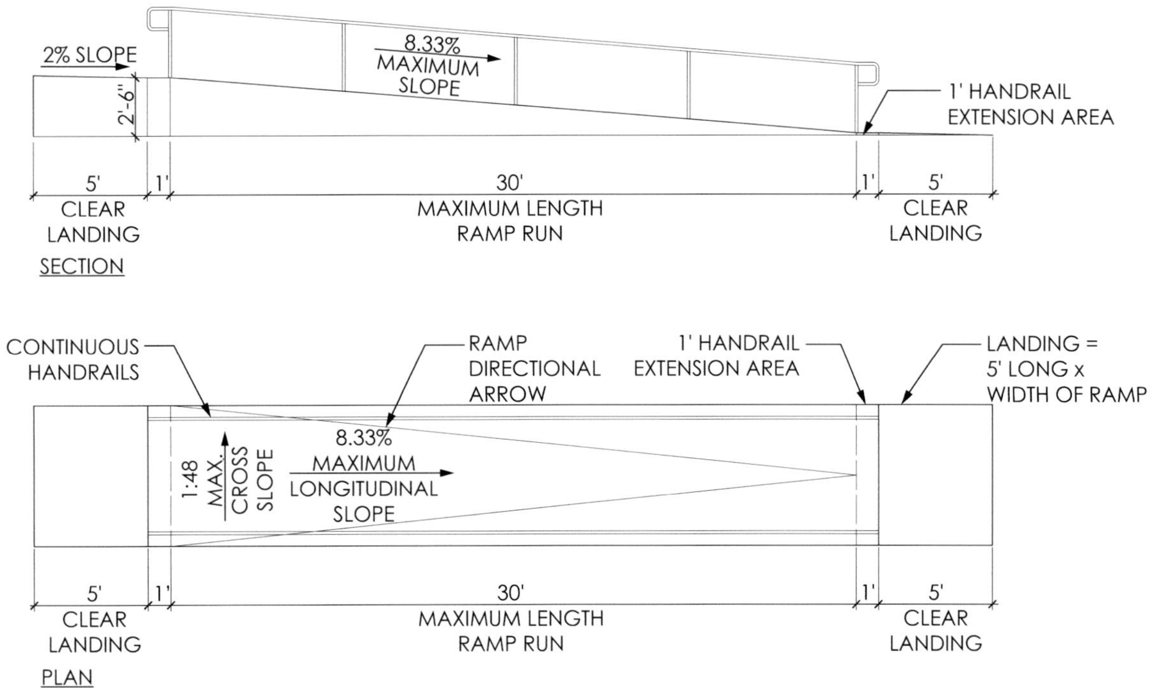

4.4–1
Ramp with rise and run labeled and 5' long landing

8.33%. The minimum length of a ramp needed to reach a given elevation is determined by the change in elevation divided by the maximum allowable slope, 8.33%. The maximum length of a ramp needed to reach a given elevation is determined by the change in elevation divided by the minimum allowable slope, 5%.

In addition to the longitudinal slope of the ramp, which may vary between the ramp runs, some ramp runs have a cross slope of 2%. These are usually on the ramp runs that are parallel to the existing topography. This means that the landings associated with these ramp runs will be warped planes. The cross slope aids in removing surface runoff from the landings, see Figure 4.4–4 for an illustration of the slope on a 5' square landing.

Ramps may be directional. A one-way ramp has a minimum width of 3' clear between handrails and a two-way ramp has a minimum width of 5' clear between handrails. Direction of the ramp should be shown with an arrow labeled with the slope for each ramp run. A ramp may be a single straight run punctuated by 5' long landings every 30' or it may double back on itself or it may have three legs at 90° to each other, see Figure 4.4–2. The style of the ramp will be determined by the amount of room on the grading problem.

Continuous handrails are required on both sides of the ramp for the entire length of any ramp runs and intermediate landings. As illustrated in Figure 4.4–3, handrails may be attached to a guard rail, a cheek-wall or to the floor of the ramp. The "clear width" of a ramp is measured from the inside of one handrail to the other. The total width of the ramp will include additional width for the cheek-wall and the handrail. Handrails must maintain a height of 34"–38" above the height of the ramp. Like stair handrails, ramp

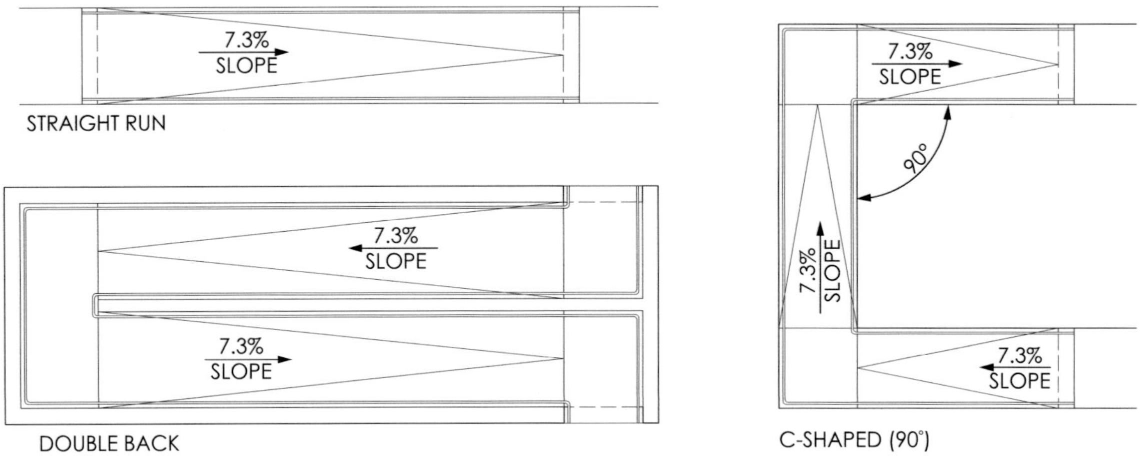

4.4–2
Straight run, double back and 90° ramps with longitudinal and cross slope labeled

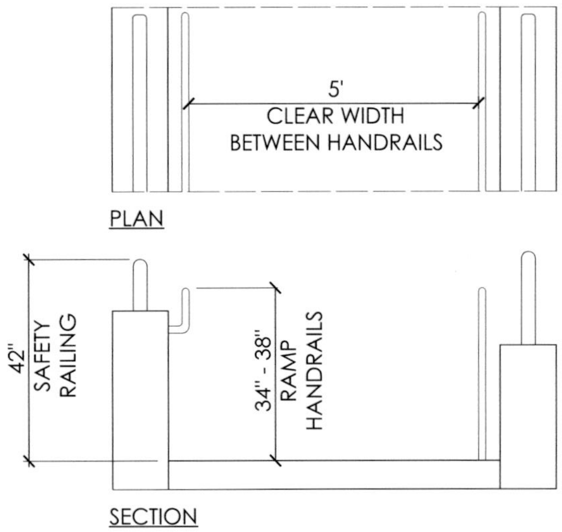

4.4–3
Plan and section view of ramp with handrail and safety rail

handrails extend past the beginning and the end of a ramp by one foot on either end. When a ramp meets an existing walk, a 1' level horizontal distance called the handrail extension is included to accommodate the end of the handrail so that it does not jut out into the existing walk.

Ramp landings are required at the beginning and end of each ramp. They should be clear of any protuberances including the handrails. Ramp landings may be combined with stair landings for a more efficient use of space.

Intermediate ramp landings have a minimum slope of 1% and a maximum slope of 1:48 (2%). On a ramp that bends the intermediate landing will be a warped plane. When setting spot elevations on a ramp landing remember that water must not settle in the corners. Calculate the slope on the landing to ensure that water drains off of the landing efficiently and there is positive drainage from all the corners, see Figure 4.4–5.

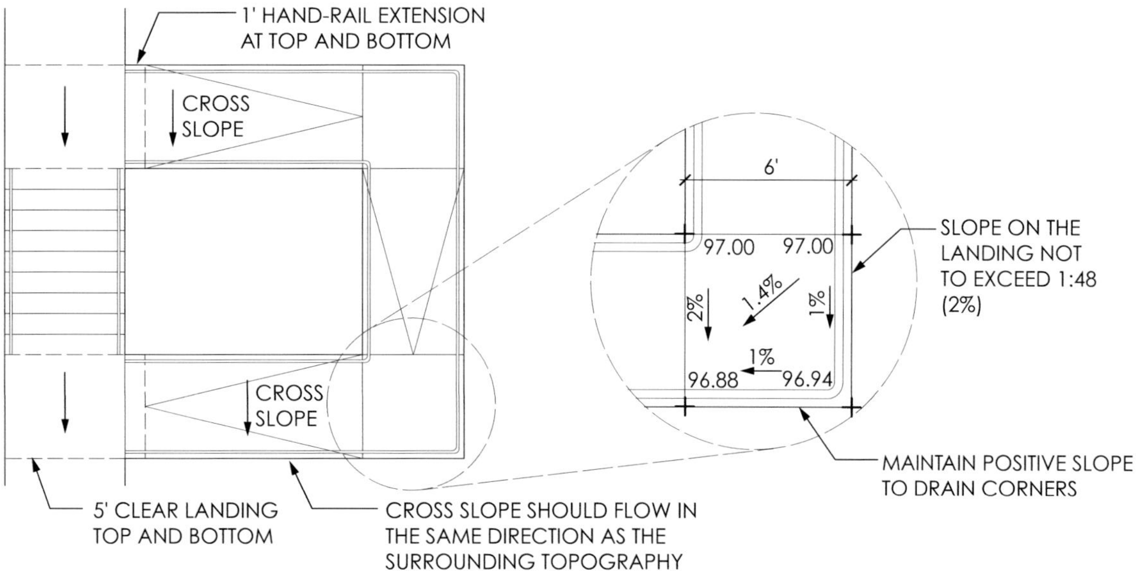

4.4–4
Stair and ramp with connections at the landings

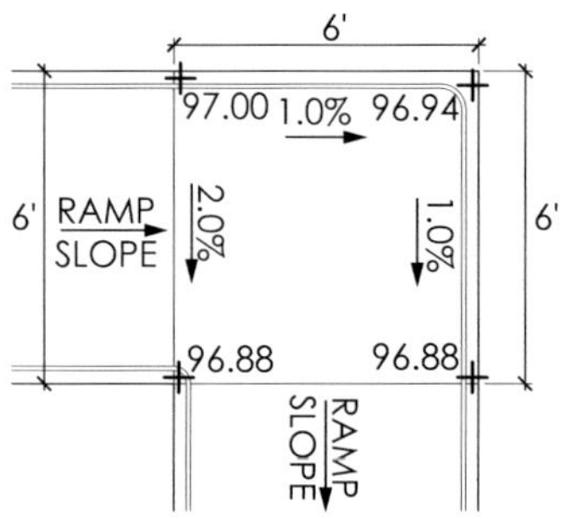

90° RAMP LANDING
1. LENGTH OF LANDING x CROSS SLOPE: 6' x 0.02 = 0.12'
2. CHANGE IN ELEVATION: 97.00' - 0.12' = 96.88'
3. TOTAL DISTANCE THAT WATER MUST TRAVEL TO ENSURE POSITIVE DRAINAGE: 6' + 6' = 12'
4. SLOPE ALONG THE OUTSIDE EDGES OF THE LANDING:
5. TOTAL VERTICAL CHANGE: 0.12'
6. TOTAL HORIZONTAL DISTANCE: 12'
7. SLOPE TO ENSURE POSITIVE DRAINAGE: 0.12'/12' = 0.01 (1.0%)

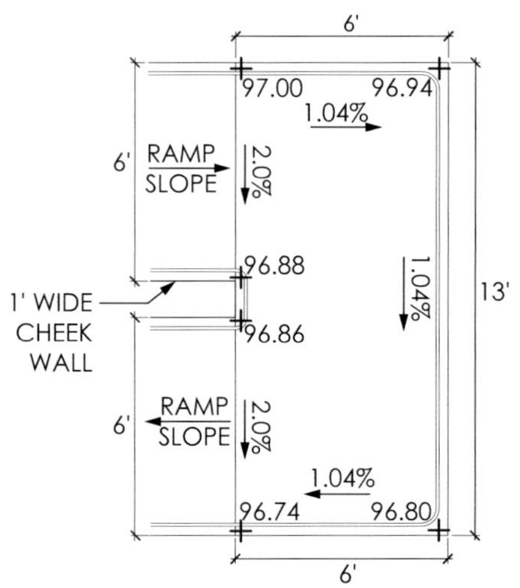

DOUBLE BACK RAMP LANDING:
1. LENGTH OF LANDING x CROSS SLOPE: 13' x 0.02 = 0.26'
2. CHANGE IN ELEVATION: 97.00' - 0.26' = 96.74'
3. TOTAL DISTANCE THAT WATER MUST TRAVEL TO ENSURE POSITIVE DRAINAGE: 6' + 13' + 6' = 25'
4. SLOPE ALONG THE OUTSIDE EDGES OF THE LANDING:
5. TOTAL VERTICAL CHANGE: 0.26'
6. TOTAL HORIZONTAL DISTANCE: 25'
7. SLOPE TO ENSURE POSITIVE DRAINAGE: 0.26'/25' = 0.0104 (1.04%)

4.4–5
Calculating slope on the ramp landings

Grading of built elements

Like stairs, ramps also have other elements that must be shown on a grading solution. The grading problem may permit a cheek-wall on a ramp, as with the stairs, if it is mentioned as part of the problem parameters it is a good idea to include it in the grading solution. If a cheek-wall is shown it usually indicates that the ramp will exceed 42" in height, the maximum height above finish grade. The cheek-wall will include a guardrail on the top. This railing is in addition to the continuous handrails.

If a ramp is nestled into the surrounding topography, proposed contour lines will be drawn across the surface of the ramp and labeled accordingly. Remember to tie the contours into the existing topography surrounding the ramp and provide any swales to move water around the ramp as needed. In some grading problems, the ramp is shown on a relatively flat site. Although tying into existing contour lines is not required, spot elevations will be required. Label spot elevations on all corners of the sloped legs of the ramp as well as at all corners of the landings. In the following example, illustrated in Figure 4.4–6, we will solve for a ramp with a change in elevation of 5'.

1 To determine the length needed for the ramp, divide the change in elevation by the maximum and minimum ramp slopes, 5'/0.0833 = 60'; 5'/0.05 = 100'. The length of the ramp will be determined by the amount of space provided on the grading problem, but it cannot be shorter than 60' or longer than 100'. We will use 60' for this example.

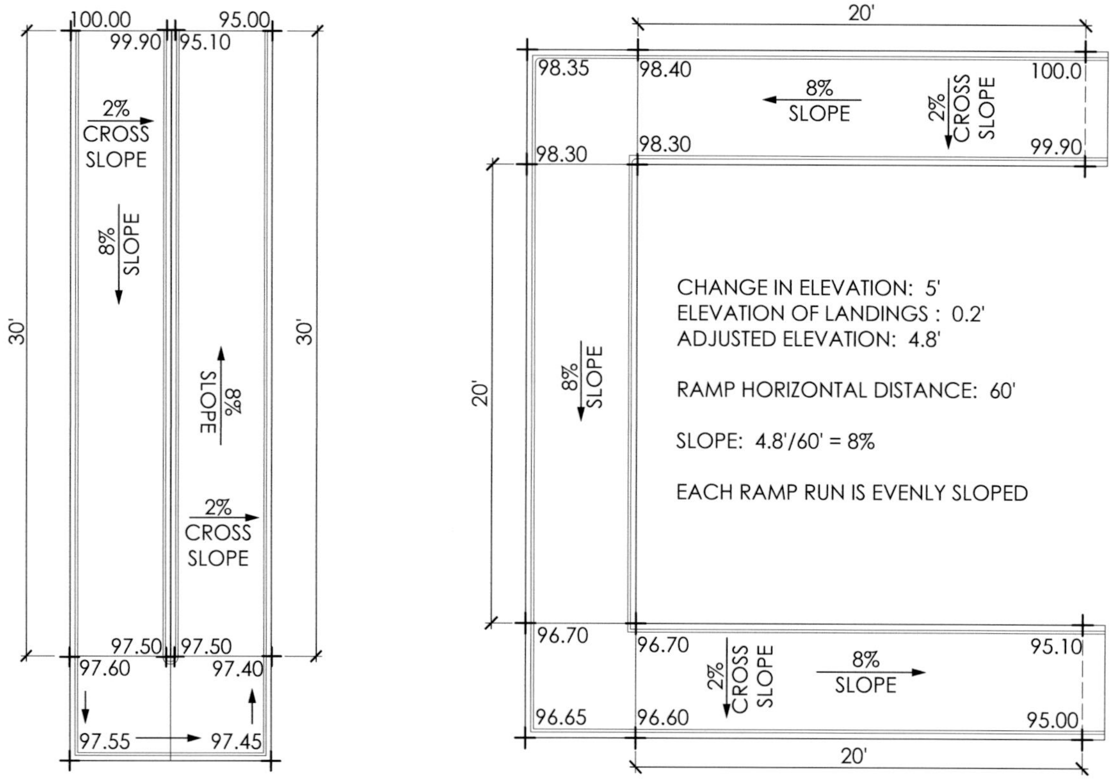

4.4–6
Evenly sloped ramps

2 Choose a style for the ramp and determine the number of landings that will be needed. A straight ramp will require a minimum of one intermediate landing and for a double back or 90° ramp, a minimum of two intermediate landings. Each 5'×5' landing is sloped at 2% to allow for drainage.
3 Using the slope formula, determine the vertical change for each intermediate landing, 5'×0.02 = 0.1'.
4 The vertical change of the landings must be subtracted from the total vertical height, similar to the stair problem, 5'−0.1' (landing 1)=4.9'; 4.9'−0.1' (landing 2)= 4.8'.
5 To determine the slope for the ramp, divide the adjusted vertical change by the length of the ramp, 4.8'/60'=0.08 or 8%.
6 Label each leg of the ramp with the calculated longitudinal slope.
7 To determine the cross slope for the ramp, multiply the width of the ramp by 2%. Apply the cross slope to the ramp according to the direction that the topography is descending, see Figure 4.4–4 as an example.
8 Show all spot elevations and provide proposed contours as needed.

The following examples show what will happen if each part of the ramp is not calculated separately and accounted for when designing a ramp. In the first example in Figure 4.4–7, the length of the landings is added to the total length of the ramp for a total length of 70', 60' for the ramp run+5' for each landing. Using the original vertical change of 5' divided by 70', you will get 7.1% as the slope for each section of the ramp.

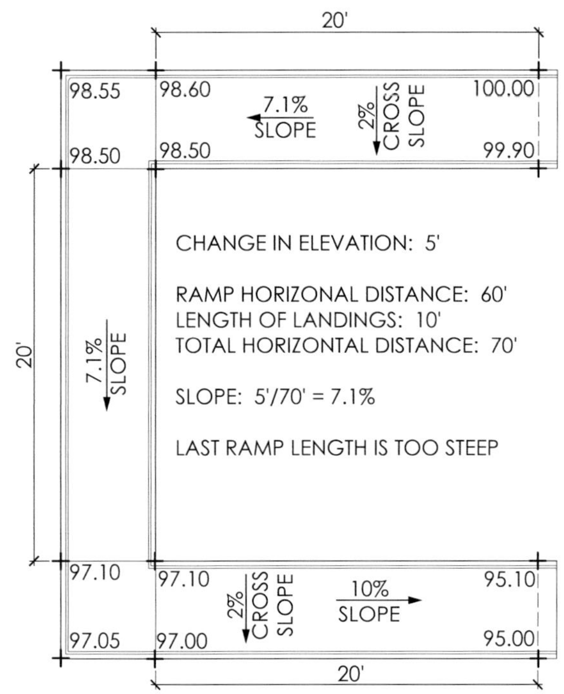

4.4–7
Ramps with uneven slopes

Calculating the spot elevations for the runs using 7.1% slope and 2% maximum slope for the intermediate landings will result in the last ramp run exceeding the maximum allowable slope for an ADA-compliant ramp.

In the second example in Figure 4.4–7, the landings are excluded from the initial ramp calculations. Using a total vertical distance of 3' for the ramp and a ramp length of 60' you calculate a slope of 5% for the ramp, the minimum allowable slope for an ADA-compliant ramp. Calculating the spot elevations for the runs using 5% slope and 2% maximum slope for the intermediate landings will result in the final ramp run being shallower than the minimum allowable slope. This last ramp run is not a ramp but a path. In most cases, having the final run of a ramp shallower than the first two runs is not a problem since a grading problem on the exam rarely asks you to provide three runs of equal slope. A problem can occur with ramps that start out with shallow slopes. The final run of the ramp may be less than the minimum slope required for an ADA-compliant ramp. If you are not careful, you may draw more pavement than is needed to solve the grading problem and/or you may draw a ramp with a slope less than 5% not realizing that a ramp signature is not needed.

This final example combines stairs and ramps using the 5' clear landing at the top and bottom of the stairs. Note that adjustments will need to be made to the calculations for the ramp based on the slope of the stair landings.

1 Calculate and solve for the stairs first. Determine the length and slope of the landing and divide the stairs and landings as needed.
2 Provide a minimum 5' long landing at the top of the stairs and provide a minimum 5' long landing at the bottom of the stairs just past the end of the stair handrails. In the example in Figure 4.4–8, the landing is 6' long at top and 7' long at the bottom. The extra foot at the bottom is to accommodate the bottom handrail from the stairs. Remember the landings for a ramp must be clear of obstructions.
3 Note the slope on the stair landings, this is the slope that will be the cross slope for the ramp. If the slope is less than 2%, for example 1.5%, make sure to calculate the cross slope for the ramp with a 1.5% slope.
4 Determine the spot elevations for the top and bottom of the landings at the beginning and end of the stairs. Calculate the total vertical change covered by the stairs including the landings, 100.0' − 94.64' = 5.36'.
5 Measure, mark, and draw two 1' wide handrail extension areas perpendicular to the stair landings. The handrail extension areas can be level or they can slope downhill at 2% slope. Because the cross slope for the ramp also applies to the handrail extension areas, it will have a maximum slope of 2% in at least one direction. Calculate the vertical change for the handrail extension area, 1' × 0.02 = 0.02'; 0.02' × 2 = 0.04'.
6 Because we are making a 90° C-shaped ramp, we will need two intermediate landings. The intermediate landings are warped planes. Measure the length of the landing and multiply it by the cross slope for the ramp. This will give you the vertical change for each landing, 6' × 0.02 = 0.12'; 0.12' × 2 = 0.24'.

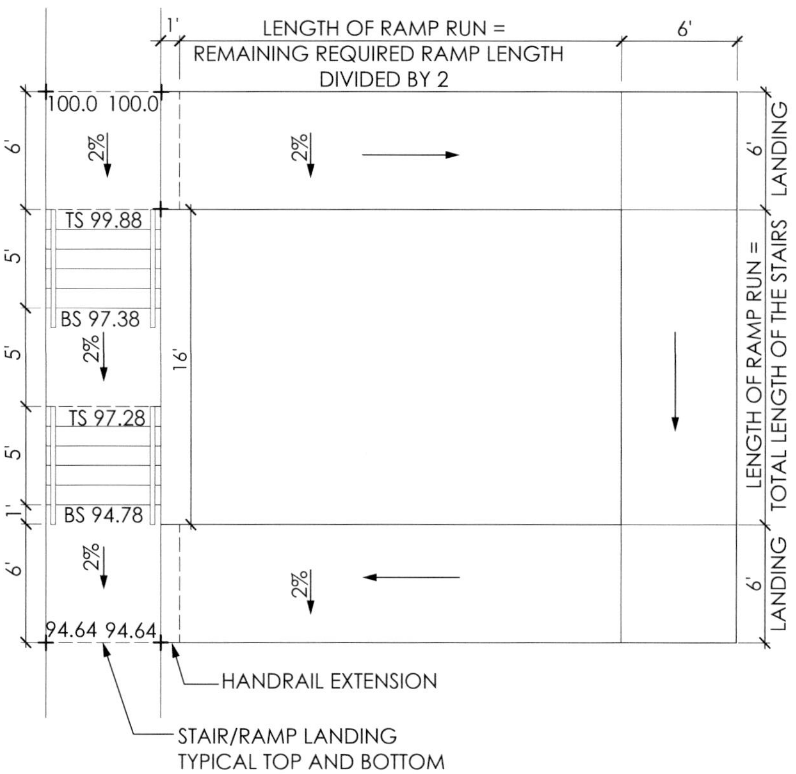

1. TOTAL VERTICAL CHANGE:
   100.0' - 94.64' = 5.36'
2. HANDRAIL EXTENSIONS:
   1' x 0.02 = 0.02'; 0.02' x 2 = 0.04'
3. INTERMEDIATE LANDINGS:
   6' x 0.02 = 0.12'; 0.12' x 2 = 0.24'
4. ADJUSTED VERTICAL CHANGE FOR TOTAL HEIGHT OF RAMP:
   5.36' - 0.04' = 5.32'
   5.32' - 0.24' = 5.08'
5. MINIMUM LENGTH OF RAMP:
   $5.08'/0.08\overline{33} = 60.98'$
6. MAXIMUM LENGTH OF RAMP:
   5.08'/0.05 = 101.6'
7. RAMP RUN PARALLEL TO STAIRS:
   16'
8. REMAINING MINIMUM REQUIRED LENGTH FOR RAMP DIVIDED BY 2:
   60.98' - 16' = 44.98'; 44.98'/2 = 22.49'
9. ROUND UP LENGTH: 23'
10. SLOPE FOR EACH RAMP RUN:
   V: 5.08'
   H: 23' + 16' + 23' = 62'
   S: 5.08'/62' = 0.08193548

4.4–8
Calculating the stair/ramp combination

7 Subtract the elevations of the handrail extension areas and the intermediate landings from the total vertical change to get the adjusted vertical change or the total height of the ramp, 5.36' − 0.04' = 5.32'; 5.32' − 0.24' = 5.08'.

8 Calculate the minimum and maximum length of ramp needed by dividing the total height of the ramp by 8.33% and 5% respectively. The ramp length that you choose should be within this range, 5.08'/0.0833 = 60.98'; 5.08'/0.05 = 101.6'.

9 You will need a minimum of three ramp runs to create the ramp. Measure the total length of the stairs including any handrail extensions at top and bottom. This is the length of the middle ramp run, 15' for stairs + 1' for handrail at the bottom of stairs = 16'.

10 Subtract the length of the middle ramp run, 16', from the minimum length of ramp, 60.98'. Divide the remaining distance in half. This is the length for the first and last ramp runs, 60.98' − 16' = 44.98'; 44.98'/2 = 22.49'.

11 Increase the length of the first and last ramp run so that the distance rounds up to the nearest whole foot. Note that you can increase the length of these two ramp runs as long as the total length of ramp is between the minimum and maximum

length of ramp that you calculated. Remember that these two ramp runs must be equal in length so that the ramp begins and ends at the handrail extensions, 22.49' rounds up to 23'.

12 Add the lengths of each ramp run, 23' + 23' + 16' = 62'.

13 Determine the slope for the full ramp from the total height of the ramp divided by the total lengths of the ramp runs, 5.08'/62' = 0.08193548. This is the slope for the ramp and will provide a uniformly-sloped ramp that meets the stair landings at the correct elevations. Do not round the slope, use the full value of the slope to generate the spot elevations for the ramp.

14 Measure, mark, and draw the ramp runs and intermediate landings starting at the handrail extension areas.

15 Calculate and label each change in elevation on the handrail extensions, ramp runs and intermediate landings. Note that it is best to calculate the elevations without rounding up or down until all the calculations have been completed and then round to the nearest hundredth.

16 Draw the continuous handrails along both sides of the ramp and around the intermediate landings. If preferred, the inside handrails for the ramp and the stairs may be joined. Figure 4.4–9 shows the completed stair ramp combination.

17 If necessary, draw any proposed contours for the ramp and tie them back to the existing topography.

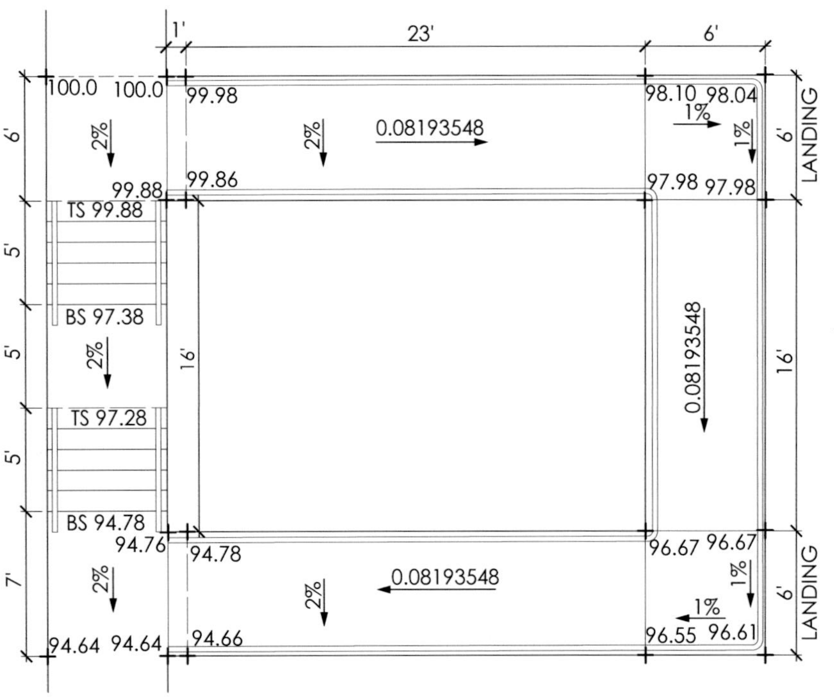

4.4–9
Stair/ramp combination with slopes and spot elevations

Creating uniformly sloped ramps simplifies finding proposed contours on the ramp. When you use the same longitudinal slope for each run of the ramp you can use the slope interval for each run of the ramp and the proposed contour lines will be evenly spaced throughout. Remember to account for the shallower slope of the landing when drawing the proposed contours.

4.5 Retaining walls

There are two types of walls in the landscape, free-standing walls and retaining walls. Free-standing walls do not support changes in the elevation of earth. Both faces of the wall are exposed to the air. In most cases, free-standing walls have similar elevations at finish grade on both sides of the wall, with variations due to drainage needs at the bottom of the wall. This section focuses on retaining walls because the elevational differences on both sides of the wall have a marked effect on topography. Unlike the previous sections in this chapter, the guidelines for retaining walls will vary greatly depending on the grading problem.

Retaining walls are used to stabilize steeply sloped areas of the landscape and provide level surfaces of different elevations on both sides of the wall. They are designed with piped drainage behind the wall below grade, and weep holes at the base to remove surface and subsurface water build up. They are also counterbalanced to prevent the wall from failing (falling forward) due to the weight of the earth behind it. In some instances, a batter may be used, particularly with dry stone or segmented walls to provide extra stability. Characteristically, one side, the front face of a retaining wall, is exposed to the air. In some cases, one side is more exposed than the other. Retaining walls are essentially very tall curbs and therefore have the same topographic signature as a curb. In plan view, contour lines travel along the face of the retaining wall and will re-emerge on the other side of the drawing.

Unlike a curb which assumes a consistent height of 6"–8", the height of a retaining wall may vary as it travels through the landscape; therefore, spot elevations are used to show the height of the wall. Spot elevations for the top and bottom of the retaining wall are called out as with stairs and curbs; top of wall, TW, and bottom of wall, BW. Spot elevations are shown at grade at the front of the wall, on top of the wall and on the finish grade behind the wall. They should be placed at intervals along the length of the wall to provide direction on whether the wall is rising, falling or level to the topography. In some instances, the top of wall will have lines that cross the top of the wall. These are not proposed contours; they indicate a change or step in elevation for the top of the wall. Spot elevations must be added at each level change to add clarity. Figure 4.5–1 illustrates some of the different options for retaining walls and the importance of labelling the top of wall at each break point. Depending on the width of the top of the retaining wall it may be pitched or cross sloped to one side to ensure drainage off the top of the wall. The TW elevation indicates the elevation at the apex of the wall.

The orientation of the wall in the landscape will vary. Depending on how the wall is sited in the landscape the finish grade in front of the wall may be flush with the bottom

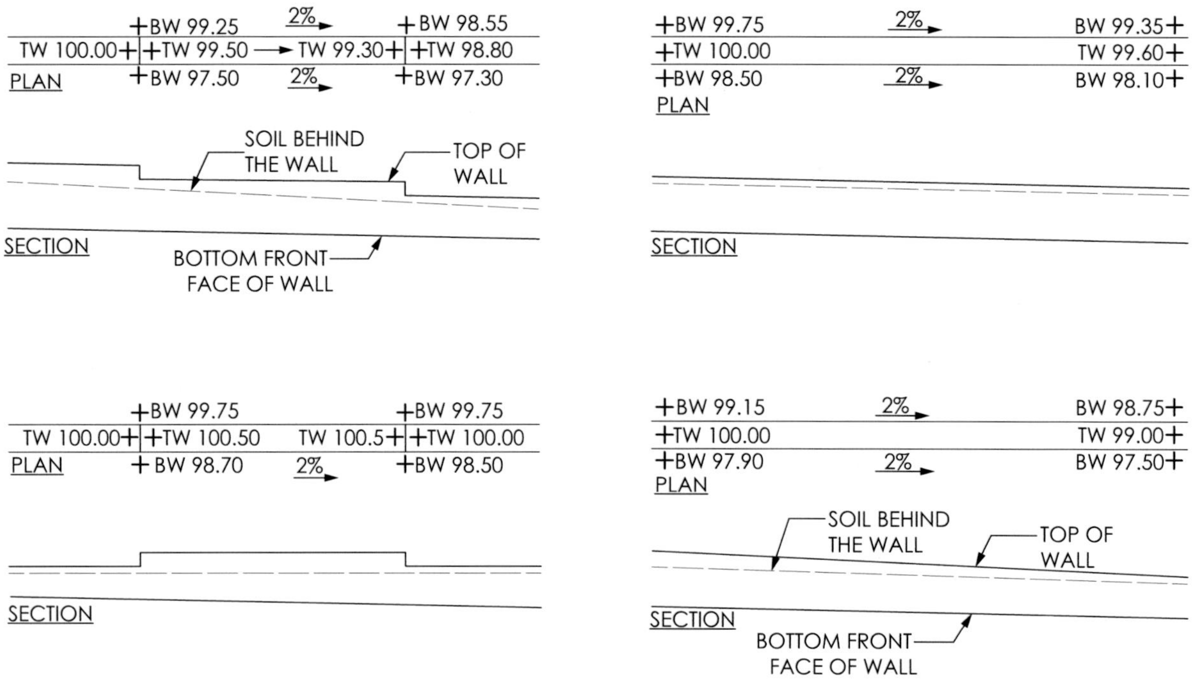

4.5–1
Examples of retaining walls labeled with spot elevations

or it may slope with the wall. The front of the wall will have minimal if any soil along its face and the grade at the back of the wall will be flush to a couple of inches lower than the top of the wall.

Retaining walls may function in myriad ways. They may hold up a terrace, they may define a planting bed, or they may surround a patio or deck as shown in Figure 4.5–2. For this reason, any retaining wall that is greater than 3' above finish grade and has pedestrian access is topped with a safety rail. The railing will vary in height depending on whether the wall is flush or partially exposed, but the total height on the pedestrian side must reach 42" as stated in the ADA requirements. Topography may either slope towards the wall or away from the wall. This decision is largely aesthetic and depends on the angle of repose of the soil being retained and the drainage system behind the wall. If you decide on a steep angle of repose behind the wall, add a shallow swale just behind the wall to prevent surface runoff from flowing over the wall.

On the exam, retaining walls are part of a larger grading problem. They are rarely used alone in a grading problem and are tied into grading patios and decks and/or split-level houses. The problem statement will provide all the parameters that must be met in the grading solution. Note, however, that because proposed contours will run along the front face of the wall, if proposed contours tie into the retaining wall on one side of the problem they must tie back to the existing contours on the opposite side of the wall.

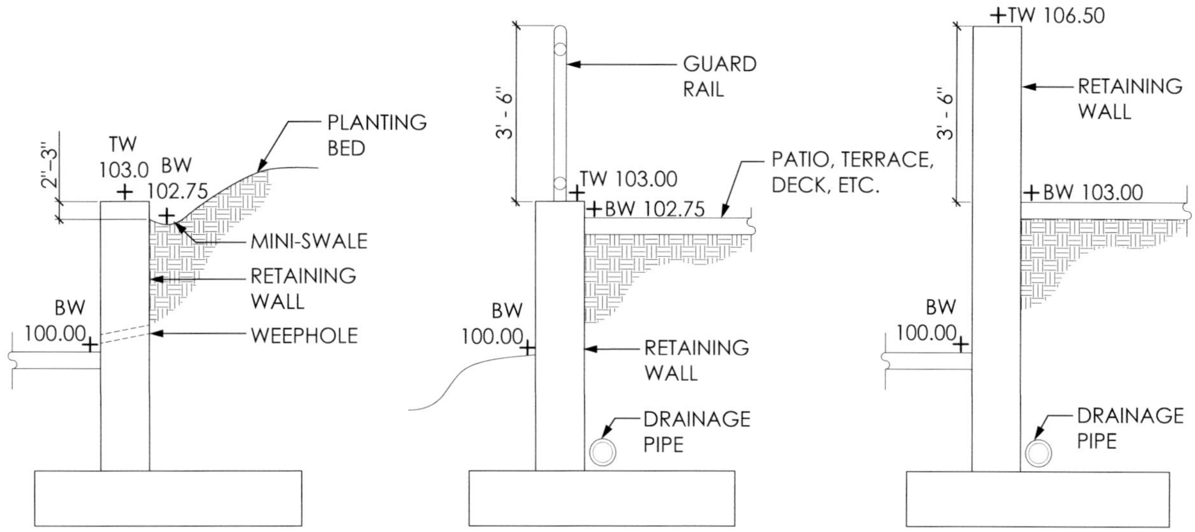

4.5–2
Retaining walls with planting bed, with patio + safety rail and with extended wall at edge

Structurally, there are two parts to a retaining wall, the arms (or legs) that bend to meet the landscape and the length of wall that does the majority of the soil retention. In some cases, the arms may be absent; however, when present they add stability. The length of the arms that bend to meet the landscape will be based on the height of the soil that needs to be retained and the maximum slope required by the grading problem. For example, if you have to retain 3' of soil and you have a maximum slope of 3:1 then the length of the arms will need to be a minimum of 9' long, 3'/0.33 = 9'.

The total length of the wall will be determined by the area of soil that needs to be retained. The length of wall may curve or undulate. Retaining walls bend into or away from the existing topography they are trying to support to provide greater stability and prevent the wall from failing. Walls with arms that bend into the topography create a "level" terrace at the top of the wall. Walls with arms that bend away from the existing topography create a "level" terrace at the bottom of the wall, see Figure 4.5–3.

There are three possibilities for the elevations along the bottom of the wall on the front face. The bottom of the wall will be level in cut, level in fill or sloping with the existing landscape. To be level in cut, the bottom of the wall will be set at the lower elevation of both corners. To be level in fill, the bottom of the wall will be set at the higher of the two elevations. A wall that slopes with the existing topography is the most common as it doesn't require additional earthwork. Proposed contours will likely tie into the front of the wall, see Figure 4.5–4.

There are two possibilities for the elevations behind the wall. The topography can slope towards the wall and surface water can flow to drains or, less likely, a saddle point can be created and surface water can be diverted into a swale around the retaining wall. Drain inlets can be placed at the corners or in the center behind the wall. The proposed

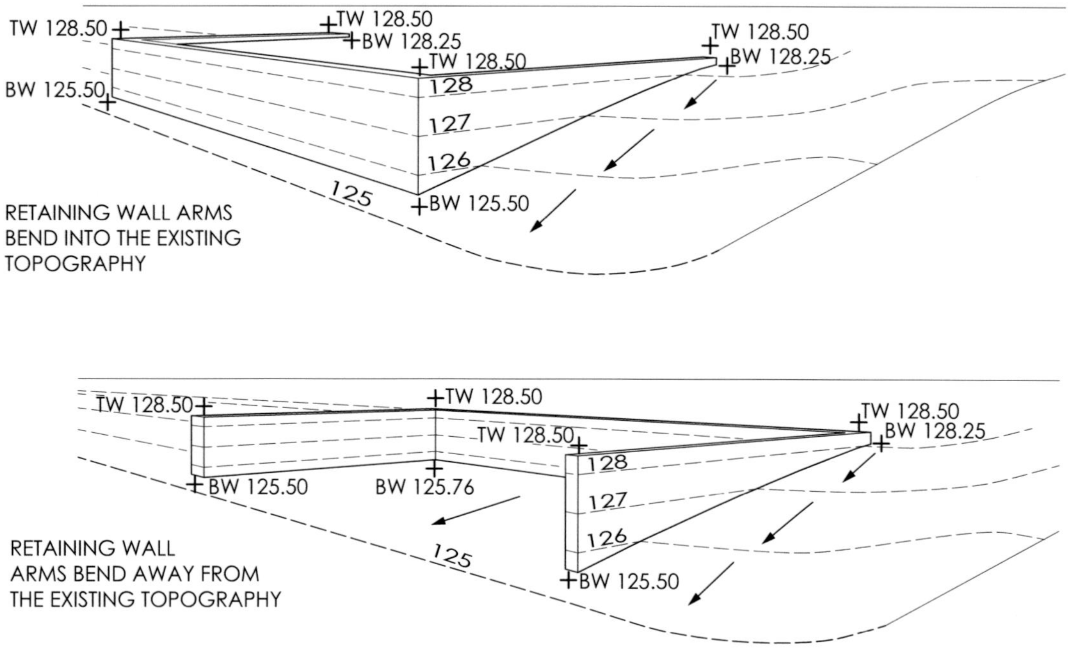

4.5–3
Retaining walls that bend into and away from the topography

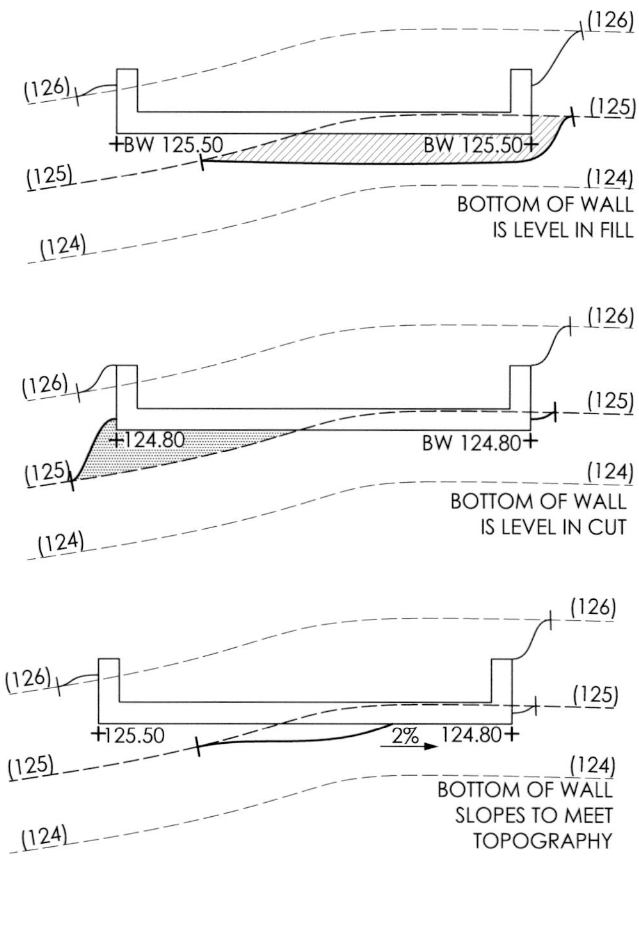

4.5–4
Retaining walls, bottom of wall elevations

topography must slope towards the drain inlets at a minimum of 2% slope. The parameters of the grading problem will usually provide the location for any drain inlet(s).

The top of the wall should be, at minimum, 1"–3" above the finish grade at the back of the wall to ensure clearance of the wall cap; to prevent water from entering the wall structure; and to prevent surface water runoff from flowing over the top of the wall. The top of the wall can be level, or it can slope as needed. The height of the wall may be uniform but the arms for the wall may slope to maintain the 1"–3" of clearance.

The example in Figure 4.5–5 shows a 3' tall wall placed at elevation 125.50' with a maximum allowable slope for the landscape at 5:1.

1 The wall is drawn with the majority of its length between the existing (124) and (125) contours. The proposed elevation at the bottom of the wall is 125.5'. This indicates that the bottom of the wall will require soil to increase the bottom elevation. The contours at the bottom right corner will be adjusted downhill. See Section Six for information about soil cut and fill.
2 Label the top (TW) and bottom (BW) front of the wall with spot elevations.
3 The wall is to be 3' tall, with finish grade behind it to be, at minimum, 3" lower than the top of the wall. Because the wall will hold back the maximum height of soil, 3', the arms of the wall must provide enough length to accommodate the 3' of soil that will be displaced. Use the maximum allowable slope for the landscape to determine the minimum length of the arms for the retaining wall, 3/0.20 = 15'.

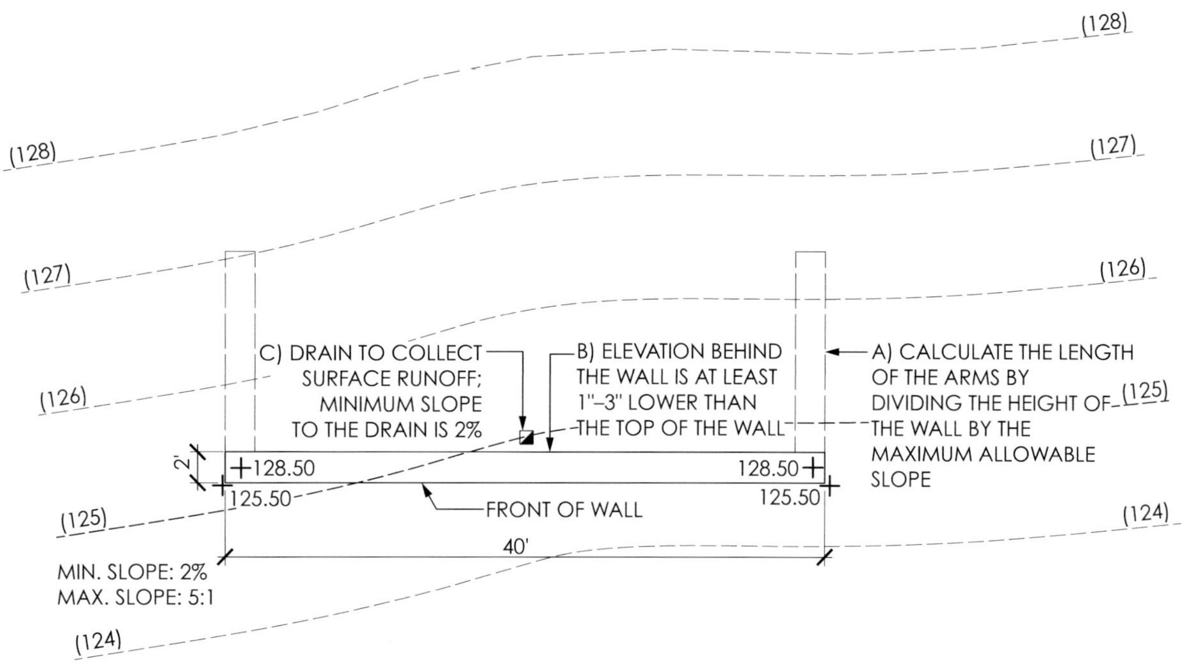

4.5–5
Retaining wall located in the landscape

4 The overall length of each arm is 15'; remember to subtract the width of the wall when drawing the length of the arms, 15'−2' width of wall = 13' of additional wall length.

5 From the bottom corners at the front of the wall interpolate using the maximum slope to find the spot elevation at the end of the arms and the whole number spot elevations along the length of each arm (see Figure 4.5–6), 15'×0.02 = 3'; 125.5' + 3' = 128.5'.

6 Label the BW elevation at the end of the retaining wall arm. Add 3" to get the top of wall elevation at each end. Note that some walls will not have caps and they can be flush with the finish grade.

7 Label the interior corner spot elevations behind the wall. They are 3" lower than the top of wall elevation, 128.25'.

8 Place a drain inlet behind the wall in the center. Slope from the corners towards the drain at 2% to find the rim elevation (RIM EL) for the drain, 18'×0.02 = 0.36'; 128.25'−0.36' = 127.89'. Label the RIM EL for the drain.

9 Check the slopes from the ends of the retaining wall arms and the existing contours to the drain inlet. Make sure the slope towards the drain is at minimum 2%. Using these slopes locate the proposed 128 whole number spot elevation.

10 These spot elevations don't tie back to the existing topography, they form a closed contour line that runs along the back of the retaining wall. Draw and label the proposed 128 closed contour line.

11 Draw and label the proposed contours along the arms of the retaining walls. Make sure that the distance between them is at the maximum allowable slope, 5:1. Make sure that water flows downhill away from the wall.

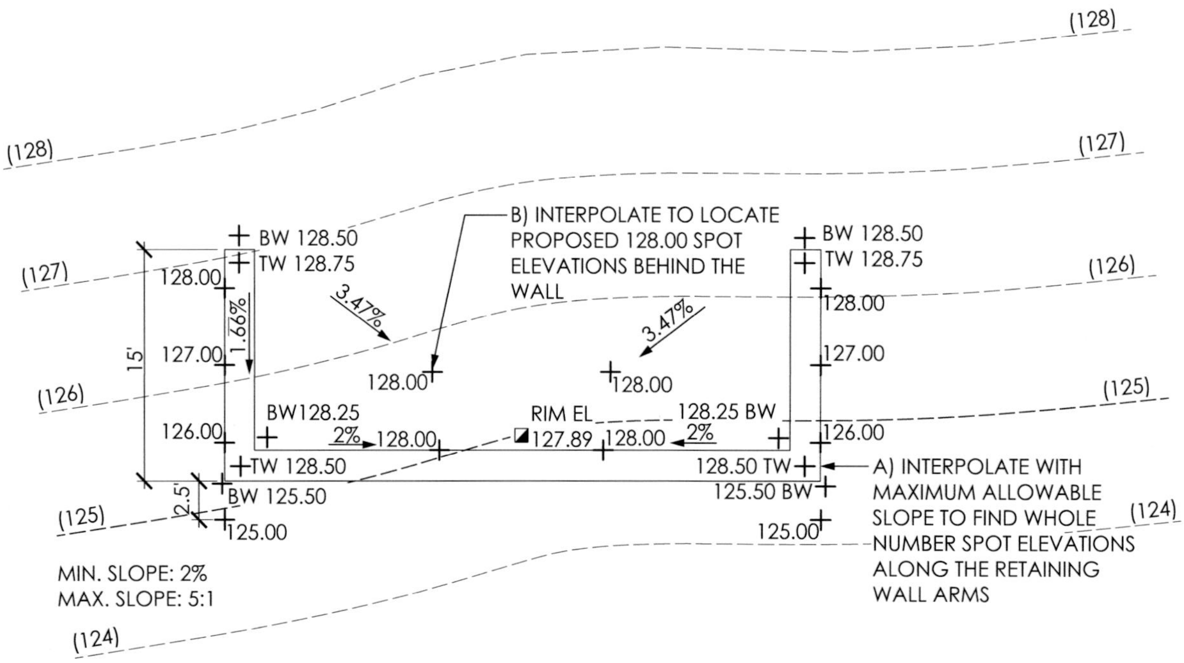

4.5–6
Retaining wall showing spot elevations

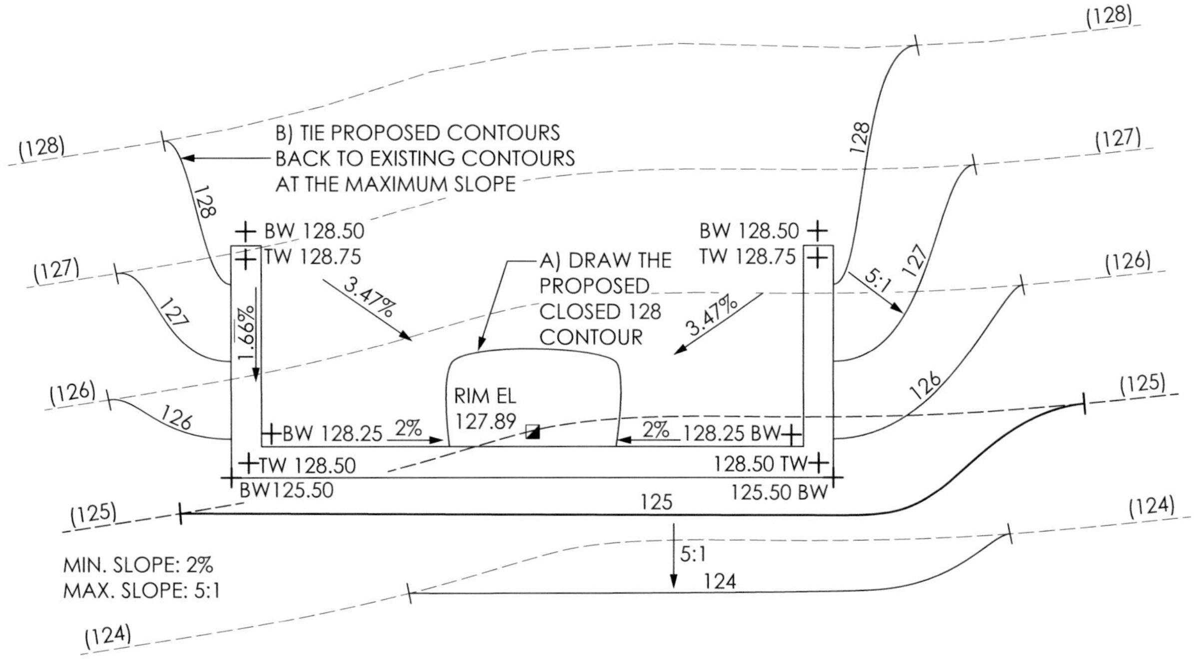

4.5–7
Retaining wall tied back to existing contours

12 Remember to adjust the contours at the bottom of the wall in front. The existing (124) and (125) contours need to be adjusted downhill at 5:1.

13 Label any slopes along the top of the wall. The completed wall illustration is highlighted in Figure 4.5–7.

The previous example showed a relatively short wall in the landscape. The following example outlined in Figure 4.5–8 is a longer curved wall with a swale behind it.

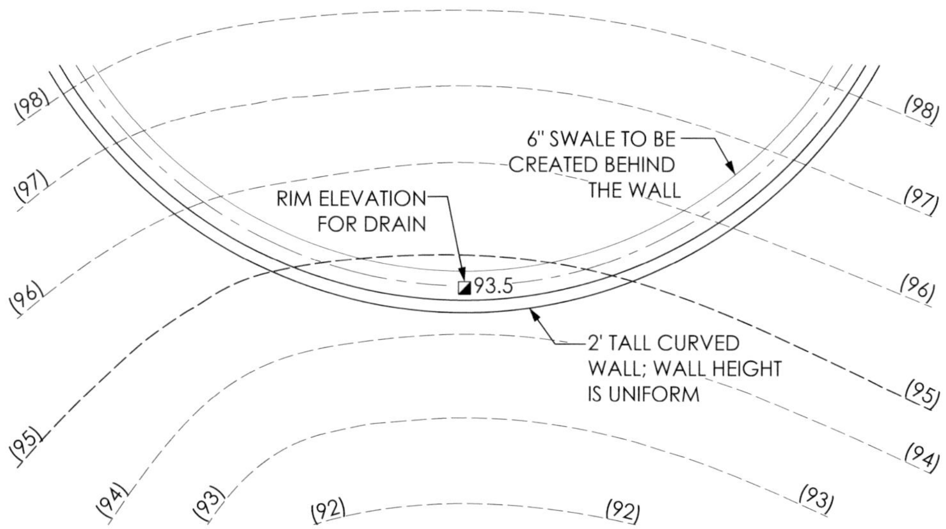

4.5–8
Curved retaining will with uniform height

A second type of retaining wall problem may require you to maintain a consistent height along sloping topography.

In this example, a known point is provided, the RIM EL for the drain inlet. A 6" deep swale needs to be added behind the wall for drainage. The top of the wall is 6" higher than the finish grade at the back of the wall and has a total height of 2' at the front of the wall. The maximum allowable slope for the landscape is 4:1.

1 Start from the known or given spot elevation and create the swale located on both sides of the drain. Use the technique shown in Figure 3.3–9 to create a 6" deep swale behind the wall.
2 After drawing the swale add 6" of height where the proposed swale contour meets the wall to get the TW spot elevation.
3 Subtract the total height of the wall from the spot elevation at the top of the wall to find the BW spot elevation at the front of the wall. Repeat this procedure for each of the proposed contours from the swale as shown in Figure 4.5–9.
4 Interpolate between the spot elevations at the bottom of the wall to find the proposed whole number spot elevations at the base of the wall.
5 Tie the proposed contours back to the existing topography at the front of the wall, maintaining at least the maximum allowable slope for the problem.
6 Adjust any proposed contours along the back of the wall to maintain at least the maximum allowable slope. Figure 4.5–10 shows the completed wall problem.

The final example uses a retaining wall to create a patio and flat lawn area for a building sited on a steeply sloping site. This is typical of the types of grading problems you may encounter professionally as well as in the exam. A 10' wide patio and a 10' wide lawn area are created for a house. There is a 1" step down from the building FFE to the patio and the front walk; and a 6" step down from the FFE to the outside corners of the

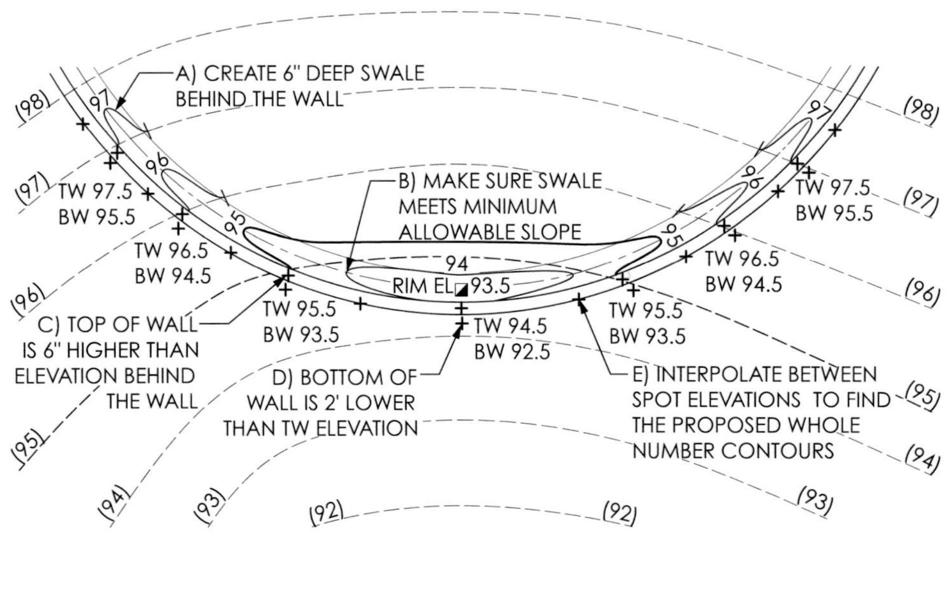

4.5–9
Curved retaining wall with 6" deep swale and TW/BW elevations

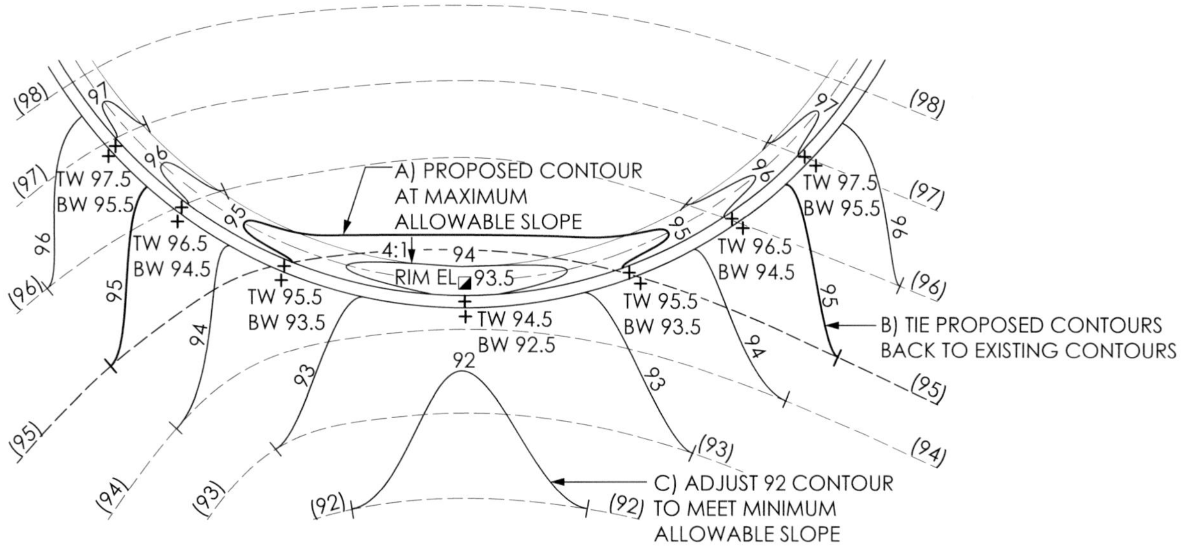

4.5–10
Curved retaining wall with proposed contours

building. The patio and the lawn area slope away from the building at 2%. A 5' buffer has been established on three sides of the building that slopes away at 2%. Although it is not a required part of this retaining wall problem there is a saddle point for a swale on the western side of the building sloping at 2% from the edge of the buffer. The retaining wall has a 1' width and is 2' higher than the patio and lawn. Note that this is the back of the wall. The height of the front of the wall will vary as displaced topography will be placed along the northern face of the wall along the wall's arm. The maximum allowable slope for the landscape is 5:1.

1. Provide the room needed for the lawn and patio area as stated in the problem statement.
2. Draw the length of the retaining wall that connects back to the building on the eastern side of the patio and lawn, 10' patio + 10' lawn + 1' width of the wall = 21' long wall. This is the front face of the wall.
3. The existing topography that travels through the house and the proposed patio and lawn, contours (98), (99), and (100), will need to be adjusted. Because there is a proposed 100 closed contour around the building and the patio, the existing (100) contour will be adjusted to the west and will not tie into the proposed retaining wall. Only the (98) and (99) contours will be supported by the retaining wall. The north arm of the retaining wall needs to support two proposed contours. Use the maximum allowable slope to determine the minimum length for the retaining wall arm, 2'/0.20 = 10'.
4. From the inside corner of the wall, calculate the spot elevations along the base of the retaining wall arm at the minimum slope, 10' − 1' width of wall = 9' for the inside length; 9' × 0.02 = 0.18'; 100.14' − 0.18' = 99.96'. Note that the elevation must slope away from the corner at the minimum slope to ensure proper drainage, see Figure 4.5–12.

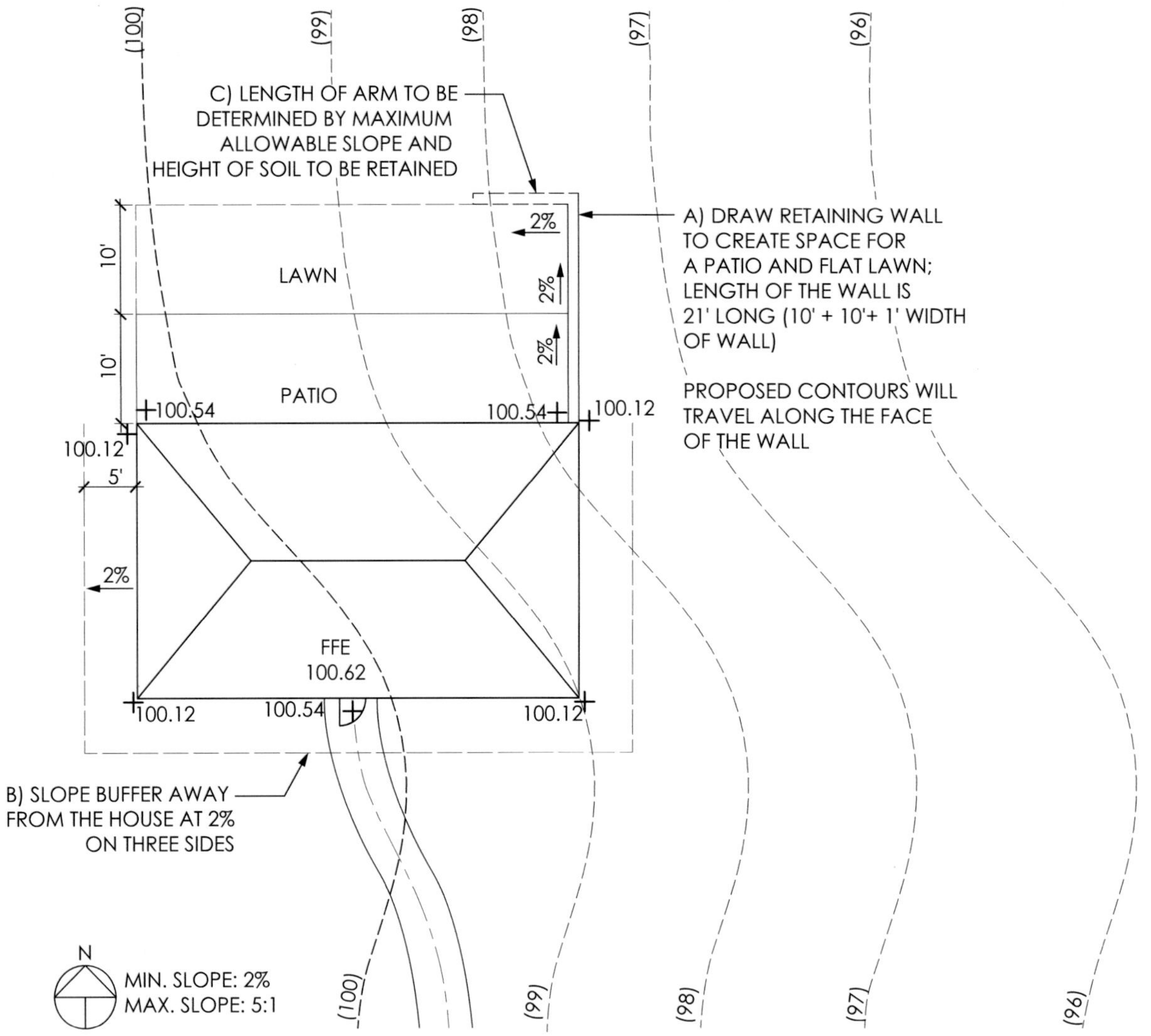

4.5–11
Retaining wall with patio and lawn

5 From the outside corner calculate the spot elevation for the northwestern corner of the retaining wall arm using the minimum slope, 1' wide wall × 0.02 = 0.02'; 99.96' – 0.02' = 99.94'.

6 Determine the bottom elevation for the northeastern corner of the retaining wall arm using the maximum allowable slope, 10' × 0.20 = 2'; 99.94' – 2' = 97.94'. Note that the full height of the retaining wall is taller than 4', TW 102.14' – BW 97.94' = 4.20'.

7 Interpolate along the eastern front face of the wall to determine the slope and the location of the whole number spot elevations, 100.12' – 97.94' = 2.18'; 2.18'/21' = 0.1038. Remember, the proposed 100 contour wraps around the full retaining wall and encircles the house.

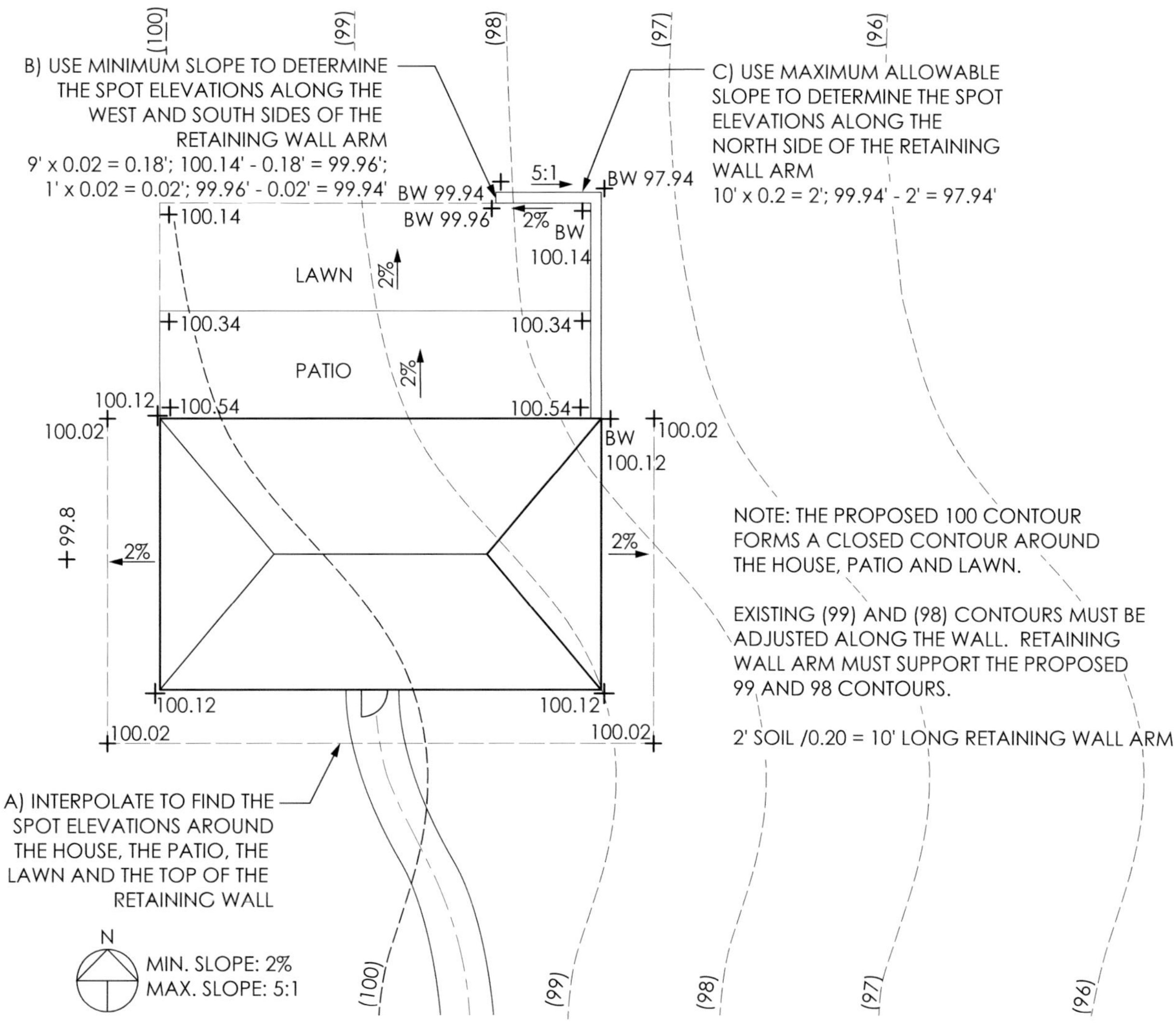

4.5–12
Finding spot elevations for the building, lawn, patio, and wall

8 Interpolate along the northern side of the retaining wall arm to determine the location of the proposed 99 and 98 whole number spot elevations using the maximum allowable slope, 5:1.
9 Label the top of the wall 2' higher than the elevation along the patio and lawn.
10 Interpolate along the southern side of the retaining wall arm to find the location of the proposed 100 contour. This will complete the closed contour around the building, retaining wall, patio, and lawn. The completed illustration is shown in Figure 4.5–13.

There are other elements in the landscape that act as retaining walls, see Figure 4.5–14. Stair cheek-walls are sometimes used as retaining walls. As with standalone retaining

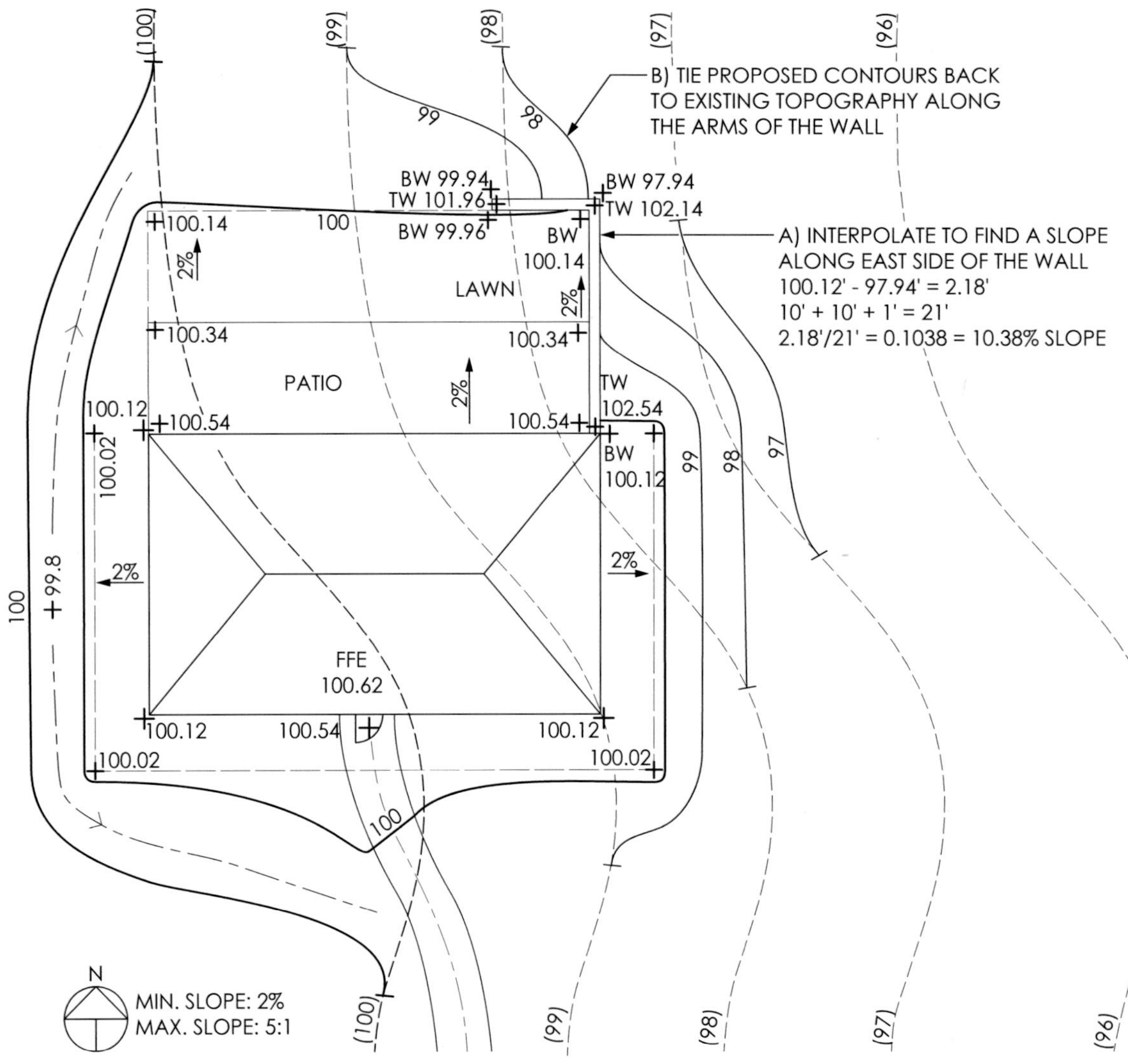

4.5–13
House with patio, lawn, and retaining wall

walls, label the top of wall (TW) and bottom of wall (BW) spot elevations. Indicate any changes in elevation with spot elevations on top of the wall. Tie the proposed contours into the cheek-walls using the maximum slope for the problem.

A split-level building has two finish floor elevations; one side is higher than the other. Soil may be retained along one or two sides of the structure. Interpolate along the length of the building to find the location of the whole number spot elevations. Remember to add a 5' buffer at the side sloping away at 2%. Swales may be needed to prevent water from flowing towards the building. The proposed contours will tie into the side of the building.

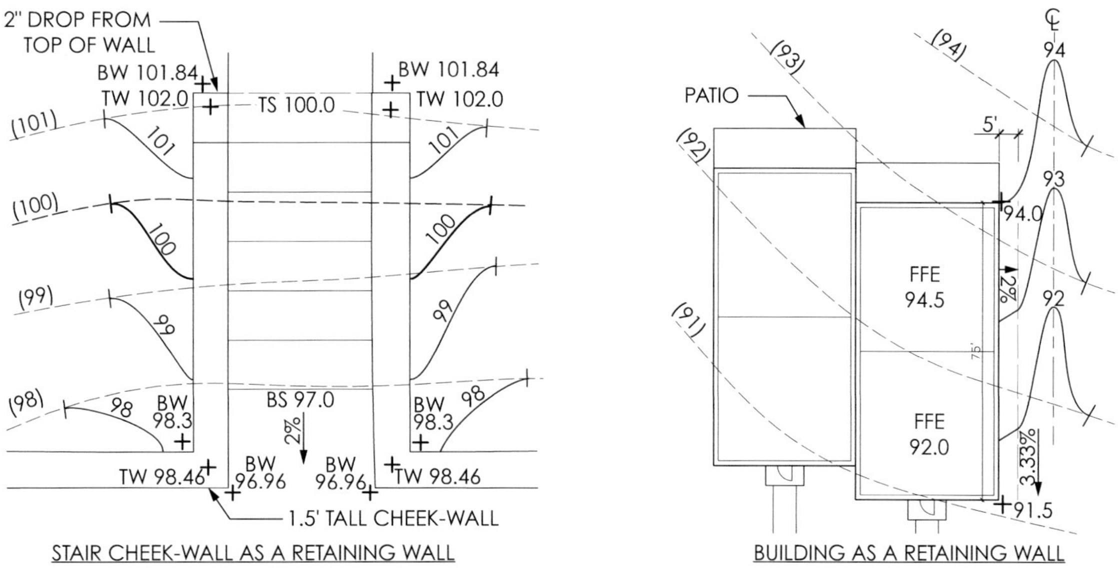

4.5–14 **Other examples of built elements that retain soil**

4.6 Summary

Roads, paths, stairs, ramps, and retaining walls are all built elements included in the landscape to enhance a design. They are more uniform in appearance than landscape elements, but the same principles of grading apply. Roads and their associated structures–curbs, sidewalks, shoulders and swales–are subject to the same longitudinal slope, but each structure may have a different cross slope. Accessible paths are closely tied to the surrounding topography and use cross slopes to create sheet drainage. Use either the slope interval of the maximum allowable slope for the path or calculate a path length and slope between a range from the minimum and maximum allowable slopes as needed.

Ramps and stairs provide the means of surmounting vertical change in areas with slopes greater than 5%. Both are subject to the ADA requirements. These requirements provide strict guidelines for landing slopes, handrail extensions, maximum ramp slopes, and stair riser heights. Understanding how to apply the requirements will allow you to answer a grading problem efficiently. Calculate the length of landing and number of risers for stairs, and length of ramp and number of landings for ramps first before drawing your solutions.

Retaining walls create vertical separation between different areas of a site for the inclusion of site features such as a terrace or a planting bed or to minimize the earthwork needed to level a site to accommodate a built element. Retaining walls have many possible solutions and few requirements; read and understand the problem statement before creating a solution that meets the parameters. For all these built elements remember to add the appropriate prefix to any required spot elevations; TC/BC for curbs, TS/BS for stairs, and TW/BW for walls.

4.7 Exercises (answers start on page 194)

4-1) Draw and grade a 6' wide path connecting Point A and Point B.

a) Path longitudinal slope is 3%; path cross slope is 2%; maximum slope is 4:1.
b) Calculate the total length of path needed.
c) Site path to avoid existing trees and connect to existing topography.

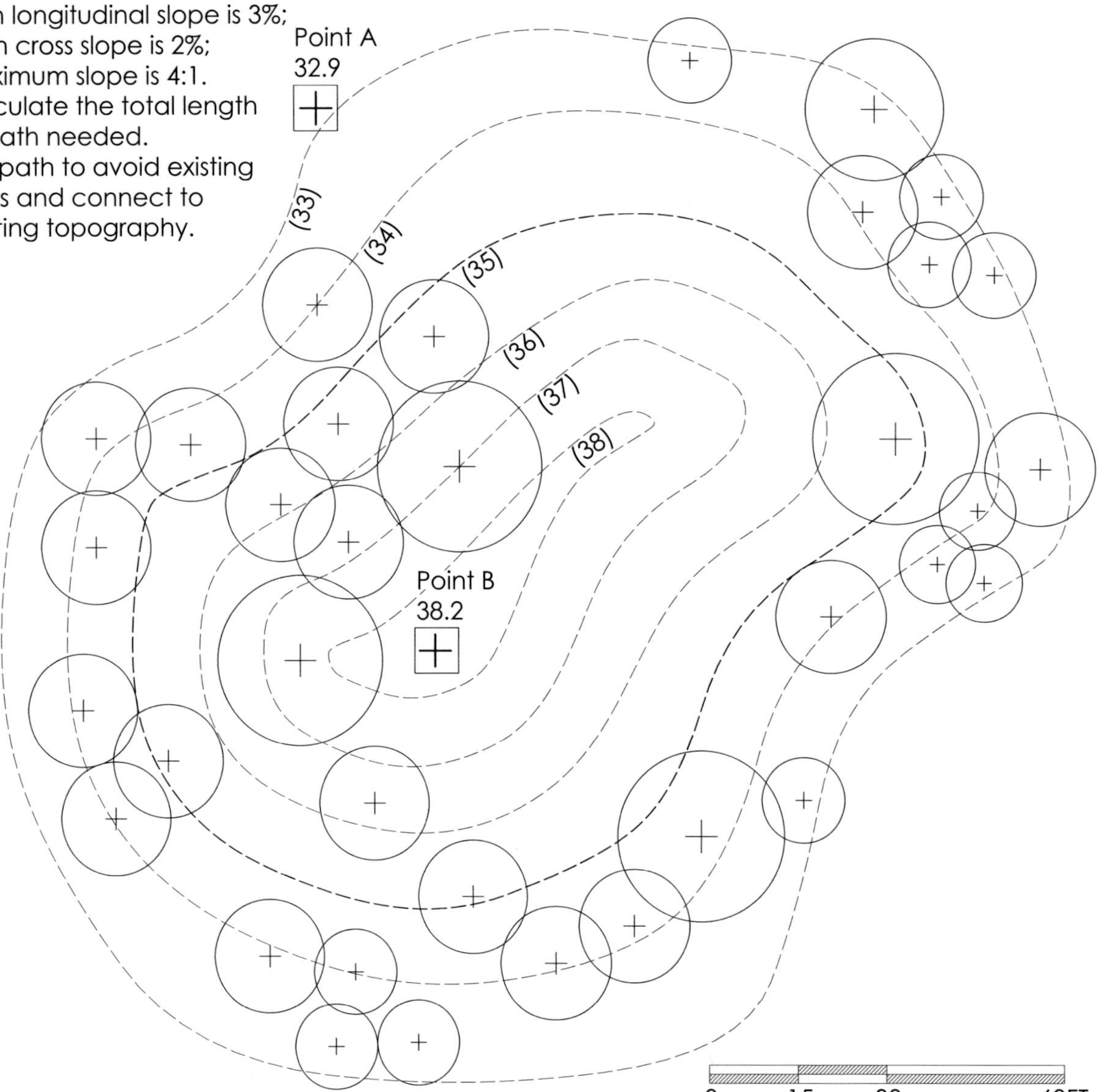

4-2) Grade the asphalt road, 6" curb and sidewalk.

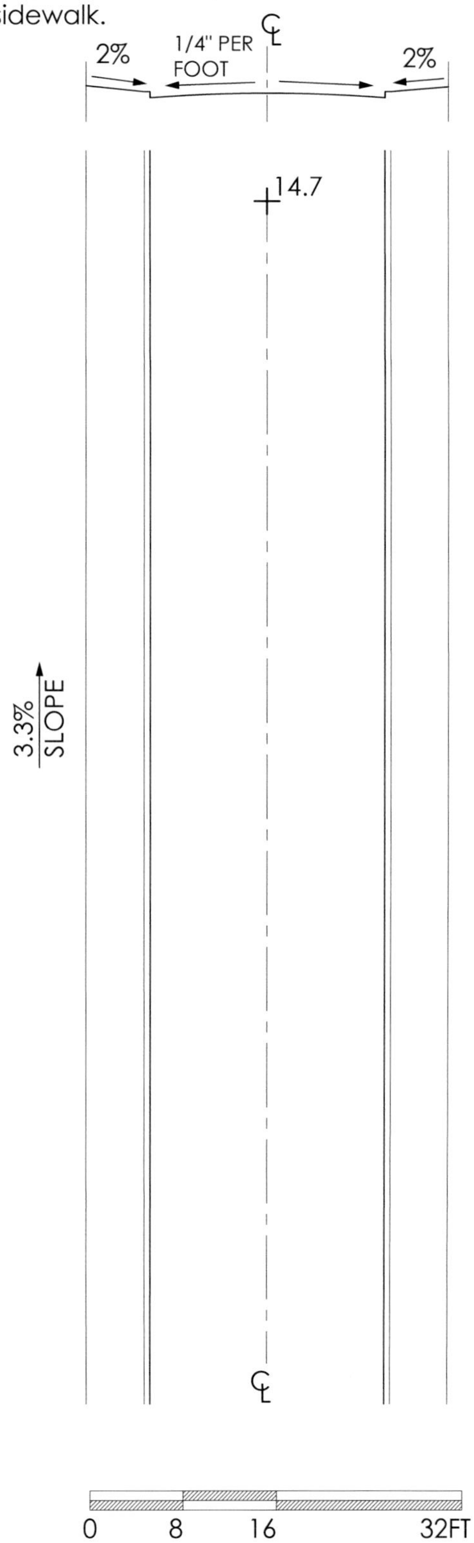

4-3) Grade the asphalt road, shoulders and swale.

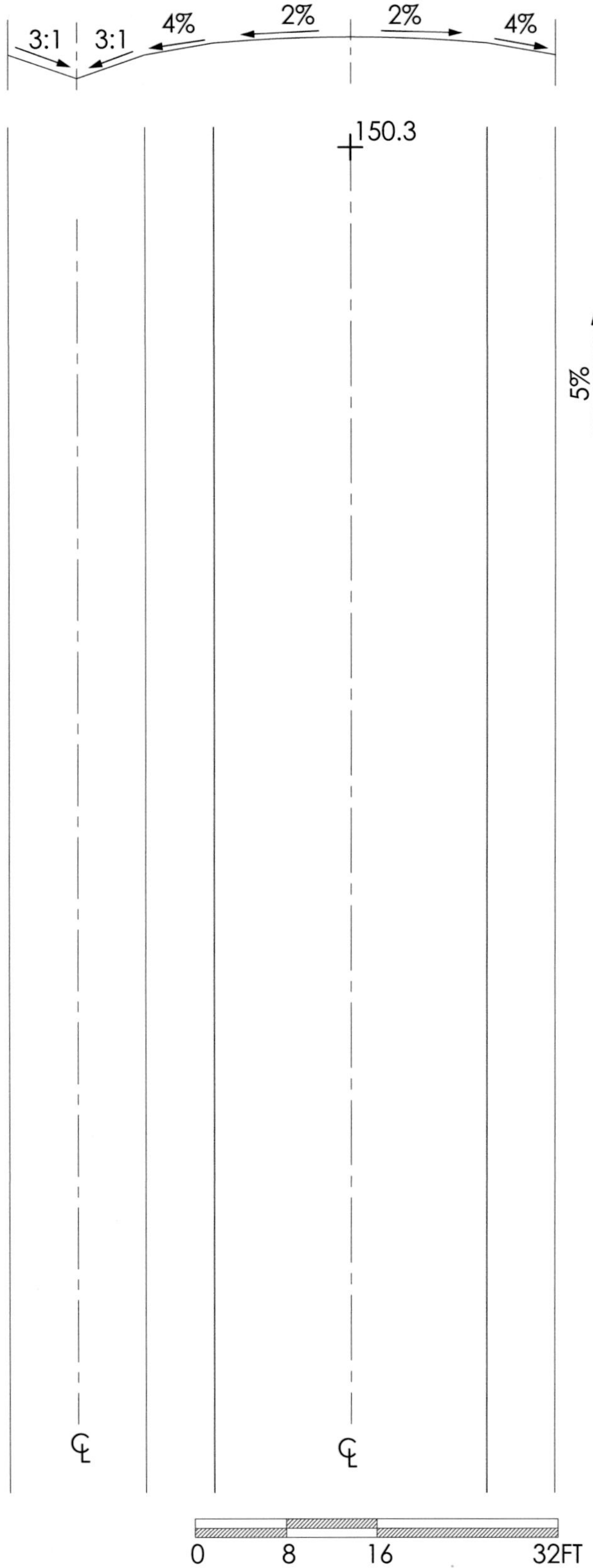

4-4) Grade the ramp with uniform slopes.

GENERAL:
A. Total change in vertical distance = 5.08'.
B. Provide critical spot elevations at 1/100th of a foot (x.xx').
C. Show required handrails.

RAMPS:
D. Label all parts of the ramp.
E. Label the slope on each leg of the ramp.
F. Provide a 2% cross slope for drainage where needed.

LANDINGS:
G. Slope on landings not to exceed 1:48 (2%).

HANDRAIL EXTENSIONS:
H. Slope handrail extension areas at 2% slope downhill.

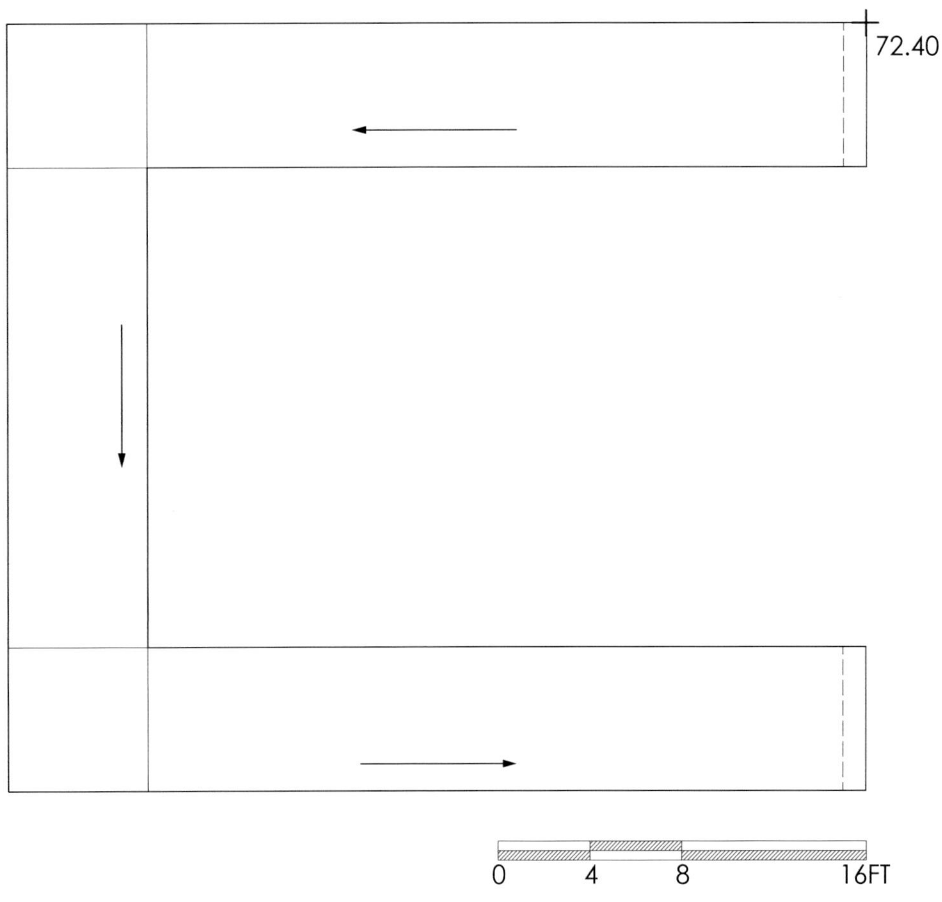

4-5) Grade stairs, cheek walls and retaining walls. Provide all spot elevations and show proposed contours.

STAIRS:

A. Create two flights of stairs with 5 steps each separated by a 5' landing.
B. Riser height = 6", Tread length =12"
C. Slope on stair landing is 2%.
D. Show all parts of the stairs.
E. Label TS and BS for each set of stairs.

CHEEK WALLS:

F. Cheek walls are 1' wide and continuous.
G. Cheek walls extend horizontally 1' above and 2' below the stairs.
H. Top of cheek walls are 2.5' higher than first TS and last BS elevations.
I. BW at the landscape side of the cheek wall is 3" lower than TW.

RETAINING WALLS:

J. 1' wide retaining wall extends perpendicular to the bottom of the cheek walls for 14' on both sides.
K. Elevation of top of retaining wall at the corner is the same as the top of the adjacent cheek wall.
L. Top of the retaining wall slopes at 2% for 14' on either side of the stairs.
M. BW at the back of the retaining wall is 3" lower than the TW.

PROPOSED CONTOURS:

N. Interpolate along cheek wall to find proposed contours.
P. Tie back to existing contours.
P. Maximum landscape slope is 3:1.
Q. Label contours.

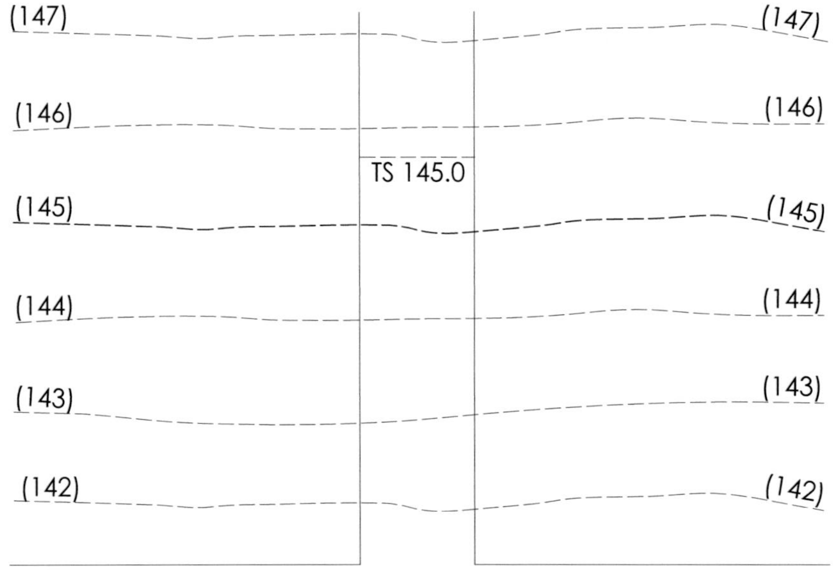

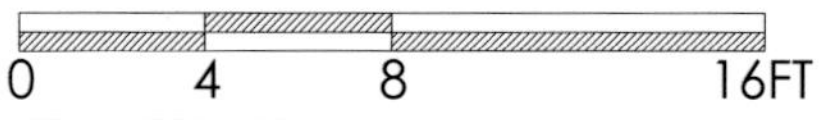

4-6) Create two flights of stairs and three equal landings from the building to the sidewalk.

a) Riser height = 6"; Tread length = 12"
b) Write the slope on the landings, show handrails as needed and provide dimensions for the stairs and landings.

DISTANCE IN FEET	STAIRS	LANDINGS	TOTAL
VERTICAL			
HORIZONTAL			

Total number of steps__________

Slope of landings__________

Length of a typical landing__________

DISTANCE IN FEET	STAIR 1	STAIR 2	TOTAL
VERTICAL			
HORIZONTAL			

DISTANCE IN FEET	LANDING 1	LANDING 2	LANDING 3	TOTAL
VERTICAL				
HORIZONTAL				

4-7) Grade the stairs and the ramp with uniform slopes. Show all parts of the stairs and ramps.

STAIRS:

A. Riser height = 6"; Tread Length = 12"
B. Overall vertical elevation for the stairs = 5'
C. Provide a 5' clear landing at the bottom and top of stairs only.
D. Stair landings slope at 2%.
E. Draw and label all parts of the stairs.

RAMPS:

F. Calculate overall vertical elevation for the ramps.
G. Slope handrail extension at 2% downhill and subtract elevation from total vertical distance.
H. Calculate minimum length of ramp needed.
I. Determine the nearest whole number higher than the minimum length of ramp.
J. Calculate and label slope for each leg of the ramp.
K. Provide a 2% cross slope for the ramp where needed.
L. Label all spot elevations on ramp.

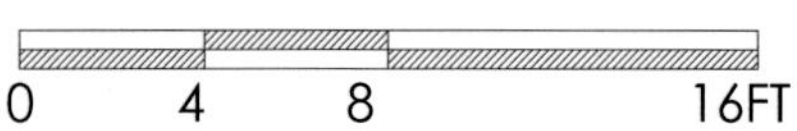

4-8) Create an ADA-compliant ramp with intermediate landings and determine the elevation at the top of the overlook deck. Topography slopes downhill from north to south. The sidewalk acts as the landing at the bottom of the ramp, no additional landings are needed. Proposed contours are not needed.

a. Grade the ramp using an 8% longitudinal slope.
b. Show all the necessary spot elevations along the ramp runs and landings.
c. Grade landings so that they drain; maximum gradient on the landings is 2%.
d. Cross slope on the middle ramp section is 2% and drains to the south.
e. Remember to add in handrail extension areas; slope on the handrail extensions is 2% downhill.
f. Show required handrails.
g. Show elevation of Overlook Deck.
h. Show elevations to the nearest 1/100th (x.xx').

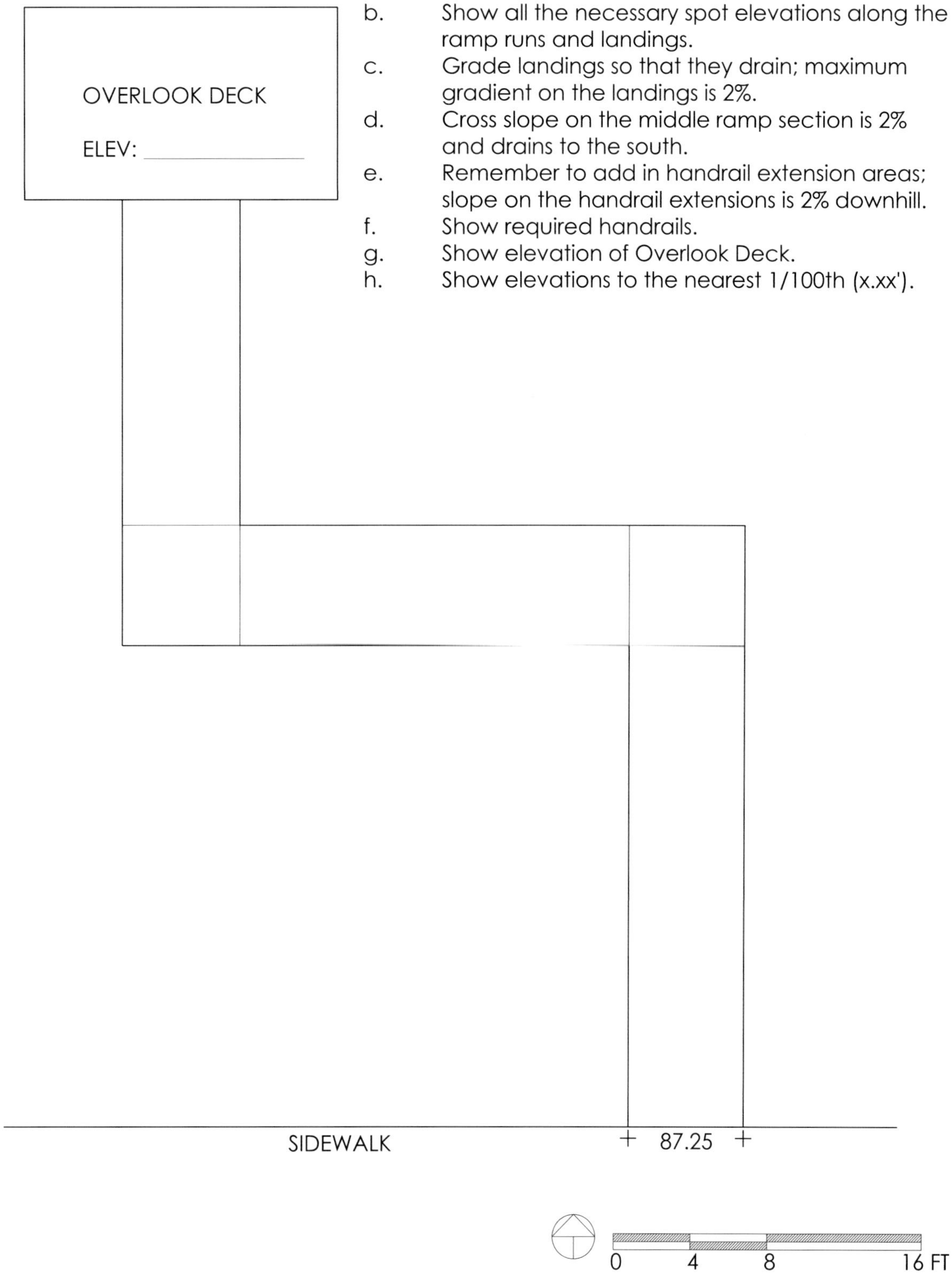

4.8 Literature cited

Harris, Charles W. and Nicholas T. Dines. 1997. *Time-Saver Standards for Landscape Architecture. 2nd Ed.* New York: McGraw-Hill.

Hastings, Jerry P. 2006. *LARE Review Section E Vignettes Grading Drainage and Stormwater Management. 2nd Ed.* Belmont, CA: Professional Publications, Inc.

Hopper, Edward J. ed. 2006. *Landscape Architectural Graphic Standards.* New York: John Wiley & Sons, Inc.

International Code Council. 2015. *International Building Code: Chapter 10 Egress, Third Printing.* Washington, DC: ICC. https://codes.iccsafe.org/content/IBC2015/chapter-10-means-of-egress.

Strom, Steven and Kurt Nathan. 1998. *Site Engineering for Landscape Architects. 3rd Ed.* New York: John Wiley & Sons, Inc.

Strom, Steven, Kurt Nathan and Jake Woland. 2013. *Site Engineering for Landscape Architects. 6th Ed.* New York: John Wiley & Sons, Inc.

United States Department of Justice. 2010. *2010 ADA Standards for Accessible Design.* Washington, DC: US Department of Justice Civil Rights Division. www.ada.gov/2010ADA standards_index.htm.

Untermann, Richard K. 1973. *Grade Easy: An Introductory Course in The Principles and Practices of Grading and Drainage.* McLean, VA: Landscape Architecture Foundation.

Section Five
Landscape and built element combinations

Understanding how to grade landscape and built elements separately is the first step to mastering grading. Knowing where to start and stop and to read topography are essential steps. Most grading problems, and any real sites, will, however, incorporate multiple elements from both landscape and hardscape grading. This chapter focuses on three common examples that might occur on the exam, but note that there are endless possibilities that occur in the professional world.

5.1 Parking lots

Warped planes are more common in the landscape than planes with a consistent uniform slope or slopes. They occur in both landscape situations, such as baseball and football fields, or in hardscape situations, such as open plazas. When grading warped planes the goal is to direct the water flow in the desired direction using a minimal slope, creating ridges or swales as needed. Surface runoff will end up in a swale or in a drain inlet. The minimum slope on landscape is 2% and the minimum slope for hardscape is 1%, although 2% is the standard and ensures positive drainage and prevents pooling of water on surfaces that are not well-graded.

The most common forms of warped planes you will encounter on the exam are parking lots and plazas. Unlike fields, plazas and parking lots have impermeable or semipermeable surfaces with minimal percolation. Surface runoff must be removed as efficiently and quickly as possible to an open or closed drainage system to prevent ponding of water. A parking lot follows the principles of a warped plane in order to drain the surface during a precipitation event. It may be sloped to the center and form a swale flowing towards a drain. If you are presented with a rim elevation (RIM EL) for a drain inlet this is the lowest point for the warped plane. All spot elevations for the plane surface are higher and must drain towards the rim elevation. If you are presented with an existing adjacent road as a starting point, interpolate along the road using the slopes provided to determine the RIM EL and proposed contours.

Although many parking lots in rural areas will allow water to flow onto the surrounding landscape, the majority of parking lots have curbs to ensure that surface runoff from the parking lot stays in the closed drainage system. Curbing creates two issues that must be addressed in the grading problem solution: proposed contours must travel along the face of the curb to tie into adjacent sidewalks or landscape similar to a road signature; and curbing adds curved structures, specifically turning radii that aid in ease of travel for vehicles; however, interpolating on these tight curves may be difficult and can be inaccurate.

The parking lot example shown in Figure 5.1–1 is for illustrative purposes, note that it is recommended, and required in most instances, that a parking lot have two entrances for improved vehicular circulation. If a parking lot has only one entrance, provide a designated area as a vehicular turnaround to avoid vehicles backing out into public rights-of-way. This example shows a parking lot with a 6" tall curb. It is a warped plane; all surface runoff will drain on the western side of the lot into a drain inlet. No water from the road or landscape must enter the parking lot. One of the slopes for the parking lot is missing and must be calculated. The minimum and maximum slopes in the parking lot are 2% and 3% respectively. It is important to visualize the parking lot problem in three stages: grade the elevations from the existing road to the edge of the driveway; the edge of the driveway to the edge of the parking lot and then the entire parking lot. Because the minimum and maximum slopes are tight, the order that you choose to grade will impact the missing slope.

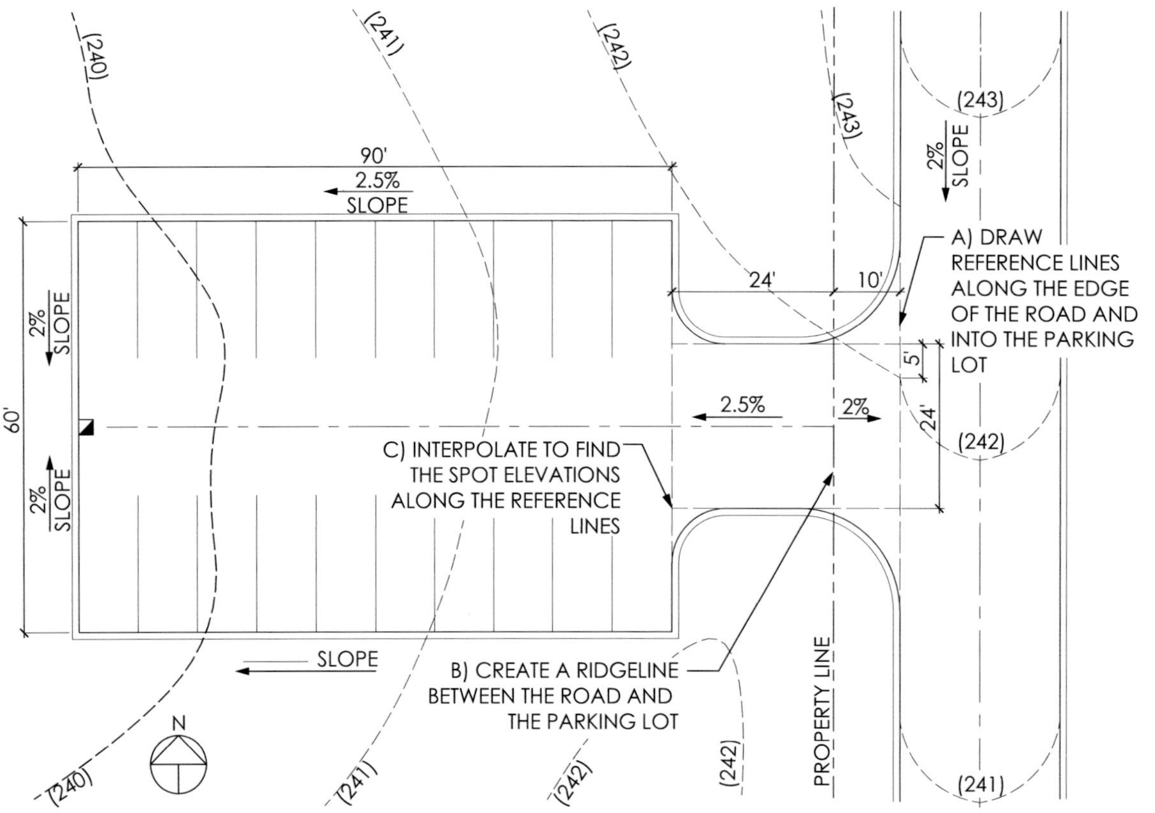

5.1–1
Parking lot and existing road with dimensions and slopes

1 Determine the slope for the existing road and any shoulder areas if not already provided.
2 Draw light reference lines perpendicular to the edge of the road into the parking lot driveway. These reference lines will be used to determine spot elevations thereby avoiding the inaccuracy of finding spot elevations along the curved curb.
3 Interpolate to find the spot elevations along the edge of the road where they intersect with the reference lines.
4 Using the slope provided, interpolate from the edge of the road up to the edge of the driveway. Note that water from the road must not cross into the parking lot and vice versa so there will be a slight ridge at the entrance to the parking lot. In this example, the ridgeline will occur on the property line.
5 Draw a light reference line from the northeastern corner of the parking lot to the southeastern corner. Make sure that all reference lines intersect at 90° angles. Interpolate to find the whole number spot elevations along the edges of the parking lot.
6 Interpolate in the driveway to find the spot elevations along the ridgeline, centerline and at each of the corners of the driveway along the reference lines, see Figure 5.1–2.
7 Interpolate from the end of the centerline to the drain inlet and calculate the RIM EL for the drain, 114' × 0.025 = 2.85'; 242.06' − 2.85' = 239.21'.

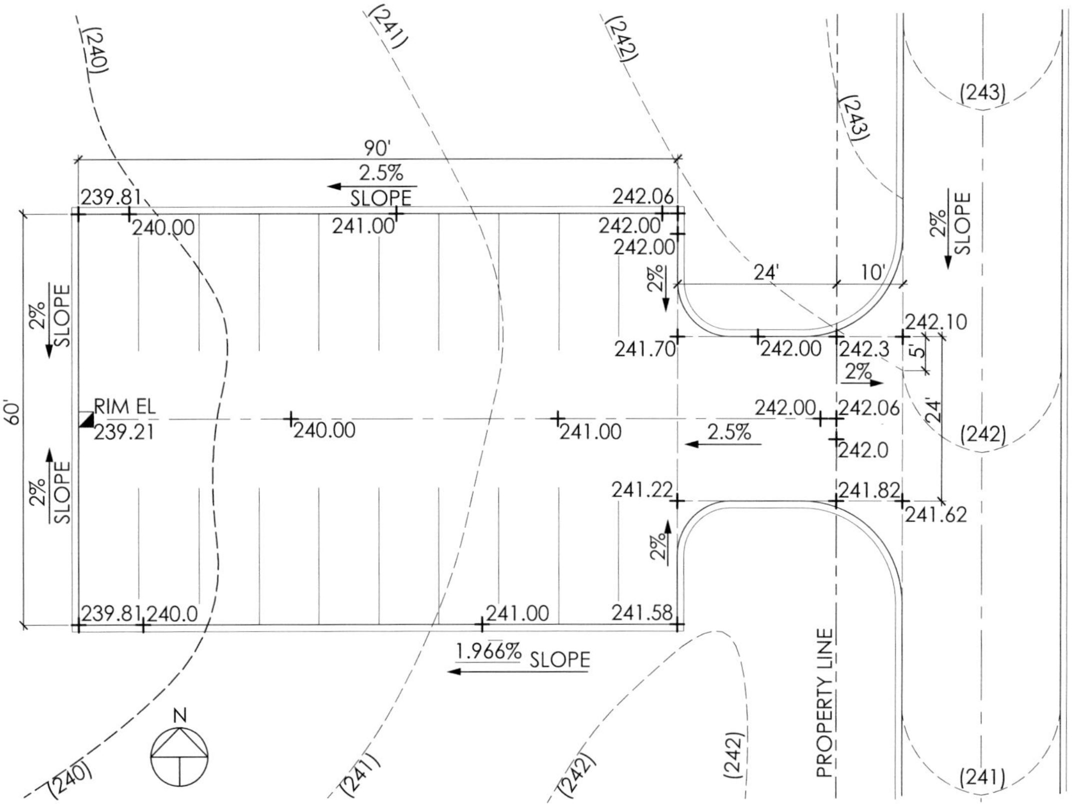

5.1–2
Parking lot showing whole number spot elevations

8 Interpolate using the provided slopes for the parking lot to find the spot elevations at the corners of the parking lot.
9 Calculate the missing slope along the southern side of the parking lot, 241.58'−239.81'=1.77'; 1.77'/90'=0.019667 or 2%.
10 After all the spot elevations have been determined for the centerline, corners and reference lines, locate the whole number spot elevations along the edges of the parking lot, driveway and property line.
11 Connect the whole number spot elevations to find the proposed contours within the parking lot. Note that the proposed 242 contour will travel along the face of the curved curb between the driveway and parking lot as shown in Figure 5.1–3.
12 Add 6" at each corner of the parking lot to find the top of curb (TC) elevation. Remember to label the TC and BC for all spot elevations at the corners.
13 Using the slopes for the parking lot interpolate to find the location of the whole number spot elevations along the top of the curb on all sides of the parking lot.
14 Draw proposed contours from the whole number spot elevations on the top of the curb to the existing topography using smooth continuous lines.

There are many working solutions to this parking lot problem, but because the range for the minimum and maximum slope is very tight and slopes for the parking lot are provided; the solutions are limited. Figure 5.1–4 looks at three possible scenarios. First, if

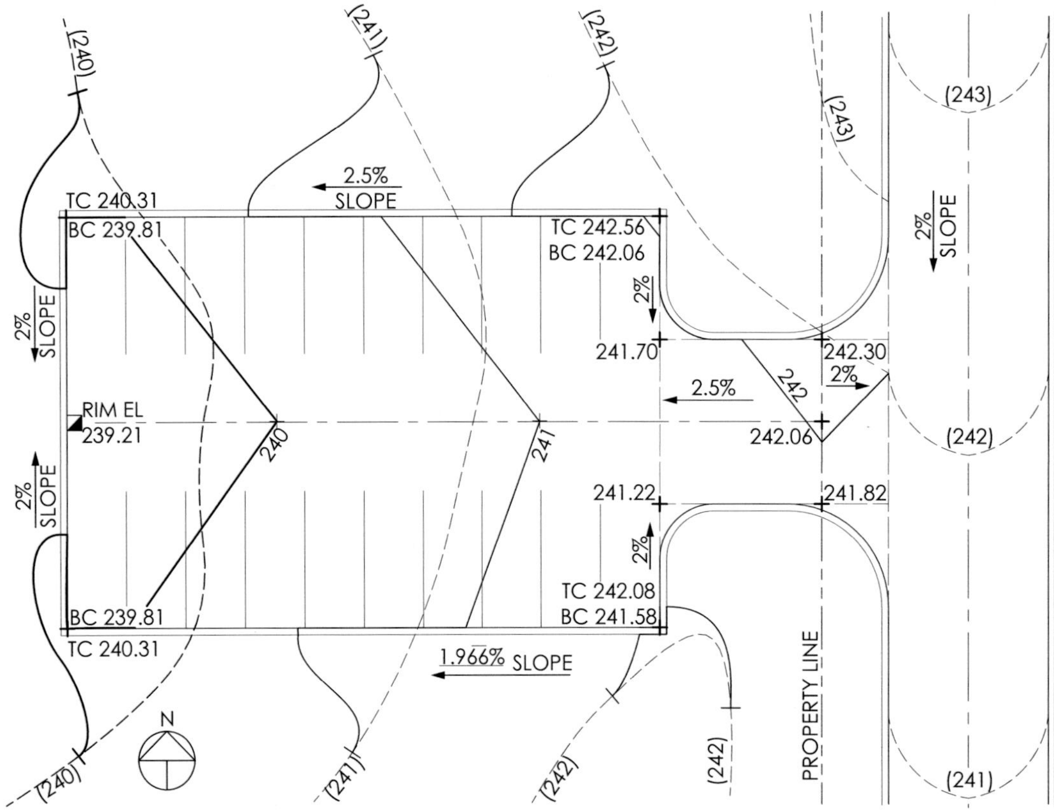

5.1–3
Finished parking lot

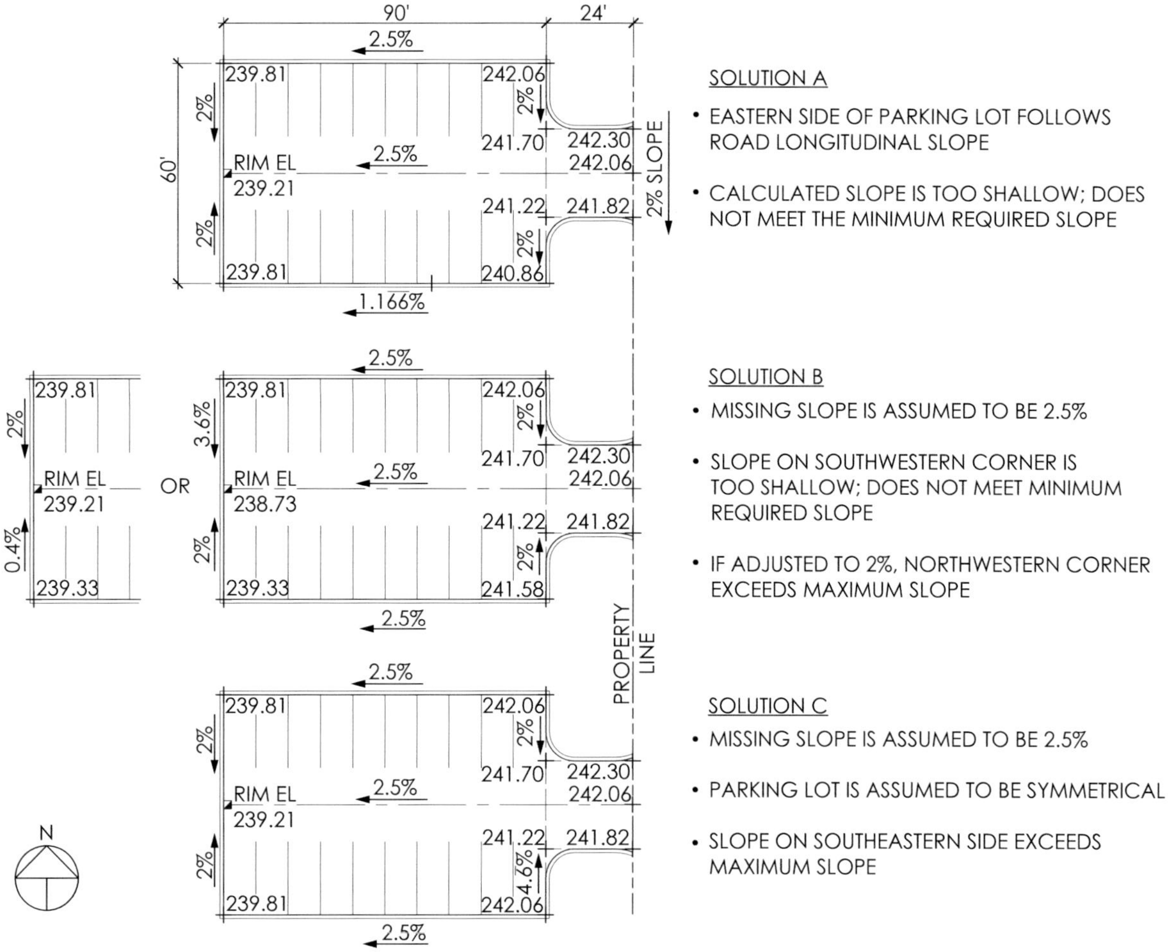

5.1–4
Possible solutions to the parking lot problem

you use the longitudinal slope from the road to calculate spot elevations along the eastern side of the parking lot the spot elevation in the southeast corner will be too low. The subsequent calculated slope for the southern side of the parking lot will be too shallow and not meet the minimum requirement as shown in Solution A.

Solution B calculates the southeastern side of the parking lot correctly using the 2% slopes for the parking lot to find the corner spot elevations. However, it assumes that the slope along the southern side is 2.5%, the same as the northern side. Calculating the spot elevation for the southwest corner using 2.5% will make it lower than the northwest corner. The slope between the southwest corner and the drain inlet is too shallow. Adjusting the slope from the southwest corner to meet the minimum of 2% by lowering the RIM EL will make the slope from the northwest corner to the drain inlet too steep.

Solution C again assumes that the slope along the southern side is 2.5%. Interpolating from the RIM EL to the northwest and southwest corners and then to the northeast and southeast corners will create a symmetrical parking lot, but the slope on the southeast side of the parking lot will exceed the maximum slope. Find the RIM EL first using the 2.5% slope, then the spot elevations in the corners using the 2% slope and lastly calculating the slope along the southern edge.

5.2 Culverts

Culverts are engineering devices, usually a pipe, that shuttles water under roads and other structures, and, in some cases, they have associated wing walls at the entrance and exit of the pipe. They are part of the open drainage system because water daylights on the other side of the structure. Culverts are minimally sloped to maintain positive water flow. Spot elevations mark the bottom of the entering and exiting sides of the culvert and it is essential to know the diameter of the pipe. The pipe will protrude on both sides of the road and may or may not have wing walls associated with it.

Solving a culvert problem requires you to grade at two different levels, finish grade and subgrade. You will have to grade a built structure, usually a road, and a landscape structure, a swale. To correctly answer the problem, you will have to work with several different slopes. Grade the structure above the culvert first, typically a road, using the road gradient and its crown. Use the spot elevations from the structure above to find the center of the bottom of the culvert pipe below; grade the pipe using the pipe slope. The swale that enters the pipe will have a longitudinal slope and a cross slope, usually the maximum landscape slope. Figure 5.2–1 shows a road in plan view with a culvert pipe underneath. Figure 5.2–2 shows the same pipe in a longitudinal cut view showing the inside of the pipe and the three slopes: maximum slope for the wing walls, longitudinal slope for the swale, and pipe slope.

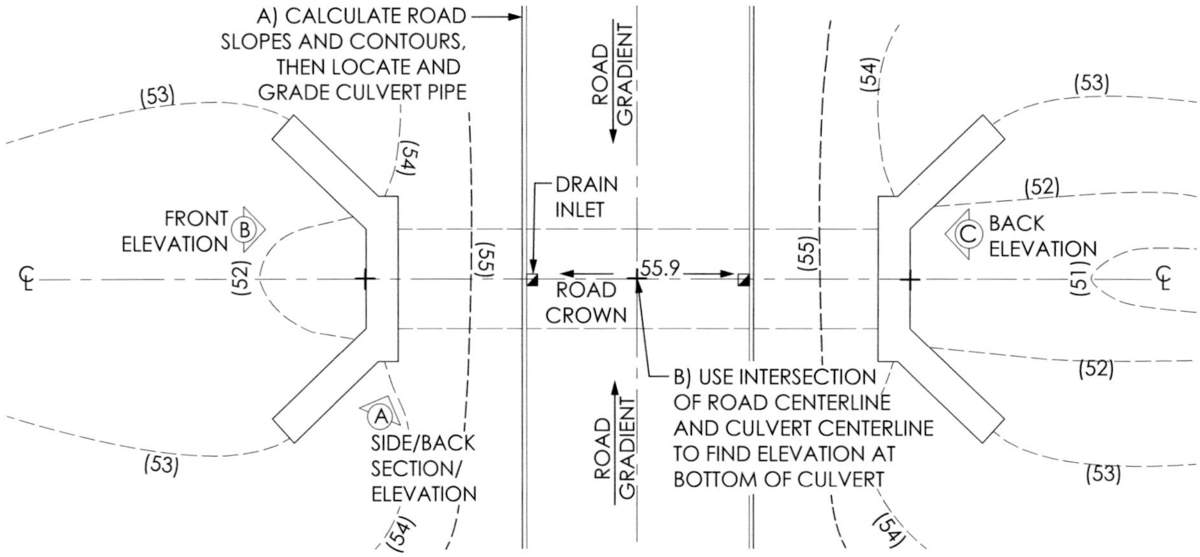

5.2–1
Plan view of road and culvert

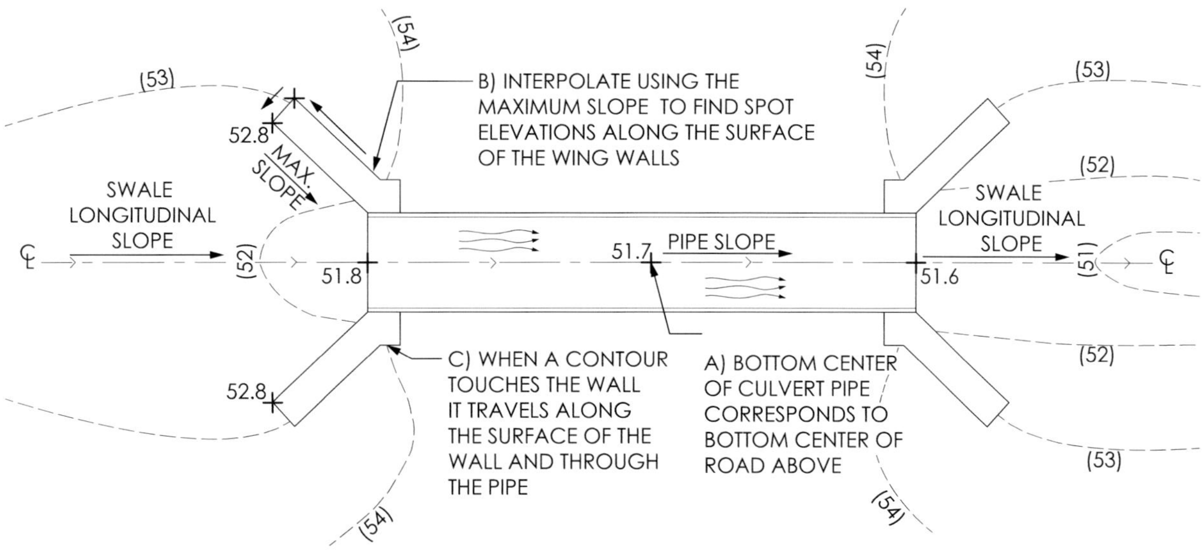

5.2–2
Longitudinal cut through culvert pipe – top view

The following example highlighted in Figure 5.2–1, shows an existing 3' diameter culvert pipe under an existing road. The wing walls at the entrance and exit to the culvert are 4' long. The maximum allowable slope for the topography is 4:1 and the pipe slope is 0.5%. Note that, in most cases, culverts are much deeper in the ground and longer, but for ease of illustration the culvert pipe is slightly more than 1' deep.

1 Determine the spot elevation along the centerline of the road where the centerline of the road above intersects with the centerline of the culvert pipe below, 55.9'.
2 From this elevation subtract the depth of soil cover over the pipe, 1', the thickness of the pipe, 0.2', and then subtract the diameter of the pipe, 3', to determine the elevation at the bottom center of the pipe, +51.7', see Figure 5.2–3, the section-elevation.
3 Using the pipe slope, 0.005 (0.5%), and the total length of the pipe, 36', interpolate to find the entering and exiting spot elevations for the pipe, 51.8' and 51.6'

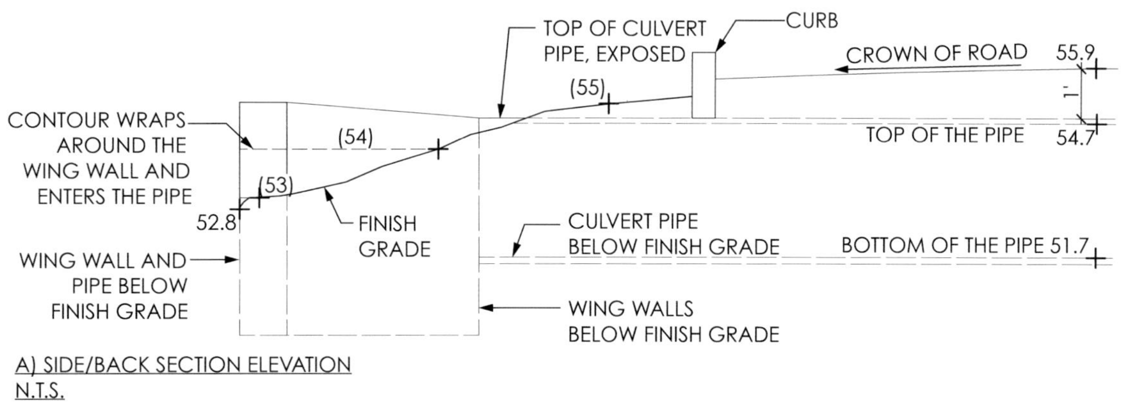

5.2–3
Longitudinal section-elevation of pipe culvert with road above

respectively. These are the invert in (INV IN) and invert out (INV OUT) elevations. In some instances, the pipe slope will not be given. If you have to determine the pipe slope you will be given a minimum or maximum slope for the problem. It is easiest to give the pipe a shallow slope which will facilitate the connection of any proposed contour lines once the swale leaves the pipe.

4 Starting at the entrance to the culvert, interpolate along the wing walls of the culvert using the maximum slope, 4:1, to determine the spot elevations at the corners of the wing walls. Remember to interpolate uphill, the spot elevation at the entrance to the culvert is a low point, see Figure 5.2–4, the front elevation.

5 Interpolate to find the whole number spot elevations on the front and the back of the wing walls, the spot elevations at each corner, and along the pipe if necessary.

6 Determine the height of the culvert at the entrance, 51.8' + 3' = 54.8'. This means that the (52), (53), and (54) contour lines will merge with the sides of the pipe and the (55) contour line will cover the pipe.

7 Once you know where and how the contour lines converge with the culvert connect the swale to the culvert.

8 The sides of the swale will connect to the culvert using the maximum allowable slope, 4:1. Calculate the longitudinal slope for the swale based on the existing contours. Subtract the INV IN from the highest contour for the swale along the centerline for the swale, 55' – 51.8' = 3.2'; 3.2'/80' = 0.04. Refer to Figure 5.2–5.

9 Repeat the same process for the exiting side of the culvert. Start at the INV OUT elevation at the exit to the culvert, 51.6'. Interpolate uphill to find the spot elevations at the corners of the wing walls using the maximum slope, 52.6'. The earth slopes up at 4:1 from the pipe exit to the edge of the wing wall, see Figure 5.2–6, the back elevation.

10 Interpolate to find the whole number contours along the front and the back of the wing walls and along the exposed pipe at 4:1.

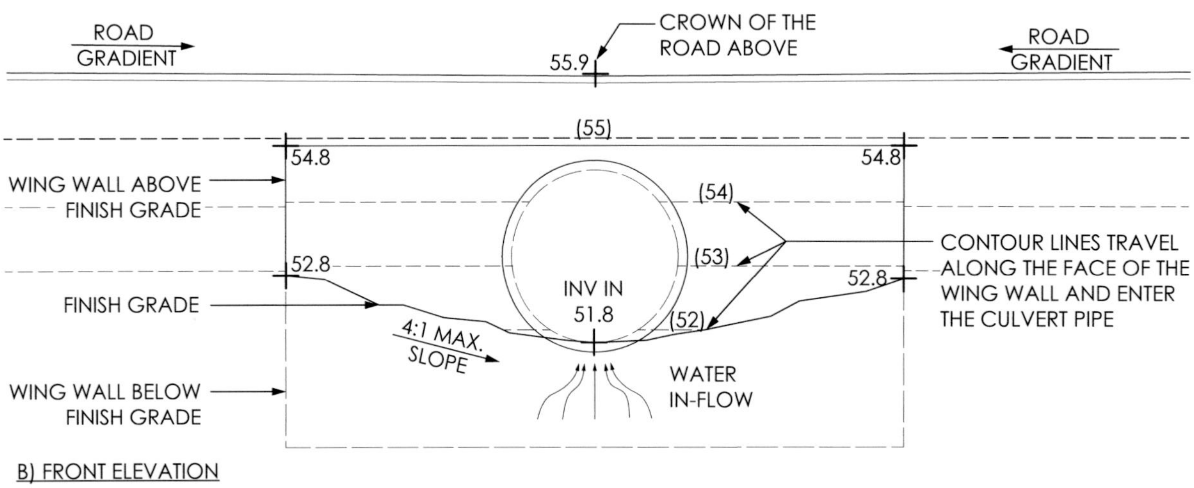

5.2–4
Culvert front elevation

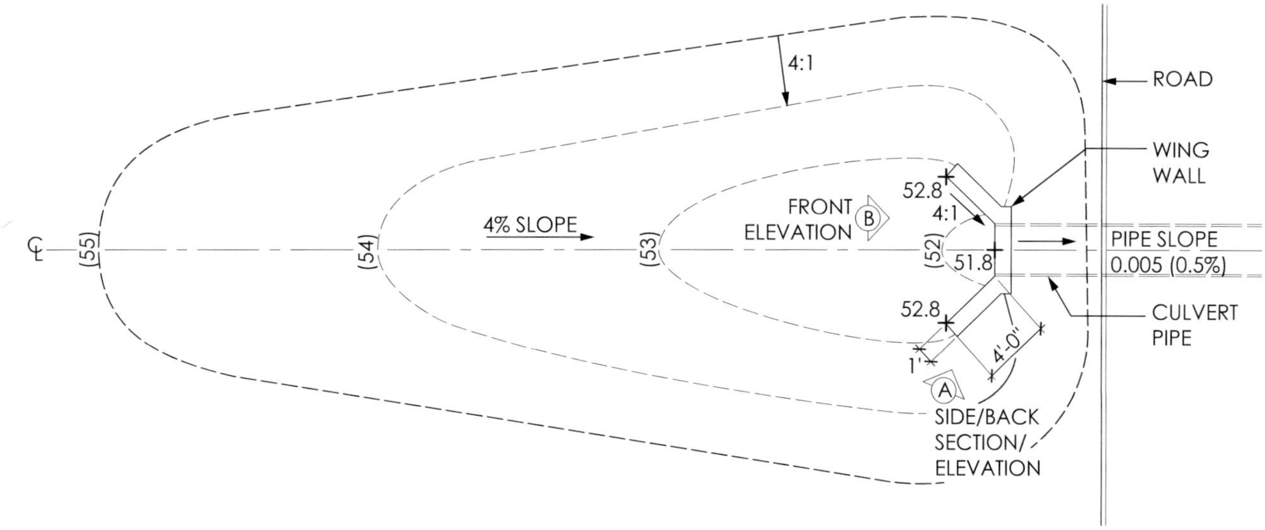

5.2–5
Swale entering pipe culvert in plan view

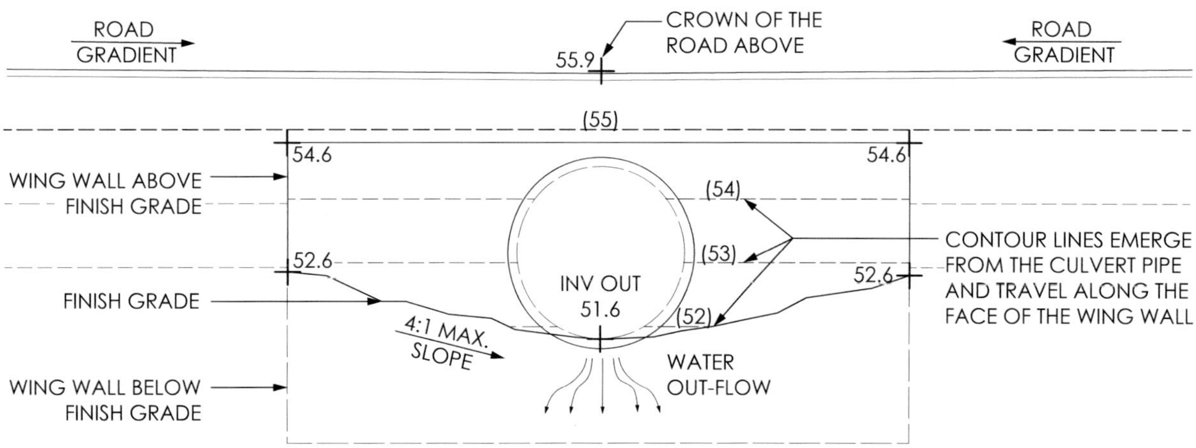

C) BACK ELEVATION

5.2–6
Culvert back elevation

11 Calculate the elevation at the top of the culvert, 51.6 + 3' = 54.6'. Contours (52), (53), and (54) will emerge from the pipe and the (55) contour covers the culvert.
12 Connect the contours to the culvert using the maximum slope. Label the contours on the uphill side and include the maximum slope so it is clear how far apart the contours are, see Figure 5.2–7.
13 Interpolate from the spot elevation on the centerline at the exit of the culvert using the swale longitudinal slope or another suitable longitudinal slope to find the location of the next contour for the swale. Make sure to provide enough room to allow the maximum allowable slope along the sides of the swale.

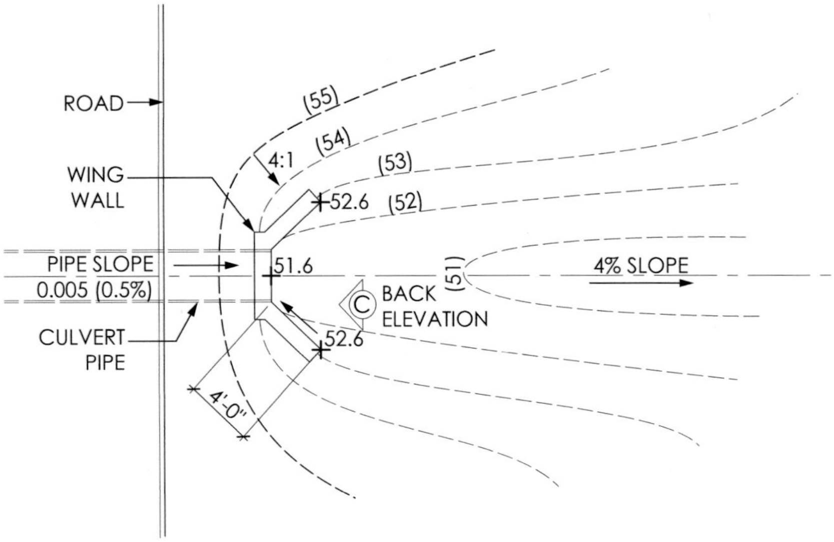

5.2–7
Swale exiting pipe culvert in plan view

5.3 Sloping berms

The last landscape/built element combination is a sloping berm. As described in Section Three, berms are used to block unwanted views in the landscape, usually a stationary object, i.e. a building. To do this requires the demarcation of a viewshed to create a berm that completely blocks the view of the structure. Sloping berms are used to block unwanted views of objects that are moving in the landscape, usually vehicles. Because the viewshed moves, you may have to create more than one berm or a sloping berm. Remember that swales may be necessary to redirect surface water that flows from the berm. The height of the berm will slope as the object moves in the landscape responding to the changing elevations in the landscape. With this in mind, you will have to make more than one line of sight section for the problem. Using a person at Point A with an eye height of 5' and an object on the road at 6' tall we will develop a sloping berm in the example highlighted in Figure 5.3–1.

1 Define the viewshed, or vision cone, if it hasn't been provided. The expanding width of the vision cone is the horizontal distance in which the top of the berm must exist. Note that proposed contours for the berm will extend past the viewshed.
2 Examine the drawing to determine any existing structures or trees that can't be moved. Identify the slope of the topography and any obvious landforms. In this example, there is a swale at contour (120), a pond at (118) and a ridge at contours (121), (124).
3 Measure the distance from the edge of the platform, Point A, to the existing contour in the road, Point B. Repeat the measurement for Points A and C. These horizontal

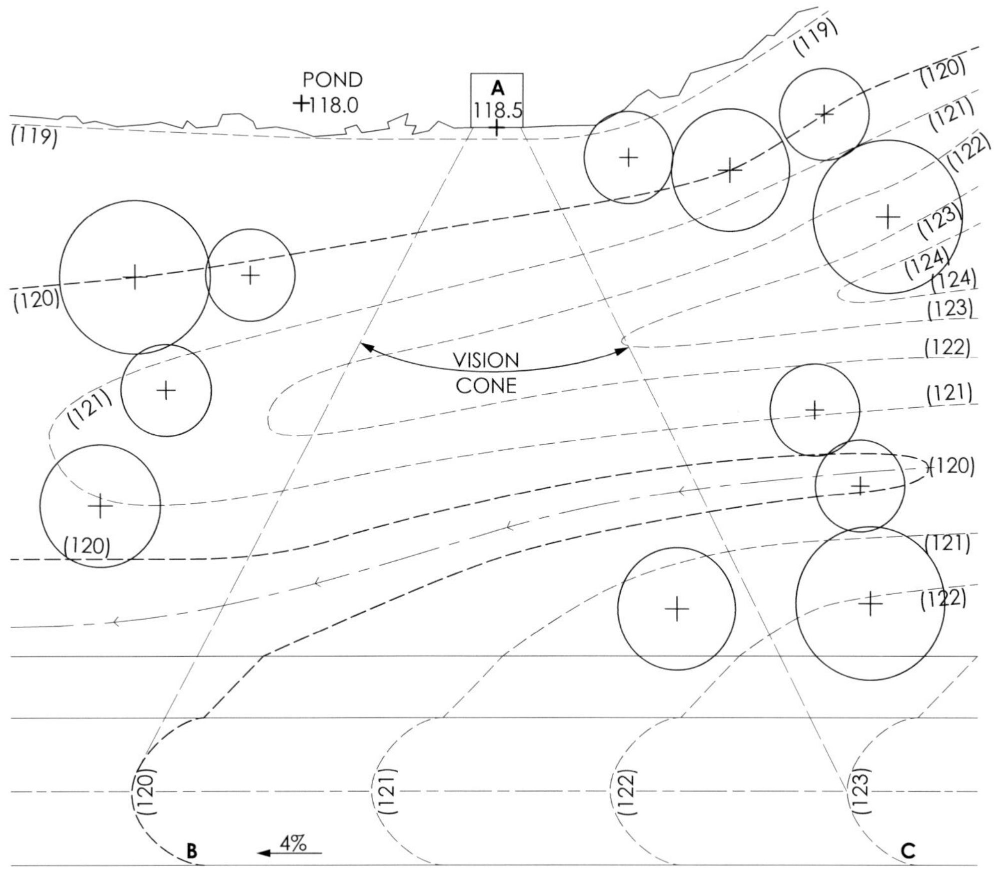

5.3–1
Vision cone for berm

distances are the baseline for your line of sight sections. Note that if the road is curved you may want to create a line of sight section for each contour along the road, see Figure 5.3–2.

4 Determine where the top of the berm should be placed. Draw a reference line, straight or curved that is aligned with the existing topography. Note that the reference line may not be the midpoint of the two lines of the viewshed. In this example, the reference line is aligned with the existing ridge.

5 Measure the distance from Point A to the intersecting points on reference lines AB and AC.

6 Draw two sections for each of the two sides of the viewshed. Each section will determine the elevation that must be covered to block the vehicles on the road. Each section corresponds to the horizontal distance of lines AB or AC.

7 Start each line of sight section at 123.5', Point A 118.5' + eye level 5' = 123.5'. End each section with the elevation of the road contour plus 6' for a vehicle traveling on the road; 120' + 6' = 126' for line AB and 123' + 6' = 129' for line AC. Refer to the sections in Figure 5.3–3.

8 Locate the horizontal distance for the reference line that bisects the vision cone. Note the distances, 40' and 51' respectively on each section.

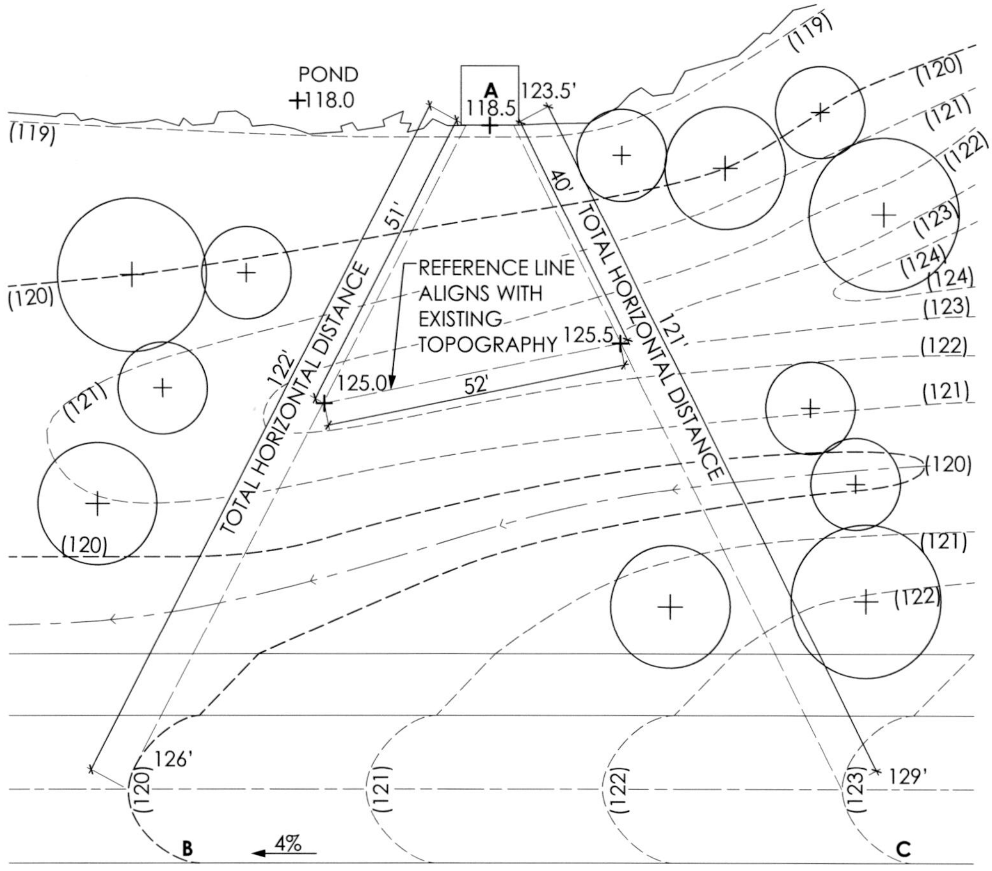

5.3–2
Determining the horizontal distance for the top of the sloping berm

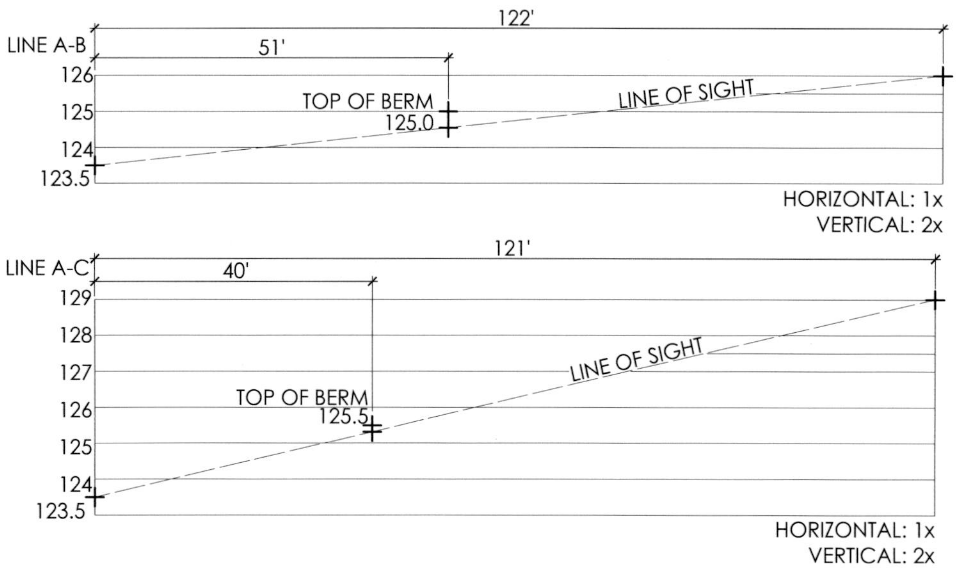

5.3–3
Line of sight sections to determine vertical elevations for the sloping berm

9 Mark the location of the spot elevation on the line of sight. Choose an elevation that is 2"–6" higher than the spot elevation on the line of sight to ensure that view of the road will be obstructed.
10 Mark the HP elevations on the plan view.
11 Find the slope between the two spot elevations by measuring the horizontal distance of the reference line, 52', and dividing it by the vertical change, 125.5' − 125' = 0.5'; 0.5'/52' = 0.0096 or 1% slope. Although the slope for this berm is minimal, other problems may have longer berms where several whole number spot elevations will need to be calculated. Interpolate along the connecting reference line to locate any whole number spot elevations at the top of the berm.
12 Use the maximum allowable slope, 4:1, to find the proposed contours around the top of the berm. Measure, mark, and draw any closed contours and tie the proposed contours back to the existing topography with smooth, continuous lines. Add any swales that are necessary.
13 Label the slopes for the berm. The completed berm problem highlighted in Figure 5.3–4 does not require any swales. The location of the berm allows runoff to enter the existing swale and pond shown in the topography.

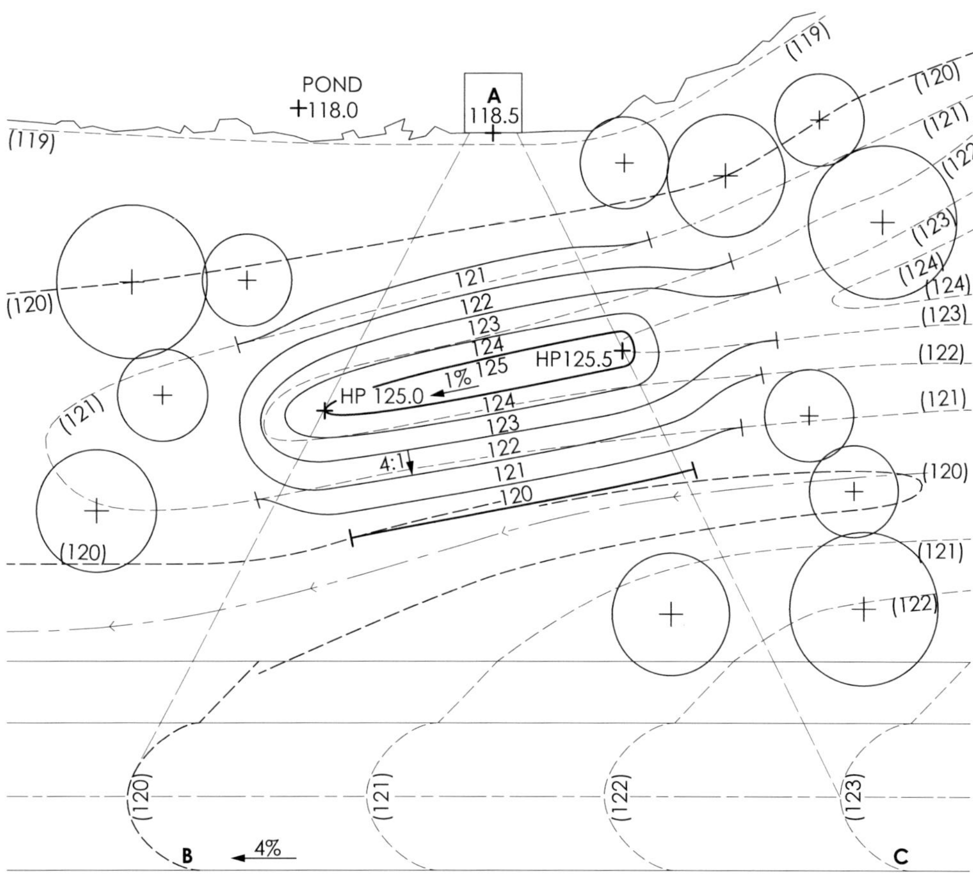

5.3–4
Sloping berm

When creating a sloping berm, it is essential to examine the topography to see where the berm should be placed. Unlike a stationary object there is more room for error. Make sure to avoid any existing trees and structures. Remember to add the vehicle height to each contour along the road. The horizontal and vertical distances for each line of sight section will vary for each contour on the road.

5.4 Summary

When solving for landscape/built element combinations it is essential to examine the grading problem or site and determine what the landforms are that will be changed and how they will affect the final grading solution. Always start at a given spot elevation and work uphill or downhill as required. In general, grade the built elements first before manipulating the landscape then tie proposed contours back into the existing topography.

Use reference lines in parking lots to avoid dealing with curved curbs and turning radii. Find spot elevations at 90° angles and interpolate uphill or downhill to corner points. Note where the property line or right-of-way line is and make sure that water from the parking lot does not flow into the adjacent property or the road. Pitch surface runoff towards a drain inlet if given or an outlet in the parking lot to a swale in the landscape. Tie proposed contours back to the existing landscape, ensuring that water does not enter the parking lot.

For culverts, find the spot elevation at the intersection of the road centerline and the culvert centerline and use it to find the spot elevation at the bottom center of the culvert pipe. Maintain the swale longitudinal slope in the landscape and the pipe slope within the culvert. Note the diameter of the culvert pipe and determine the number of contours that will pass through the pipe. Interpolate along the pipe and wing walls to find where the swale contours will tie into the built structure at the entrance and exit for the culvert.

Lastly for sloping berms, remember to add the eye height of a person on the viewing platform and the height of the vehicle to each contour on the road that falls within the viewshed. Create a section for each contour line and determine the vertical elevation for the berm. Make sure that the berm is placed in the landscape so that it ties in easily to existing landforms. Create swales if needed to remove surface runoff from the berm.

5.5 Exercises (answers start on page 202)

5-1) Grade the asphalt road and parking lot.

a) Start at the spot elevation provided and grade the road with the gradients provided.
b) Runoff from the road must not drain into the parking lot.
c) All hardscape surface runoff from the parking lot must drain into the inlet provided.
d) Minimum slope is 2%, maximum slope is 4:1 and curb height is 6".
e) Grade the landscape island in the middle into a mini-ridge.
f) Show TC/BC for each spot elevation in the parking lot.

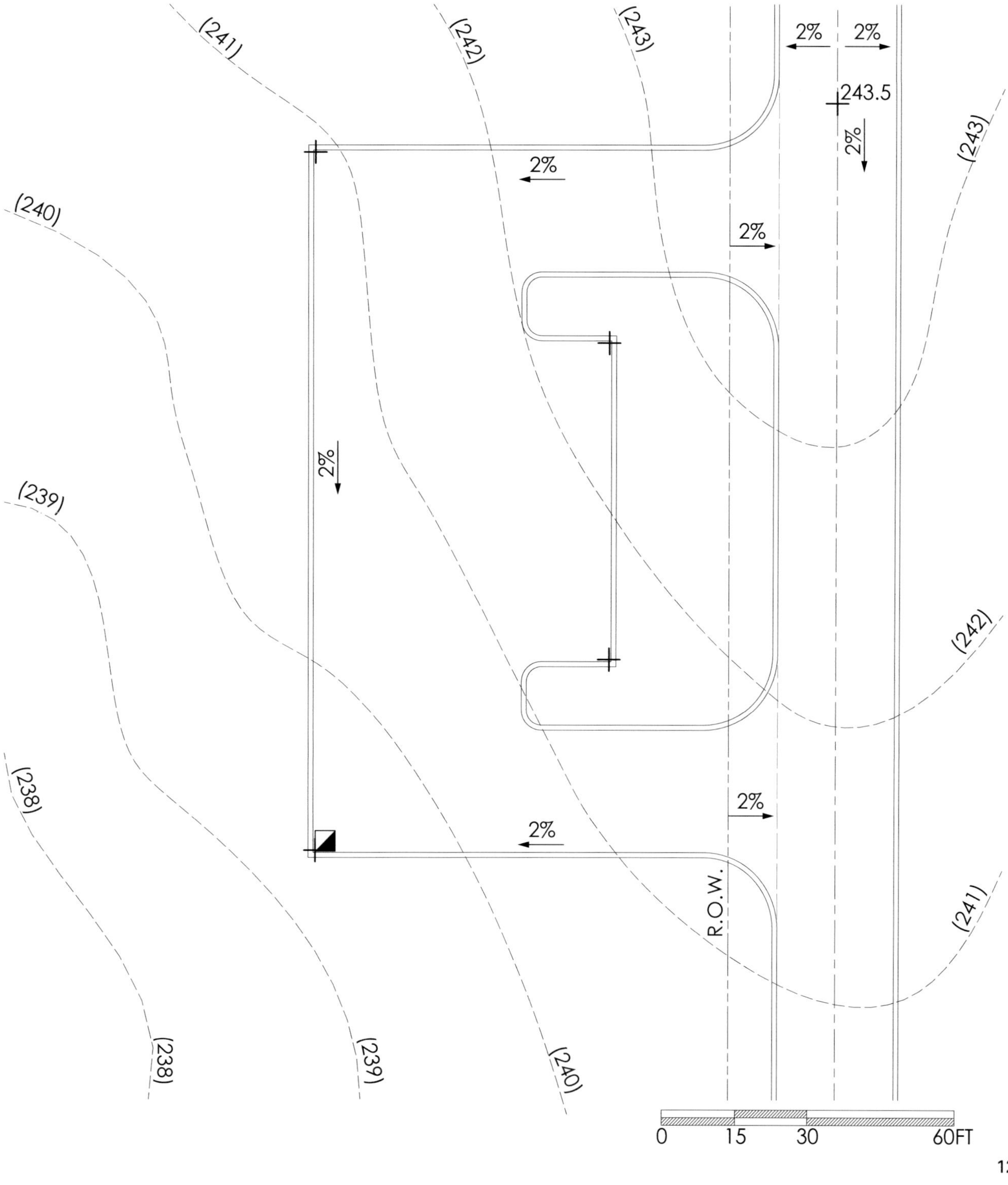

5-2) Grade the asphalt road, shoulders, swales and culvert pipe.

a) Crown of the road is 2%. Shoulder slopes at 2% away from the road. Swale along the side of the road is 6" deep; maximum slope is 5:1.

b) The pipe culvert slope is 0.5% (0.005).

c) Bottom of the pipe culvert is 2' lower than the spot elevation for the road.

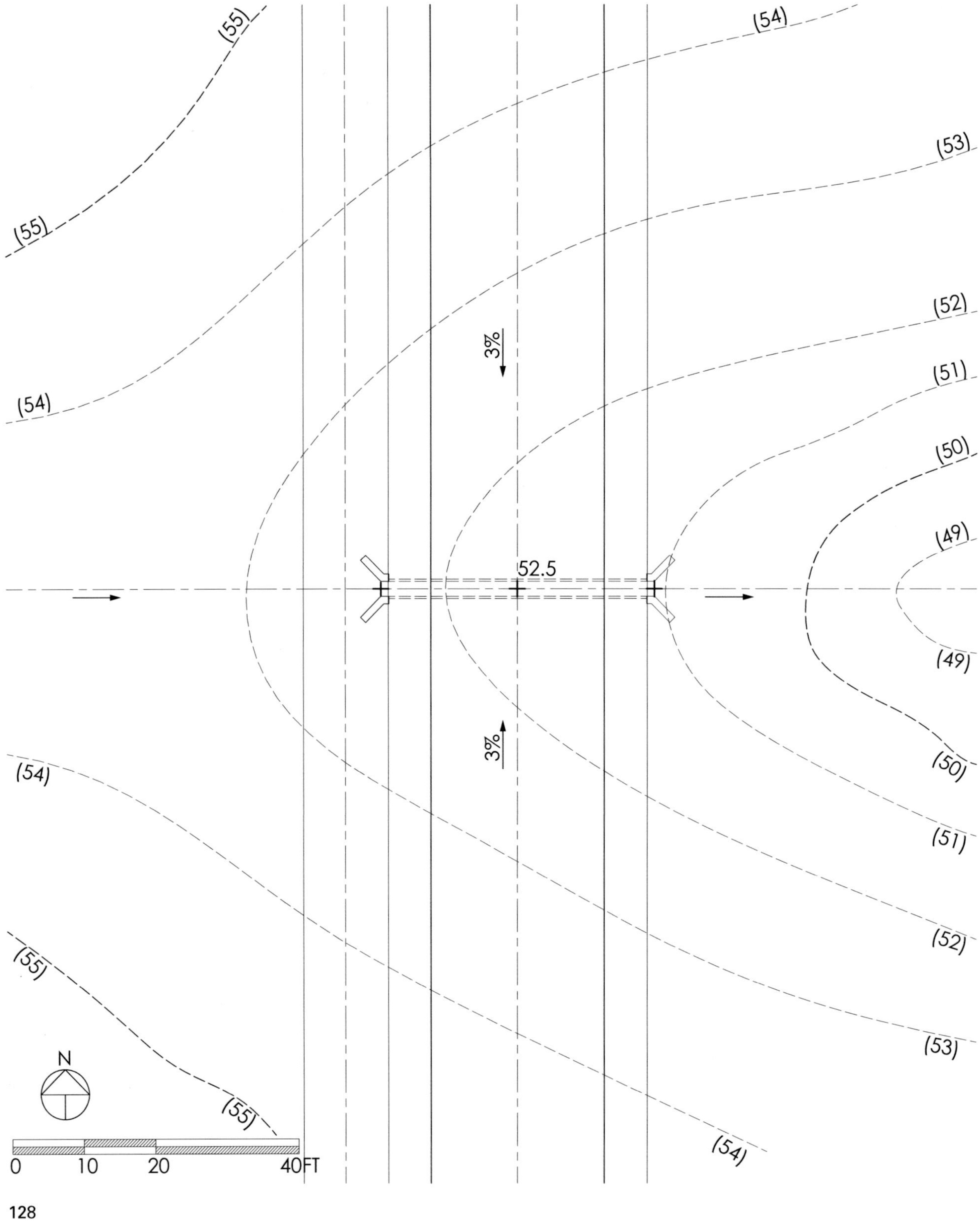

5-3) Draw a sloping berm.
  a) Measure the distance from Point A to Points B and C and create sections to find the height of the berm.
  b) Choose a distance on each line of the viewshed that corresponds to the topography.
  c) Label the spot elevations and slope for the berm and draw proposed contours; maximum slope is 4:1; minimum slope is 2%.
  d) No swales are needed; water may flow onto existing topography.
  e) Eye height at Point A is 5' and vehicle height on the road is 6'.

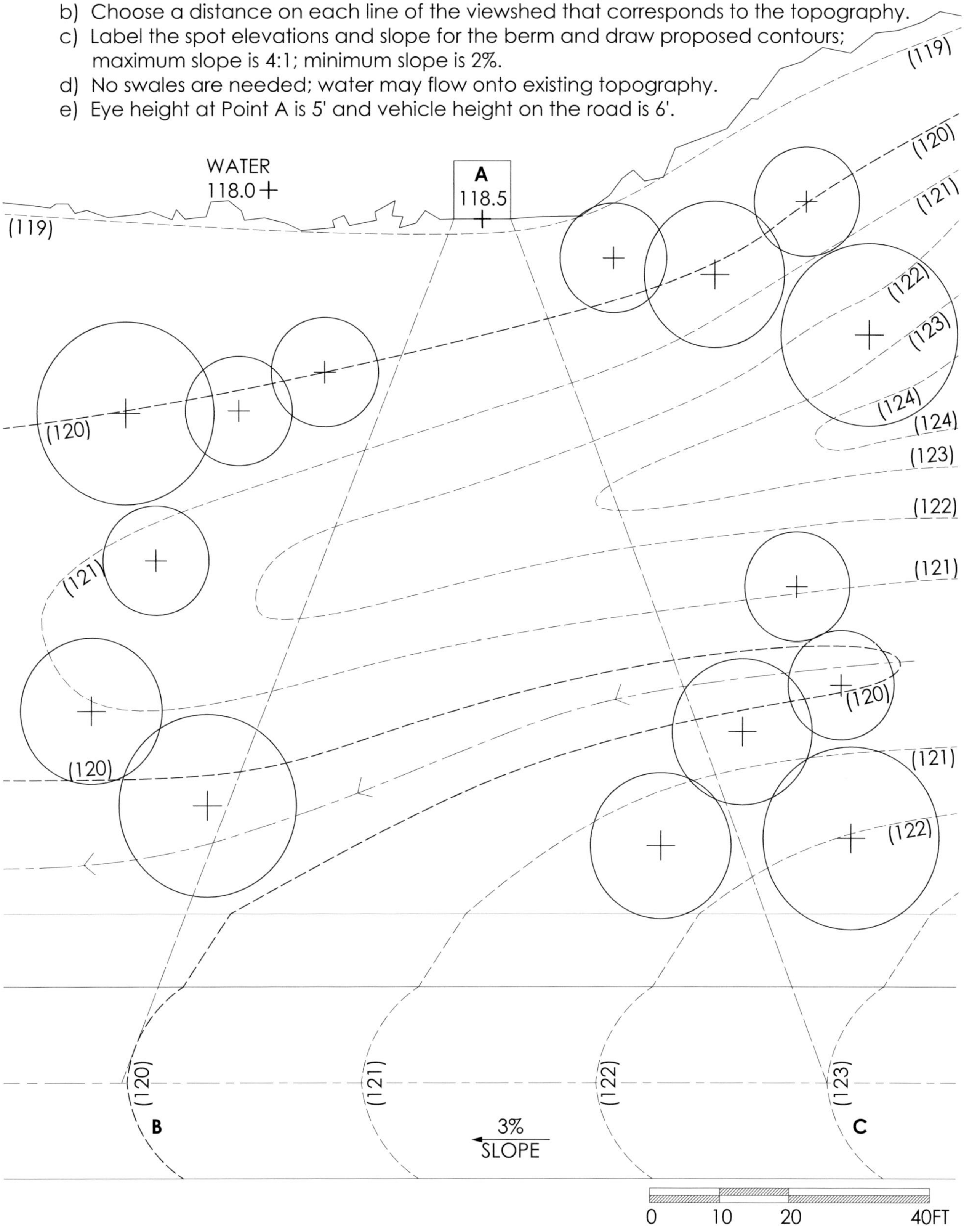

5.6 Literature cited

Harris, Charles W. and Nicholas T. Dines. 1997. *Time-Saver Standards for Landscape Architecture. 2nd Ed.* New York: McGraw-Hill.

Hopper, Edward J. ed. 2006. *Landscape Architectural Graphic Standards.* New York: John Wiley & Sons, Inc.

Strom, Steven and Kurt Nathan. 1998. *Site Engineering for Landscape Architects. 3rd Ed.* New York: John Wiley & Sons, Inc.

Strom, Steven, Kurt Nathan and Jake Woland. 2013. *Site Engineering for Landscape Architects. 6th Ed.* New York: John Wiley & Sons, Inc.

Untermann, Richard K. 1973. *Grade Easy: An Introductory Course in The Principles and Practices of Grading and Drainage*. McLean, VA: Landscape Architecture Foundation.

Section Six
Other concepts that are good to know

In addition to testing your understanding of grading, drainage and stormwater management, CLARB tries to determine if you know how to successfully incorporate data from other methodologies used in landscape architecture and from related disciplines, such as architecture, civil engineering, and regulatory agencies.

6.1 Calculating FFE

For grading problems with architectural elements, a common part of the problem statement is providing the finish floor elevation (FFE) for a building before grading the surrounding site elements. Options for determining the FFE are based on the problem parameters; sometimes an initial spot elevation is provided, sometimes the placement of the FFE is based on the existing topography and/or other geographic elements. These problems usually provide hints.

If an initial spot elevation is provided at a known point along a built structure such as a road or sidewalk, use interpolation to determine the unknown spot elevations from the initial spot elevation to the threshold of the building using any slopes provided as part of the problem statement or any slopes that you have to determine based on the existing topography. When you reach the threshold of the building, add 1" (0.08') to the spot elevation for the FFE unless the parameters say the FFE is flush. Once the FFE is established complete the grading solution. An example of this process is illustrated in Figure 6.1–1.

As a second example, you may be asked to determine a reasonable elevation for a building pad on a steep slope. The object is to minimize soil cut and fill. The easiest method is finding a midpoint elevation between the highest and lowest spot elevations of the building pad as it sits in the existing topography.

1. Determine the existing spot elevations for the highest and lowest corners of the pad by drawing a perpendicular line through the corners of the pad to the nearest contour lines, see Figure 6.1–2.
2. Determine the slope for each line based on the total horizontal distance between the contour lines and the contour interval, 1'/6.5' = 0.15384 and 1'/8.5' = 0.11764'.
3. Interpolate to find the spot elevation for the corners of the building pad, 6' × 0.15384 = 0.923'; 44' − 0.923' = 43.08'. 7' × 0.11764 = 0.824'; 39' + 0.824' = 39.82'.

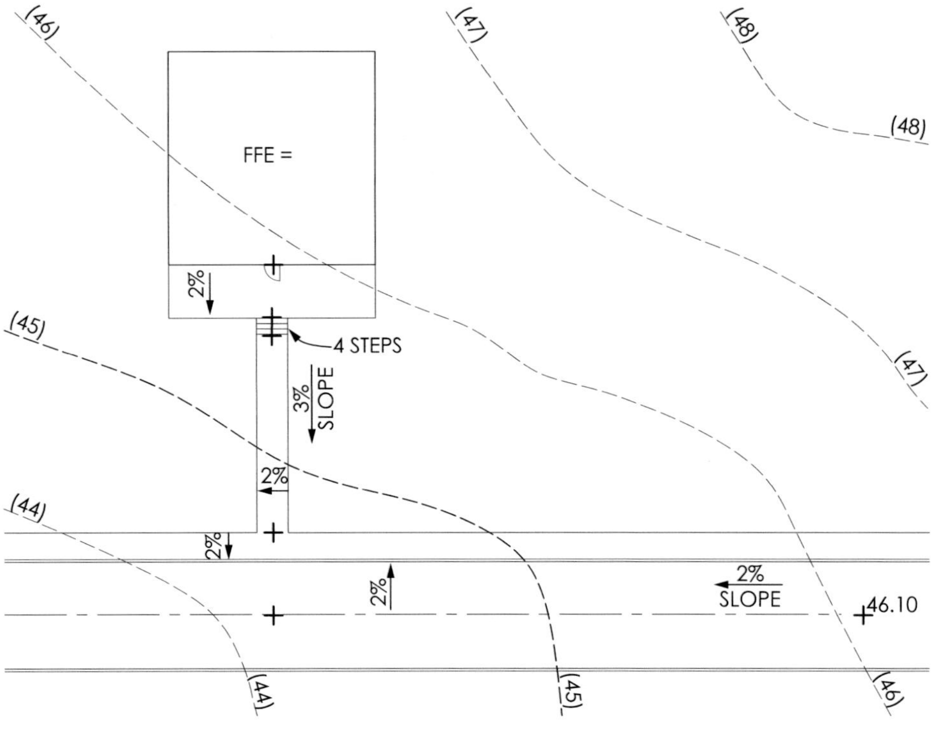

6.1–1
Figuring FFE from spot elevation on the road

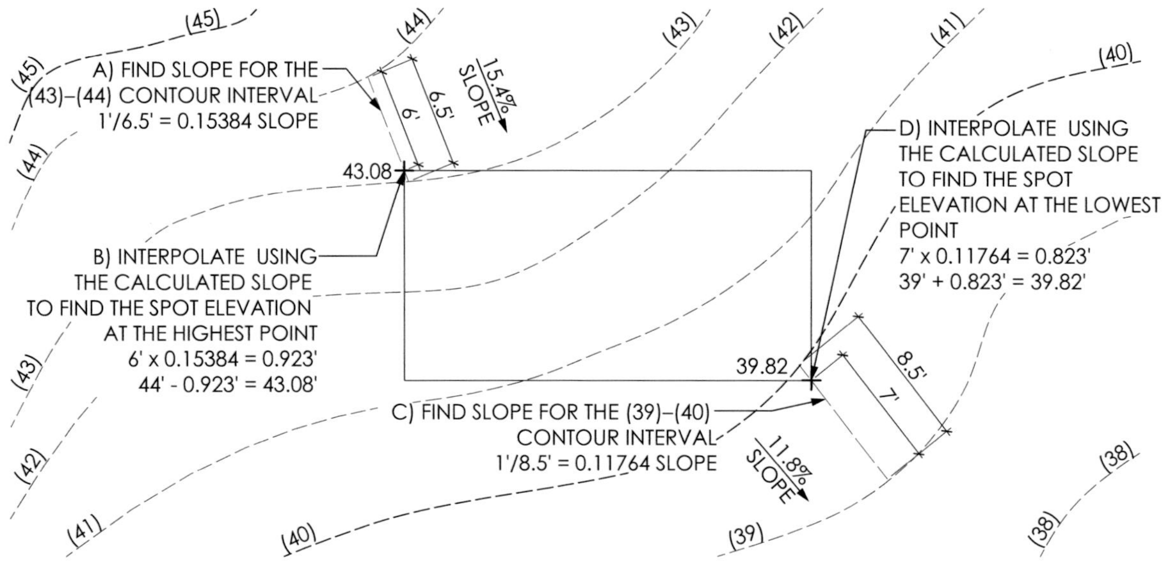

6.1–2
Interpolate to get corner spot elevations

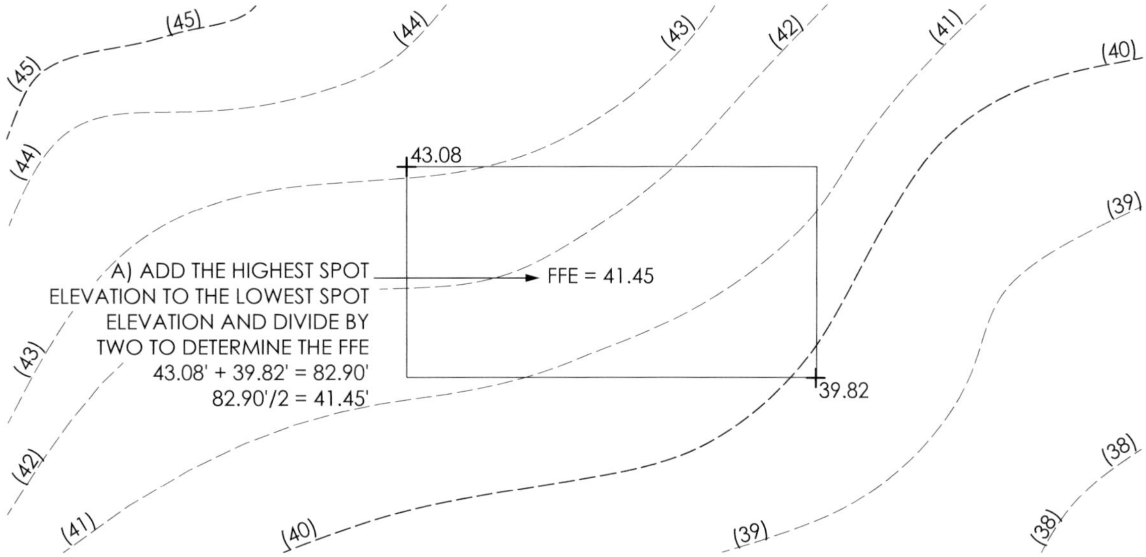

6.1–3
Finding FFE

4 Add the spot elevations at each corner together and divide by 2 to determine an appropriate midpoint elevation, 43.08' + 39.82' = 82.90'; 82.90'/2 = 41.45' (see Figure 6.1-3). Check the elevation to determine if you will have ample room around it to create any drainage structures, such as swales that are needed to solve the grading problem. Raise or lower the elevation by 3"–6" increments if needed.

On a large-scale watershed problem, you may be required to locate the building pad and provide a suitable FFE. For example, a building pad must be sited in an area with slopes less than 5%. First, determine the areas that are steeper than 5% and exclude them. Second, in the areas that remain, remove any that don't comply with the parameters outlined in the problem statement. Finally, remove those areas that don't provide adequate drainage around the structure or don't allow you to tie back into the existing topography at the maximum slope allowed by the problem.

6.2 Cut and fill

Minimizing earthwork as mentioned in previous sections is one of the goals of correctly siting building structures. Calculating soil volumes is the domain of civil engineering but recognizing what is cut and what is fill aids in understanding how as designers we are manipulating existing topography. It is crucial to be able to read topography and understand which proposed contours are in cut and which in fill for soil estimation.

Cut requires removing existing soil, it is shown as proposed contours that are uphill from their existing contour lines. An example of a landscape element that is predominantly cut is a swale. Fill adds soil volume to the site. It is shown as proposed contours that are downhill from their existing contour lines. An example of a landscape element that is predominantly fill is a berm. Generally minimizing the amount of cut and fill for a site or trying to maintain a balance between the amount of cut and fill is recommended for live projects. Figure 6.2–1 illustrates a proposed pad showing sheet drainage where

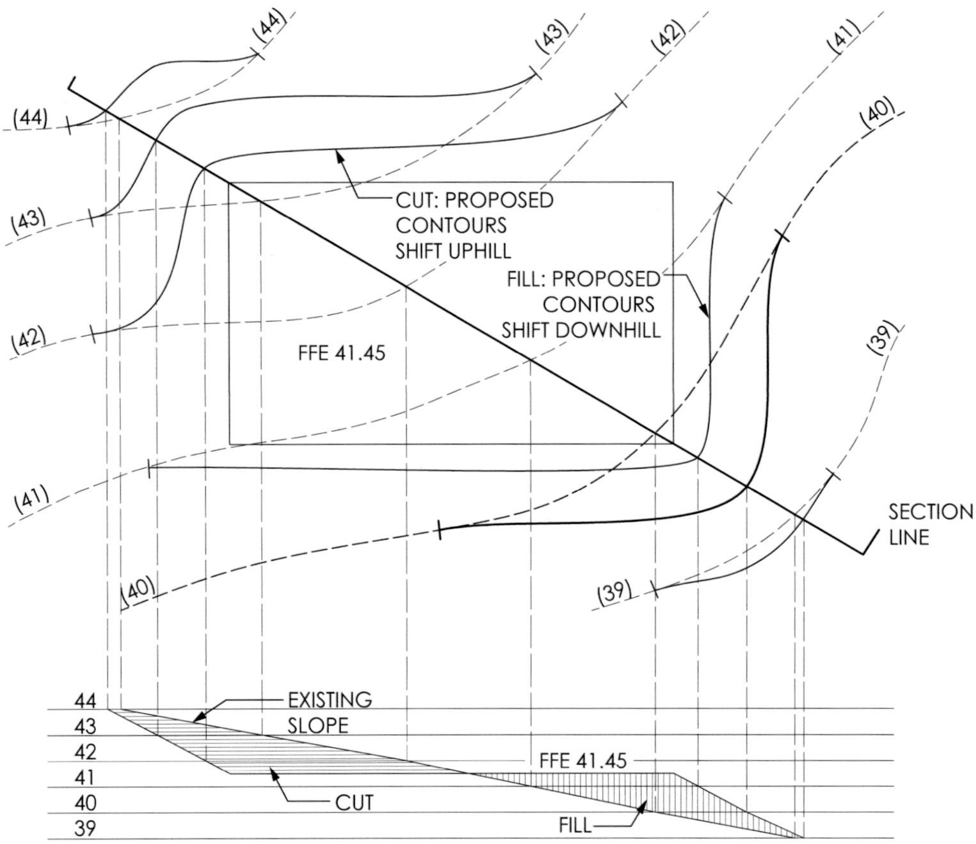

6.2–1
Cut and fill shown with proposed contours

cut and fill are roughly equal. The recommended reading provides detailed explanations of soil volume calculations for linear projects, open green field projects and dense urban projects.

6.3 Pipes

In addition to figuring out how architectural elements will fit into a grading solution you may also encounter existing utility pipes. Generally, stormwater pipes are civil engineering structures which cannot be easily moved and need to be addressed if they appear in a grading problem. Existing pipes may be shown underneath existing topography with instructions that state that sufficient cover must be maintained over the pipe. Large structures should not be placed over existing pipes. Before proceeding to the grading solution, the type of piping and the depth of the pipe must be ascertained.

There are two types of utility pipes that collect surface runoff. Subsurface drainage pipes are part of the open drainage system. They collect excess groundwater and surface water that percolates into the soil. These pipes, as illustrated in Figure 6.3–1, are placed under playing fields and playgrounds with permeable surfaces that require a flat finish grade with a slope less than 2% or behind retaining walls and under planted areas to quickly drain a site after a rain event. Subsurface drainage pipes are perforated, wrapped in filter fabric and laid perforated side down in a herringbone or grid pattern

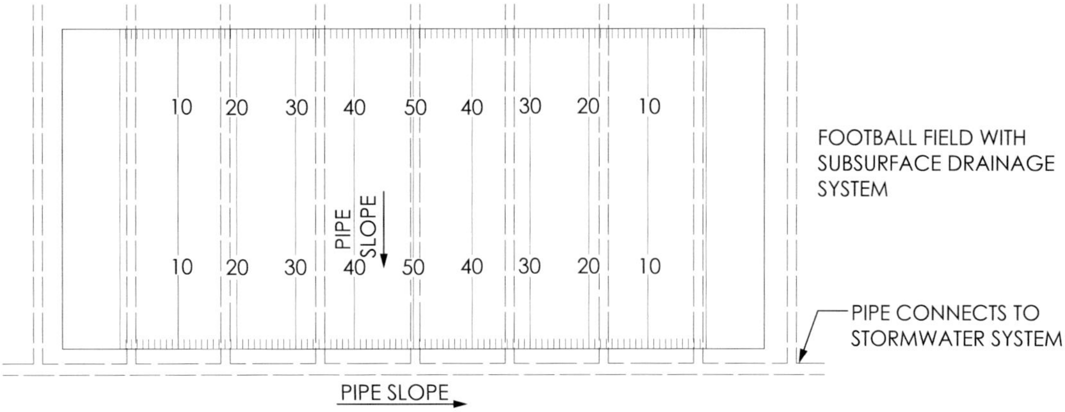

6.3–1
Subsurface drainage of a football field

with pipe sizes that increase from one end of the field to the other. Water holding capacity of the soil will determine how many pipes are needed and how close together pipes should be. The pipes may be connected to an open source of water like a pond or stream or they may be tied to a stormwater drainage system. Subsurface drainage pipes are not as deep nor as big as pipes in the stormwater drainage system and they may be sloped in different directions than finish grade.

Stormwater drainage pipes are part of the closed drainage system. They collect stormwater surface runoff from landscape and hardscape areas. The pipes are solid; no groundwater can enter the system. These pipes are dimensionally larger and found at greater depths than subsurface drainage pipes. These are the utility pipes most commonly found in landscape problems on the exam. In addition to pipe type and depth it is useful to know how to size pipe and determine pipe velocity and runoff rates. Refer to the recommended reading for an in-depth review of pipe sizing and the Rational method, Q=CiA, for calculating stormwater flow rates and volumes.

To determine the depth of an existing stormwater pipe in the landscape, we need to know all the variables associated with the stormwater pipe: the invert elevation, which is located at the inside base of the pipe where it connects to a catch basin or area drain; the diameter of the pipe, which provides the total width of the pipe; and the slope of the pipe. It is important to note that pipe slopes are, in general, more shallow than topographic slopes and pipes may slope in different directions than finish grade. We also need to know which way water is flowing over the existing topography above the pipe.

1 Interpolate spot elevations along the length of pipe where it intersects with existing contour lines using the pipe slope starting at the invert elevation, see Figure 6.3–2. This indicates what the spot elevation is at the bottom of the pipe where the height of the topography has changed by 1' vertically. Although the topography may have changed by 1', the pipe depth may have only changed 2"–3".

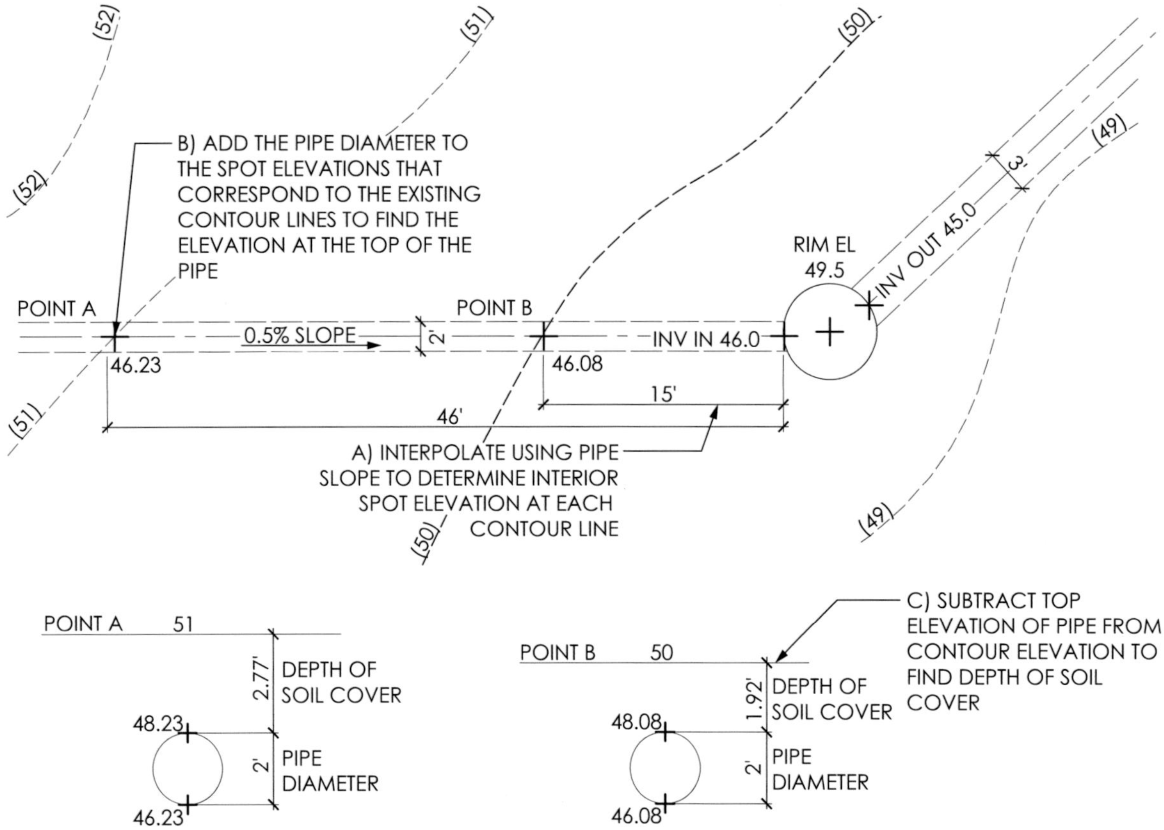

6.3–2
Pipe with existing topography and catch basin

2 Add the pipe diameter to the spot elevations along the length of the pipe. This is the elevation for the top of the pipe.
3 Subtract each of the top of the pipe spot elevations from the contour line that is directly above it to determine how much existing soil is covering the pipe. It will also indicate whether the slope of the land and the slope of the pipe are fairly constant or if the pipe is descending or ascending in comparison to the soil that covers it.

If the existing cover is 3' and the problem requires at least 1.5' of cover over the pipe, then 1.5' of soil can be moved before the pipe will have insufficient coverage. This information is most useful when placing swales over existing pipelines.

In order to determine the direction of stormwater flow in a piped system it is helpful to have the invert elevations for any pipes that are shown on the grading problem and the pipe diameters. The depth of a catch basin will vary depending on how many pipes are connected to it. Along the depth of the catch basin there can be more than one piped connection flowing to the drain. The lowest pipe is always flowing away from the drain. To determine which pipes are flowing to the drain and which are flowing away, determine the depth of the pipe attachments using the invert elevations. An invert in (INV IN) is a spot elevation that is flowing to a catch basin and an invert out (INV OUT) is a spot elevation that is flowing away from a catch basin. Invert out elevations are always lower than invert in elevations to maintain positive water flow through the pipes.

Another hint to determine which way water is flowing through the piped system is to determine the diameter of the pipes. A pipe that is flowing away from a catch basin is often bigger than the pipes that are flowing to a catch basin. The pipe flowing away from the drain will have both stormwater entering from the drain above and water piped into the catch basin from one or more pipes and thus will carry more stormwater than any individual pipe entering the system. The pipe diameters and the invert elevation for the pipes along the depth of a catch basin together will help identify the direction of water flowing through the closed drainage system. To find the INV OUT subtract the difference in diameter size from the lowest pipe that is flowing to the catch basin. As shown in Figure 6.3–3, the INV IN elevation of the 8" diameter pipe is 45.33', the pipe diameter for the water flowing out is 15". The difference in diameter size is 7" (0.5833'). Subtract 0.58' from the INV IN to get the INV OUT for the 15" diameter pipe, 45.33' − 0.58' = 44.75'.

Some grading problems will require that a pipe which daylights at a stream, channel or pond prevents backflow from the stream from entering into the closed drainage system. Backflow of stormwater into the drainage system is more commonly called surcharge. Knowing or determining the slope of the pipe will enable you to determine the outfall from a pipe. In order to maintain a positive flow and prevent surcharge, make sure that the invert out elevation for the pipe is at least 6" above the high-water mark for the stream or channel. This elevation, above the water level, more commonly called freeboard, should be at minimum 6" above the elevation of any pond, river or wastewater treatment facility.

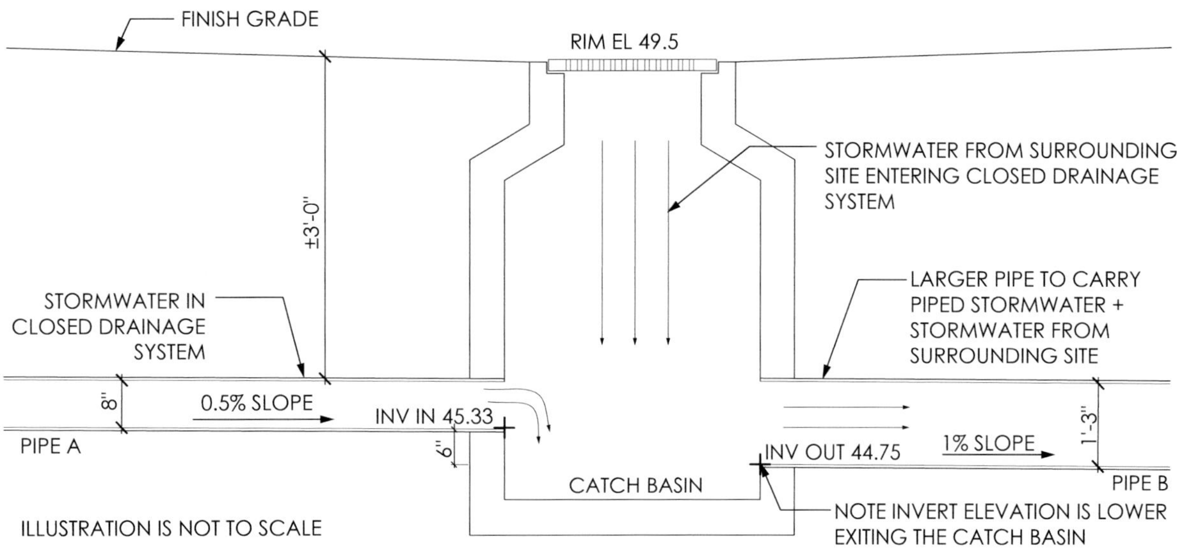

6.3–3
Catch basin showing pipe in and pipe out

6.4 Horizontal and vertical curves

Roads are not straight lines; they curve horizontally and vertically through the landscape. A horizontal curve is part of site layout determining the orientation of the road through the landscape, and a vertical curve is part of site grading determining the vertical alignment of the road to the existing landscape.

Horizontal road alignment is the site layout for a road. It considers the existing topography and vegetation, avoiding areas that require extensive cut and fill. The road is laid out in plan view using traverses, straight tangential lines with directional bearings to north, south, east, and west. A traverse is the centerline for a road. Once the traverses have been laid out, horizontal curves are added. Figure 6.4–1 shows two horizontal curves and their associated traverses.

Horizontal curves are true arcs with continuous radii. Curve sizes are assigned based on sight distance and proposed road speed. Refer to the literature cited for information about road design speeds. After the traverse and its curves have been designed road stations are added and the width of the road, shoulder and associated swales are added.

Stationing divides a proposed road or an existing road segment into 50' or 100' sections. Unlike most grading problems that require spot elevations throughout, because road layouts are longer than typical sites, spot elevations are assigned at 100' stations. Stations are assigned at the beginning of the traverse, at every 100' feet, at the beginning and end of each horizontal curve, and at the end of the road. Stations are written with a plus sign in the middle. For example, 1+00 equals 100', 5+25 equals 525' and 10+00 equals 1000'. Stations indicate horizontal distance but will have spot elevations denoting vertical change associated with them. Stations that indicate the beginning and the end of the curves are called the point of curvature and point of tangency, respectively. These special stations plus the end of the road station are noted to the decimal foot. For example, 160.91' as the start of a horizontal curve is 1+60.91'. Stations must be

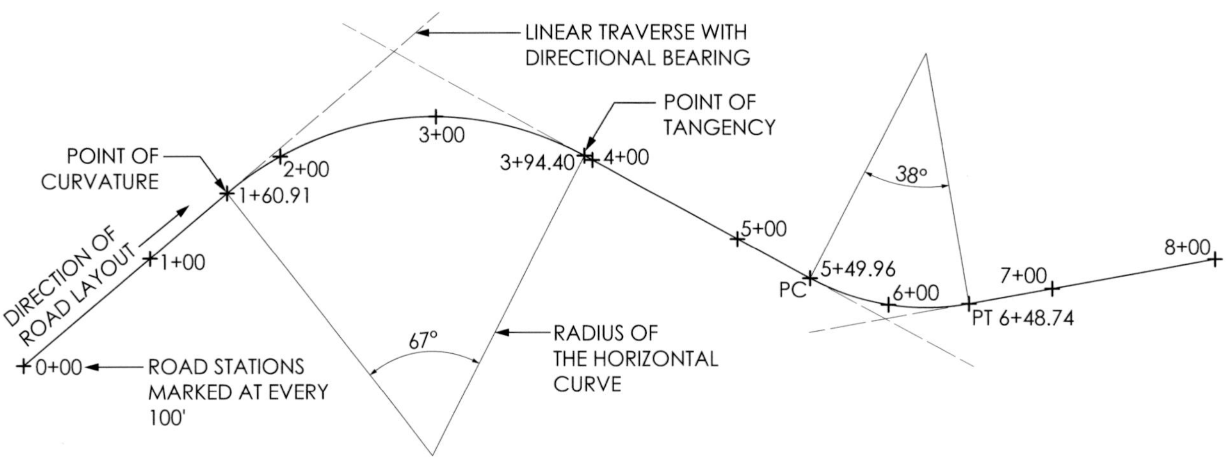

6.4–1
Road layout showing horizontal curve and stationing

assigned before vertical road alignment can be done. Stationing is also used to estimate cut and fill on linear project sites such as roads and utility lines.

Vertical road alignment also starts with a series of straight lines. These tangential lines are road gradients, longitudinal road slopes. The elevation changes for the horizontal road layout are surveyed in a road profile, a longitudinal section. Road gradients are applied that relate to the surveyed landscape.

Roads never form sharp peaks or valleys, they always curve, so when road gradients change slope from shallower to steeper or change direction from ascending to descending a vertical curve needs to be applied to maintain a smooth continuous road.

Unlike horizontal curves, vertical curves are not true arcs, they are parabolic. One side of the curve may descend or ascend at a steeper or shallower gradient than the other side. The goal of vertical alignment is to create peak or sag curves that create continuous flow, maintain sight distance so that a car on the road traveling at speed has enough distance to slow or stop, and the driver's view is not obstructed.

To create a peak or sag curve, an offset is calculated from the existing topography and proposed road gradient at each station point. The subsequent vertical curve is plotted from the offsets. The gradients for the vertical curve will be shallower than either the first or second gradient for the curve. Once the curve has been drawn, a LP or HP is calculated. Figure 6.4–2 illustrates a sag vertical curve and how it affects the proposed contours for a road.

Horizontal and vertical curves don't have to correspond to each other in the landscape. A road can have one horizontal curve and two or more vertical curves and vice versa. The horizontal road alignment sets the layout through the landscape and the vertical curve adjusts the existing landscape into smooth continuous curves depending on the surrounding topography.

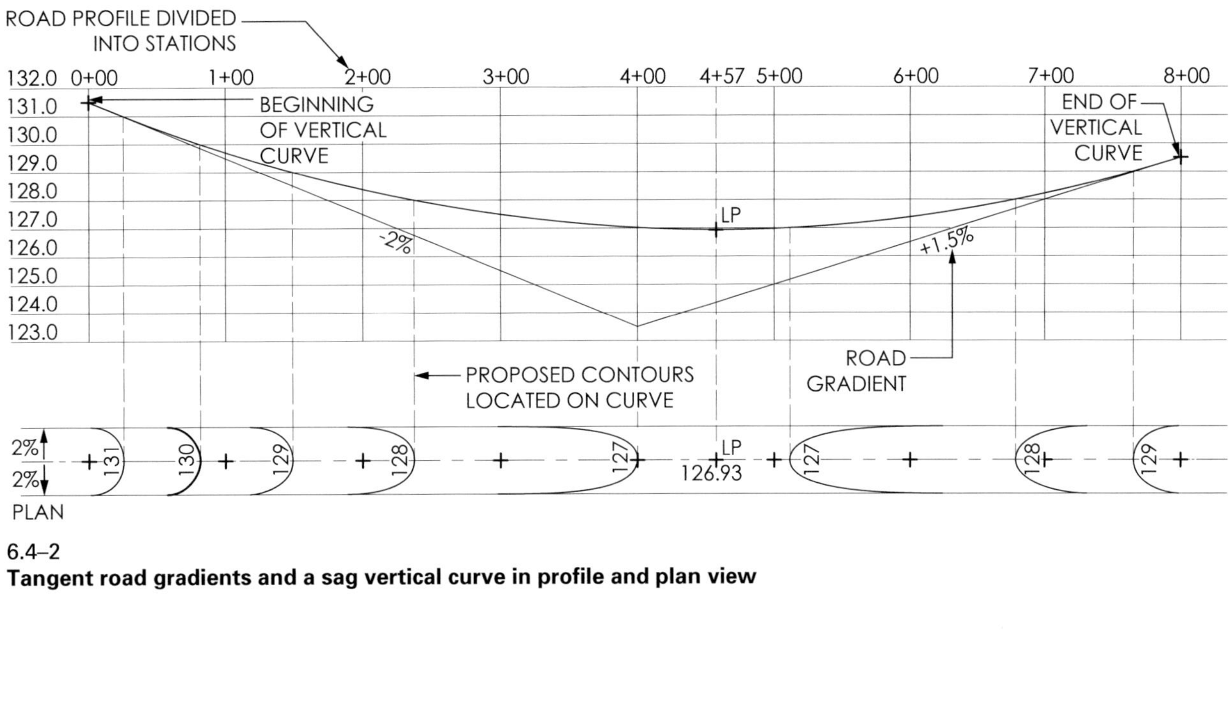

6.4–2

Tangent road gradients and a sag vertical curve in profile and plan view

Other concepts that are good to know

The recommended reading has detailed descriptions of how to solve for horizontal and vertical curves. If a road problem includes a vertical curve, it will likely include a road profile that shows the associated curve and LP or HP. Note that, in plan view, the proposed contours will elongate as they get closer to the center of the curve; they will not be evenly located along the length of the curve, and the gradients will be shallower than the linear part of the road on either side of the curve. Make sure to use the appropriate cross slopes for the crown, shoulder and swales if provided to completely grade the proposed road and tie it back into the existing topography.

6.5 Landscape grading standards

Finally, the recommended reading also mentions standard slopes used for landscape projects. Minimum and maximum slopes refer to what will allow a site to efficiently drain water. On a landscape site, because some stormwater percolates into the soil, the minimum slope to ensure positive drainage is recognized as 2%. On a hardscape surface, the minimum is recognized as 2%, but can go as low as 1% for live projects. For hardscape slopes less than 1%, consider using subsurface drainage to ensure that surface water does not pool. The minimum 2% requirement for hardscape surfaces ensures positive drainage allowing for differences in surface types (concrete paver vs. asphalt) and construction installation techniques. If a minimum slope is not explicitly stated use 2% for cross slopes for paths, road crowns and shoulders.

Maximum slopes will vary by the problem statement but will not exceed 3:1. This slope is recognized as the maximum allowable slope for planting turf grasses; the minimum vegetative cover required to prevent soil erosion on a slope. The maximum slope for a problem will vary based on type of soil. It will not exceed the angle of repose for the soil. The angle of repose is the angle at which a given soil naturally settles. A clay-based soil has a steeper angle of repose than a sandy soil

Other standards that you will be expected to know are national ADA requirements; i.e., understanding how to make a site ADA-compliant, specifically knowing the minimum and maximum slopes for ADA-compliant walks, ramps and curb ramps; and understanding how to apply the recommended guidelines to stairs and ramps including handrails, landings, safety railings and fall height. Refer to www.ada.gov for the most current standards and further information about accessibility guidelines

6.6 Summary

Understanding how other methodologies, disciplines and guidelines interact with grading, drainage and stormwater management is part of the successful completion of a grading vignette. Learn how to calculate FFE either by using given spot elevations and slopes or by estimating it using existing topography.

Determine how existing utilities will affect the grading problem and address these before creating a solution. Calculate the depth of soil over an existing utility, to understand how much soil can be moved or if you should avoid placing any built structures over the utility.

Know the difference between subsurface drainage pipes and stormwater drainage pipes and the conditions in which you may encounter them. Practice reading topography to clearly identify which landscape structures are in cut and which in fill. These are a helpful check with any soil estimation calculations that may be necessary.

Understand the difference between horizontal and vertical road alignment. Horizontal curves are part of the layout of roads and they are connected to vertical curves that show how proposed roads will be graded into the landscape. Familiarize yourself with stationing and road profiles and how they will affect the road grading.

Lastly, know your national standards not just your state, county or local standards. Understand how to apply the ADA requirements to your grading solution. Learn to recognize when these standards are not applied correctly as well. Refer to the recommended reading for information about common standards for grading sidewalks, roads, etc.

6.7 Exercises (answers start on page 206)

6-1) Find the spot elevations along the length of the pipes and determine the depth of soil covering the pipes.

a) Calculate the missing slope.

b) Maintain a minimum depth of 2' of cover over the pipes.

c) Provide INV OUT at the catch basin and at the end of the culvert; and TW elevation.

d) Freeboard is 6" above the stream elevation.

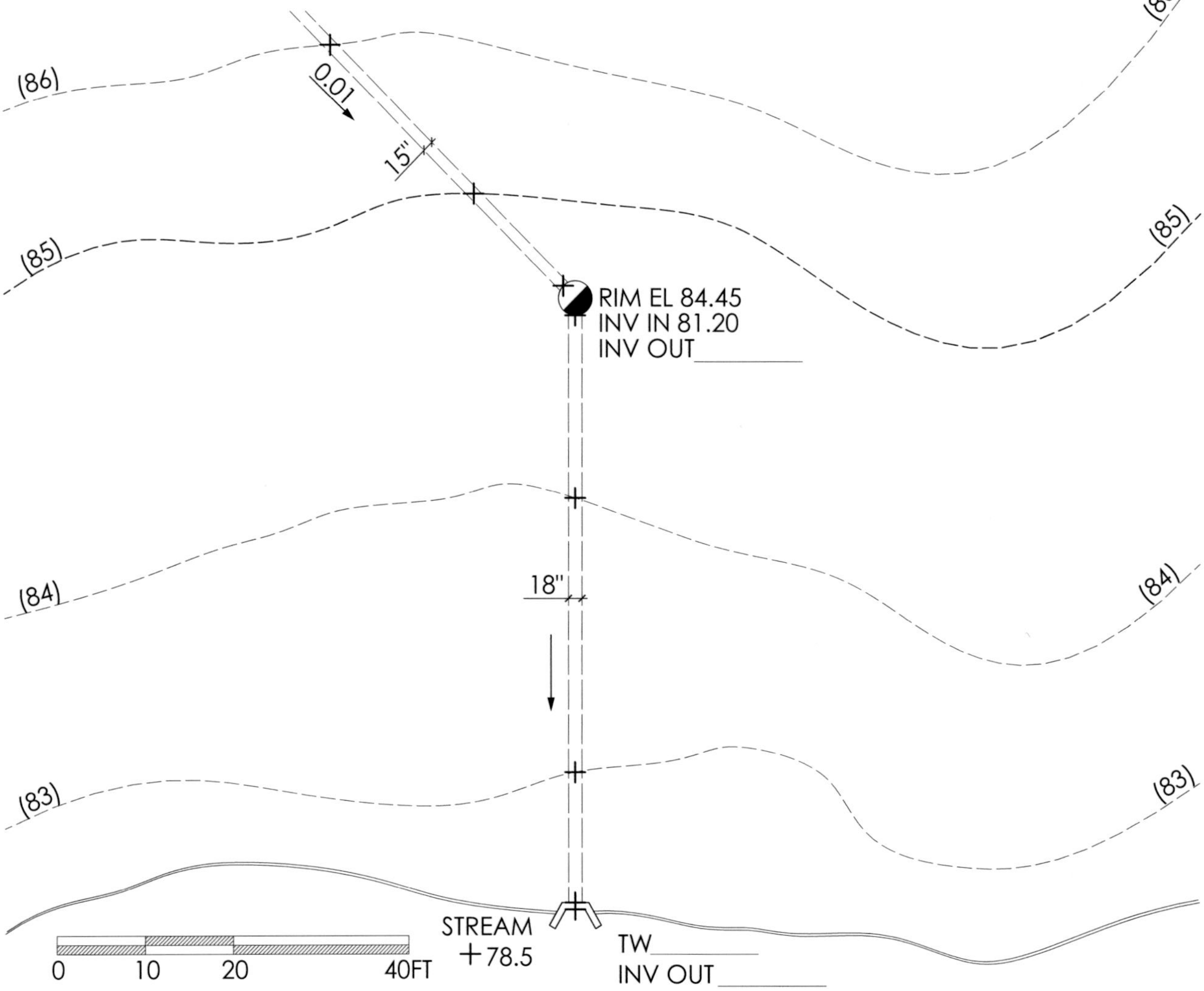

6-2) Find the FFE and create drainage around the building pad.

a) Longitudinal slope on the eastern swale is 3%; longitudinal slope on the western swale is 4.5%. Maximum slope is 4:1.
b) Slope away from building pad at 2% for the first 5'.
c) HPS is 5' from the edge of 5' clear area and slopes at 2%.
d) Provide RIM ELs for the drain inlets.

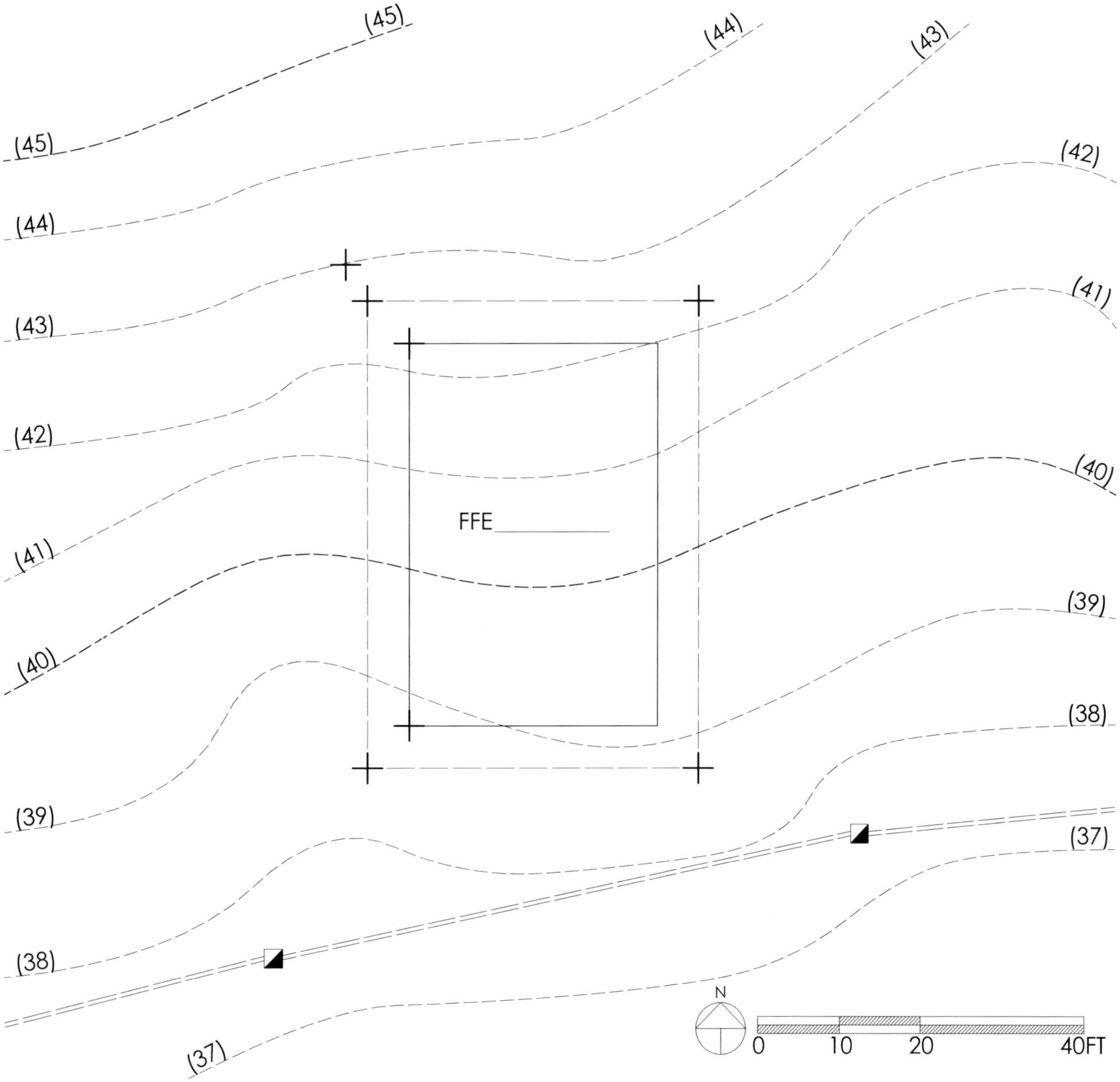

6-3) Create a swale.

a) Maintain a depth of 2' over the existing pipe.
b) Longitudinal slope of the swale is 3.5%.
c) Maximum slope is 5:1.
d) Provide slope of existing pipe.
e) Provide depth of cover where swale crosses the pipe.

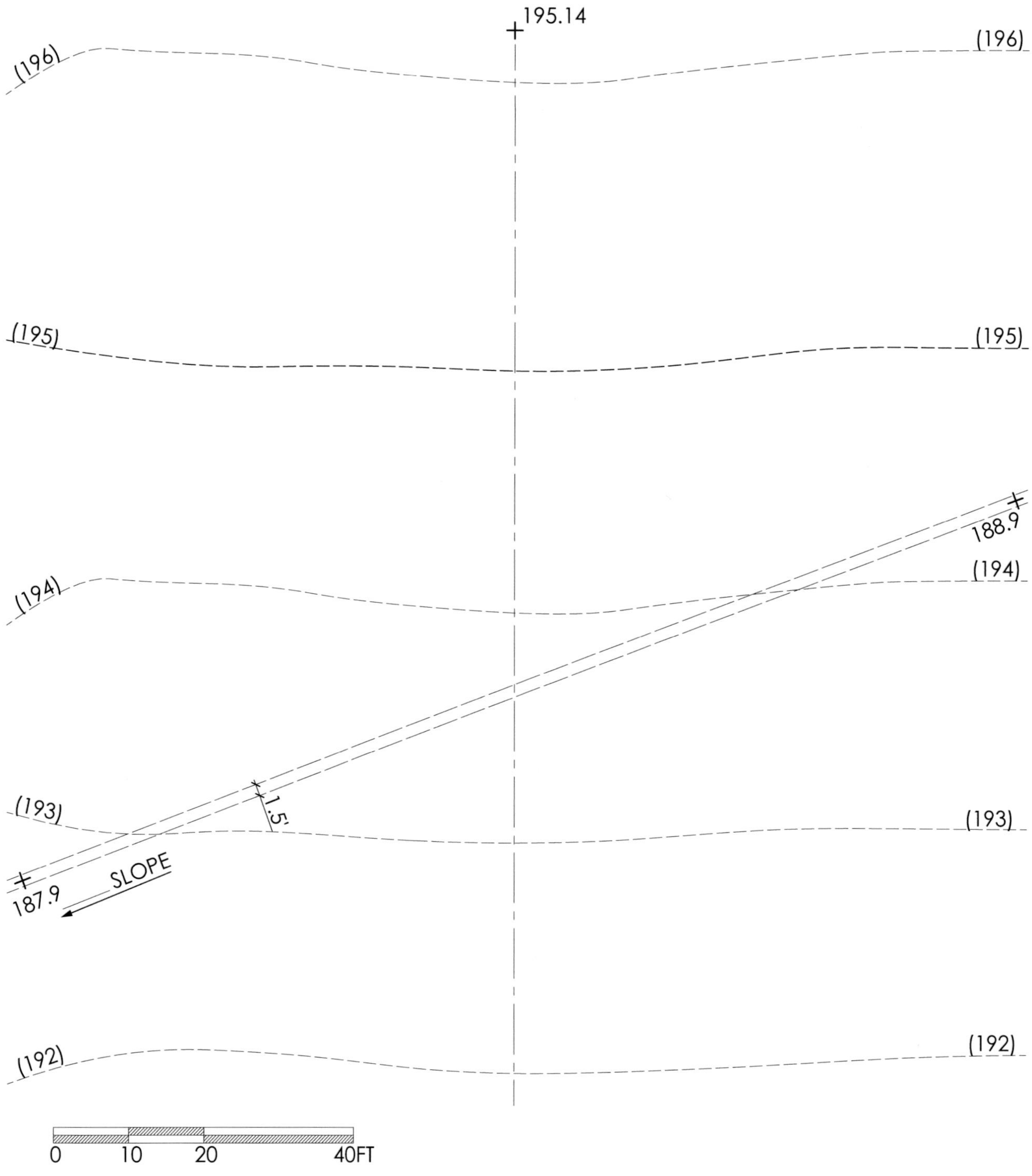

6.8 Literature cited

Harris, Charles W. and Nicholas T. Dines. 1997. *Time-Saver Standards for Landscape Architecture. 2nd Ed.* New York: McGraw-Hill.

Hopper, Edward J. ed. 2006. *Landscape Architectural Graphic Standards.* New York: John Wiley & Sons, Inc.

Strom, Steven and Kurt Nathan. 1998. *Site Engineering for Landscape Architects. 3rd Ed.* New York: John Wiley & Sons, Inc.

Strom, Steven, Kurt Nathan and Jake Woland. 2013. *Site Engineering for Landscape Architects. 6th Ed.* New York: John Wiley & Sons, Inc.

United States Department of Justice. 2010. *2010 ADA Standards for Accessible Design.* Washington, DC: US DOJ Civil Rights Division. www.ada.gov/2010ADAstandards_index.htm.

United States Department of Transportation 2014. *Safety Effects of Horizontal Curve and Grade Combinations on Rural Two-Lane Highways.* McLean, VA: US DOT Federal Highway Administration. www.fhwa.dot.gov/publications/research/safety/13077/13077.pdf.

Untermann, Richard K. 1973. *Grade Easy: An Introductory Course in The Principles and Practices of Grading and Drainage.* McLean, VA: Landscape Architecture Foundation.

Section Seven
Taking the examination

7.1 A note about the computer-based test

In the years since the first edition of this book was published, the LARE has been completely converted to a computer-based test. There are now four sections instead of five. The fourth section, Grading, Drainage and Construction Documentation, includes the concepts covered in this study guide. Test takers now have the opportunity to demonstrate basic competency in site grading not just in four vignettes but over a series of smaller problems with varied design and complexity during a 4-hour online test. Although the format has changed, the test still requires understanding and properly applying slope calculations and interpolation, identifying topography, knowing how to grade landscape and built elements, and incorporating the work of different disciplines and national standards into your grading solutions. The CLARB website (www.clarb.org) has a comprehensive section on the different question formats on the test.

Note that this is a professional licensing examination and is not similar to other standardized computer tests that you may have taken to get into college or graduate school. These questions are very specific to the profession, they require an understanding of the subject matter, the profession and related disciplines. Process of elimination and other tricks that are used for other computer tests will not work for this examination.

Currently, the test is offered three times a year and there are no available online practice exams. CLARB and ASLA (the American Society of Landscape Architects) have provided examples of the types of questions that you will encounter on the exam on their websites and CLARB has videos that show the computer format that is used to administer the test so that you can familiarize yourself prior to the test date. The best way to study for this exam is still hard copy vignettes and study guides such as this one.

7.2 Scheduling your life before the exam

Most designers shy away from site grading because the concepts involve more math than design. The best way to conquer the apprehension caused by math problems is through practice; repetition of the simple subjects such as finding slopes and interpolation and, more importantly, doing as many grading vignettes as possible. Most practice vignettes are similar and the LARE is constantly changing so there is no way to anticipate what the grading problems will be in the exam. Know, however, that if you understand the basic principles of site grading and how to solve the individual landscape elements the problems will be doable.

This study guide was compiled to help you readily gain access to and practice many of the topics involved in site grading, however, it is not meant to be an exhaustive guide and it should be used in conjunction with the study materials recommended by CLARB (www.clarb.org) and ASLA (www.asla.org). Careful and thorough preparation for the examination is the key to a successful outcome. In addition to the exercises and vignettes in this book, there are free vignette problems available on the ASLA website. Workbooks provided by PPI are the standard study guides with several practice grading vignettes and there are exercises and vignettes in *Site Engineering* by Strom, Nathan and Woland. Many landscape architecture schools have created their own vignettes some of which are posted on the internet.

My best advice for studying is to approach the test from different angles. Individually, do at least one vignette problem every day for at least two months, alternate between an hour in the morning before work, an hour during your lunch break or an hour before leaving work. Do a problem every day even if you're tired, stressed or excited; being able to focus on the basics of grading, even at your lowest levels of concentration, will only help you during the test when you're down to the last hour. Once you have the basics of a vignette problem down and you understand where you've made mistakes, do the problem again in a few weeks and time yourself doing it. Accuracy under time constraints will help you get through the exam and give you enough time left over to make corrections.

It is also extremely effective to join a study group. Each person will bring their knowledge to the group, providing insight into subjects which may be unfamiliar. If you choose to join or form a study group, meet regularly. A simple setup is to assign a practice vignette each session and review the answers together to determine where your most common errors occur.

Lastly, take advantage of all the review sessions offered by your local ASLA chapter. The review sessions are thorough, and, in some cases, you may be able to obtain discounted copies of some of the recommended study materials. If they provide a mock test; do it. There is no greater preparation for how test day will feel.

Every person takes a different approach to the test, some people want to take the graphic sections first so they can get them out of the way, and some take the sections in order. My advice for those taking Section 4 as their first exam, familiarize yourself with some of the other concepts of site engineering, such as pipe sizing, cut and fill, and vertical and horizontal road alignment. These topics are covered briefly in Section Six and more thoroughly in the recommended reading. Understanding how these concepts tie into site grading and drainage will only enhance your chances of success. It is also wise to familiarize yourself with synonyms for common civil engineering elements so that you are not caught off guard by common landscape elements called by an unfamiliar name.

Remember too that this exam tests your knowledge of construction documentation, including construction detailing and methodology, paving and wall sections, planting details, etc. Even if you are confident in your site grading skills make sure you are equally prepared to tackle construction documentation. You will not be able to pass the exam if you only know one of the subject matters well.

7.3 Test-taking tips

The purpose of the LARE is to test basic grading competency. CLARB does not expect you to get everything correct, but because the vignettes are so specific there is usually one answer they are looking for and your answer will fall within a range of acceptable answers for the problem.

1 Read the problem statement and identify what elements you'll have to provide in the grading solution.
2 Note the orientation of the vignette and the north arrow.
3 Note the dimensions of the elements in the problem. Because this is an international exam, CLARB uses a universal scale to avoid imperial and metric unit calculations.
4 Determine the direction that the existing topography slopes and get a rough idea of the total slope(s) for the existing topography. Identify any specific existing landscape features, ponds, trees, ridges, swales, setbacks, culverts, etc. to be protected or avoided in your selection of the correct answer.
5 Review any key information or exhibits that accompany the plan or problem statement; any charts with added information that needs to be addressed or accounted for prior to choosing a solution.
6 Reread the problem statement and make sure you understand everything that is being asked. Do only what is asked.
7 Visualize what each element should look like. Rough out a solution on scratch paper; keep it simple, avoid over-designing.
8 At the very end of each problem go back and review every element carefully. Include the correct naming conventions for special spot elevations (TW, BW, RIM EL, etc.) Make sure to include all required spot elevations. Make sure the answer is complete as the program will not accept partial answers.
9 Be mindful of your time, try not to get hung up on any one problem. If you're having difficulty tag the problem you're working on for review and go back once you've done the other problems.

Each question will give you more than enough space to provide a solution to the problem. If the space seems very large compared with your solution, go back and reread the question to make sure you've met all the parameters in the problem statement, make sure you have accounted for all of the surface water runoff on the site.

When stressed we all do things that we would not otherwise do, even after a month or two of studying, so when reviewing your drawings make sure that your roads are ridges, unless specifically stated otherwise in the problem, and your swales point uphill. Make sure every contour line is accounted for and that two adjacent proposed contour lines don't tie back into the same existing contour. Make sure each contour is facing the correct direction to ensure drainage, each line is labeled on the uphill side and every fifth contour line is darker.

In the LARE Orientation Information booklet, CLARB provides a list of all the required items you can bring to the exam. Double-check the list of prohibited items provided in the Orientation booklet to ensure that you are bringing only those things approved by CLARB. My recommendation is as follows:

1 Wear a simple, non-computerized watch so that you can be aware of the time, but not something that will distract you or any other test takers.

7.4 Identifying topographic signatures

While studying for the exam, I found that knowing my topographic signatures for each grading concept made the test more understandable. If I knew ahead of time what the end result should look like, I could guide myself to a correct answer. As designers we are taught to think and interact visually. We speak to each other through sketches and even the roughest of sketches have meaning. Using our ability to obtain visual meaning from sketches we can generalize site elements and make the process of grading less mathematical and more design oriented. A topographic signature is a generic symbol or abbreviated note of how a typical graded landscape element would look in plan view. For example, a retention pond will have at least 1'–2' of depth and it will be accompanied by a dam or embankment. Adding specificity for a particular problem, i.e. maximum side slopes at 3:1, will help remind you of what the grading solution must include. With topographic signatures, you can break down a complex grading problem into its smaller elements and in so doing determine which element to tackle first. They should be fairly generic and closely mimic your thought triggers. Like sketches, the symbols should be specific to the given grading problem and have meaning.

The topographic signatures in Figure 7.4–1 are generic sketches I used for the common site elements both natural and man-made. When solving a grading problem it is helpful to note each topographic signature that will be needed to solve a problem. This will ensure that you don't omit any of the problem parameters and will help you to visualize what the resultant grading solution should look like.

If you have difficulty visualizing the site elements on a grading plan even after you've created recognizable topographic signatures, use your orthographic drawing tools. The best way to understand any site existing and proposed is to do some quick line sections on top of each other to understand the before and after of your design. Precision is not necessary; these sections are only giving you the rise and fall of the earth. These sections should be exaggerated 2–5x vertical to 1x horizontal in order to understand clearly what the design is doing and what your proposed intent is.

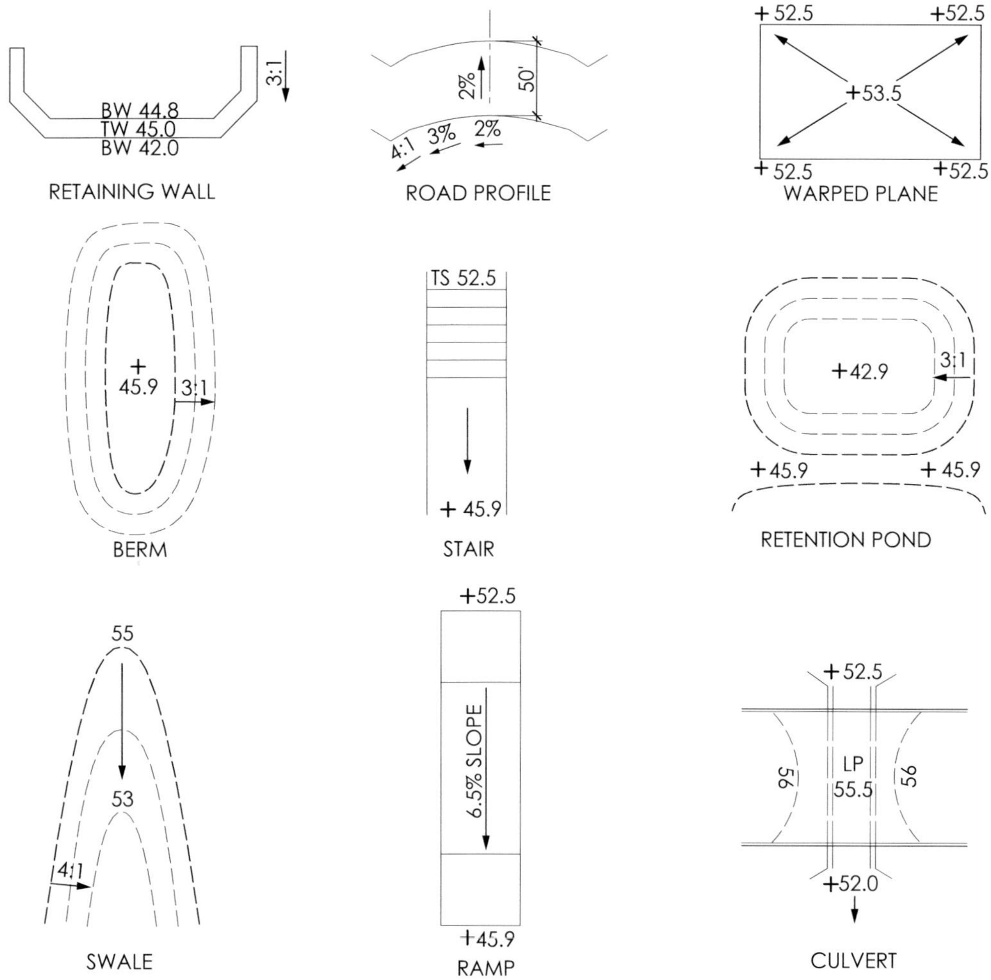

7.4–1
Examples of sketches that can be used as quick visual triggers

7.5 Where to start

For every grading vignette, there is a hierarchy of processes that must be followed in order to solve the problem. The key is to identify each of the landscape elements that must be solved. Create a list of these elements and then, based on the given parameters of the problem, organize them in hierarchical order based on the grading concept with the most known variables first. Solving each element on the hierarchical list systematically creates additional variables needed to solve the next element on the list. Generally, solve the man-made built elements before the natural landscape drainage elements. Solve all the other disciplines' (architecture and engineering) elements before tackling the landscape grading. A typical list will probably follow this order:

1 From the problem statement, identify the different elements that need to be included in the grading solution.
2 Provide any necessary spot elevations using slope and interpolation.

3 Determine any potential conflicts with underground and existing structures that must remain.
4 Grade any plane surfaces (building pads, fields, or parking lots).
5 Grade any built elements (stairs, ramps, berms, roads, paths, retaining walls).
6 Tie into any drainage structures (swales, culverts, retention ponds, drain inlets).

Of course, this list is malleable and will change based on the variables of the grading problem, but it is a useful starting point when doing practice vignettes. The more vignettes you do, the more comfortable you will be recognizing the order for completing a grading vignette on the exam.

For each vignette you do, think about it logically. For example, in order to provide a swale around a building, you must first grade the building pad; in order to provide a retention pond you have to draw a swale to attach to the pond. Start from the highest elevation on the problem statement and work your way down to the lowest spot elevation for the solution. This last point has more to do with the spot elevations than the orientation of the existing topography or the design within the problem.

Lastly, although this information is written with practice vignettes and the LARE examination in mind, know that the information in this study guide applies to real-world landscape grading problems. Landscape architecture is not a conceptual profession; it is a rigorous discipline that affects the public's health, safety and welfare and as a professional it is your responsibility to understand how your designs will be applied and implemented. Landscape grading vignettes are snapshots of the kind of work we as practitioners are expected to be able to do in order to get a design built. Grading plans are the way we communicate our designs to the landscape contractors who will build our designs. Using this study guide and doing practice vignettes will help you understand landscape grading better and ultimately make you a stronger designer.

7.6 Recommended reading

In addition to solving practice grading vignettes as a way to study for the exam, here is a list of recommended reading for greater depth and comprehension. Also refer to the CLARB website for their list of recommended reading. As important as solving vignettes is, it is equally important to understand the concepts behind what you are doing.

Site Engineering, 6th Edition – by Steven Strom, Kurt Nathan and Jake Woland – this edition also has a separate workbook with some grading vignettes and covers information about soil volume calculations, pipe sizing and horizontal and vertical road alignment.

Landscape Architectural Graphic Standards – edited by Leonard Hopper – a great and thorough study aid for all sections of the examination.

Time Savers Standards for Landscape Architecture – by Charles Harris and Nicholas Dines – has the basics for all landscape architecture topics.

LARE Review, Section E Vignettes: Grading, Drainage, and Stormwater Management – by Jerry P. Hastings – the starting point for all grading vignettes.

7.7 Literature cited

American Society of Landscape Architects. 2012. *ASLA Technical Workshop LARE Prep Section 4, Handouts*. Phoenix, Arizona: ASLA. www.asla.org/uploadedFiles/CMS/Become_a_LA/LARE_Content/Section 4 Materials 1.pdf.

American Society of Landscape Architects. 2012. *ASLA Technical Workshop LARE Prep Section 4, Post Session Handouts.* Phoenix, Arizona: ASLA. www.asla.org/uploadedFiles/CMS/Become_a_LA/LARE_Content/Section 4 Materials 2.pdf.

American Society of Landscape Architects. 2014. *SAMPLE EXAM – Landscape Architect Registration Examination Section 4 – Grading, Drainage and Construction Documentation*. Washington, DC: ASLA. www.asla.org/uploadedFiles/CMS/Become_a_LA/LARE_Content/Section-4-Grading-Drainage-Construction-Documentation-Sample-Questions.pdf.

Council of Landscape Architectural Registration Boards. 2017. *L.A.R.E. Orientation Guide*. Fairfax, Virginia: CLARB. www.clarb.org/docs/default-source/take-the-exam/lareorientation guide.pdf.

Section Eight
Grading vignettes

The problems that follow incorporate several landscape and built elements that need to be graded. Remember that most grading problems have more than one correct answer. What is important is that you follow the steps to get a correct answer. A solution to each is located in the back of the book.

Grading vignettes

Grading vignette 1 (answer on pages 209–210)

A gazebo has been placed at the top of a hill. Provide an accessible concrete path from the gazebo to the existing sidewalk.

1. From the FFE for the gazebo a 6' long ramp has been placed at 8.33%.
2. Longitudinal slope for the path is 3% minimum and 4% maximum.
3. Path cross slope is 2%.
4. Minimum slope on land is 2%; maximum slope is 4:1.
5. Label all proposed contours.
6. Water may drain across the path; no drainage swales are needed.
7. Existing trees are to remain.

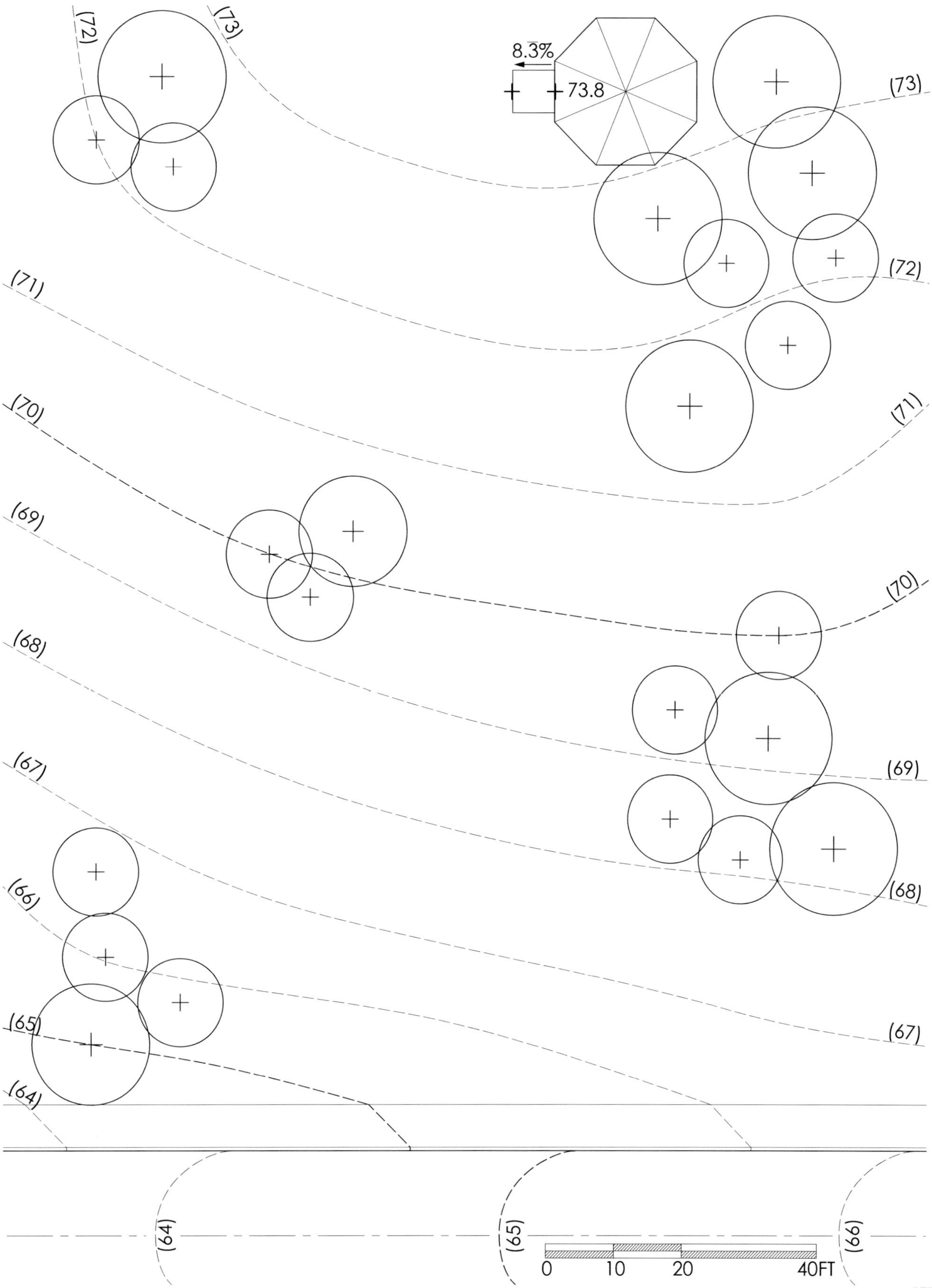
(72)
(73)
8.3%
73.8
(73)
(72)
(71)
(71)
(70)
(70)
(69)
(69)
(68)
(68)
(67)
(67)
(66)
(65)
(64)
(64)
(65)
(66)
0
10
20
40FT

Grading vignette 2 (answer on pages 211–213)

Grade a ramp and stairs for the proposed office building.

1. Courtyard in front of the building is 1" lower than the building FFE.
2. Courtyard slopes evenly away from the building at 2% gradient towards the stairs.
3. The bottom of the stairs is 3' lower than the top of the stairs.
4. Landing at the bottom of the stairs slopes evenly towards the existing pier at 2% gradient.
5. Determine the longitudinal slope for the ramp and grade the ramp runs with a uniform slope.
6. Cross slope on the ramp runs is 2% max.
7. Show the handrails for the ramp and all the critical spot elevations.
8. Round spot elevations to the nearest hundredth (1/100) x.xx'.
9. Proposed contours are not necessary.

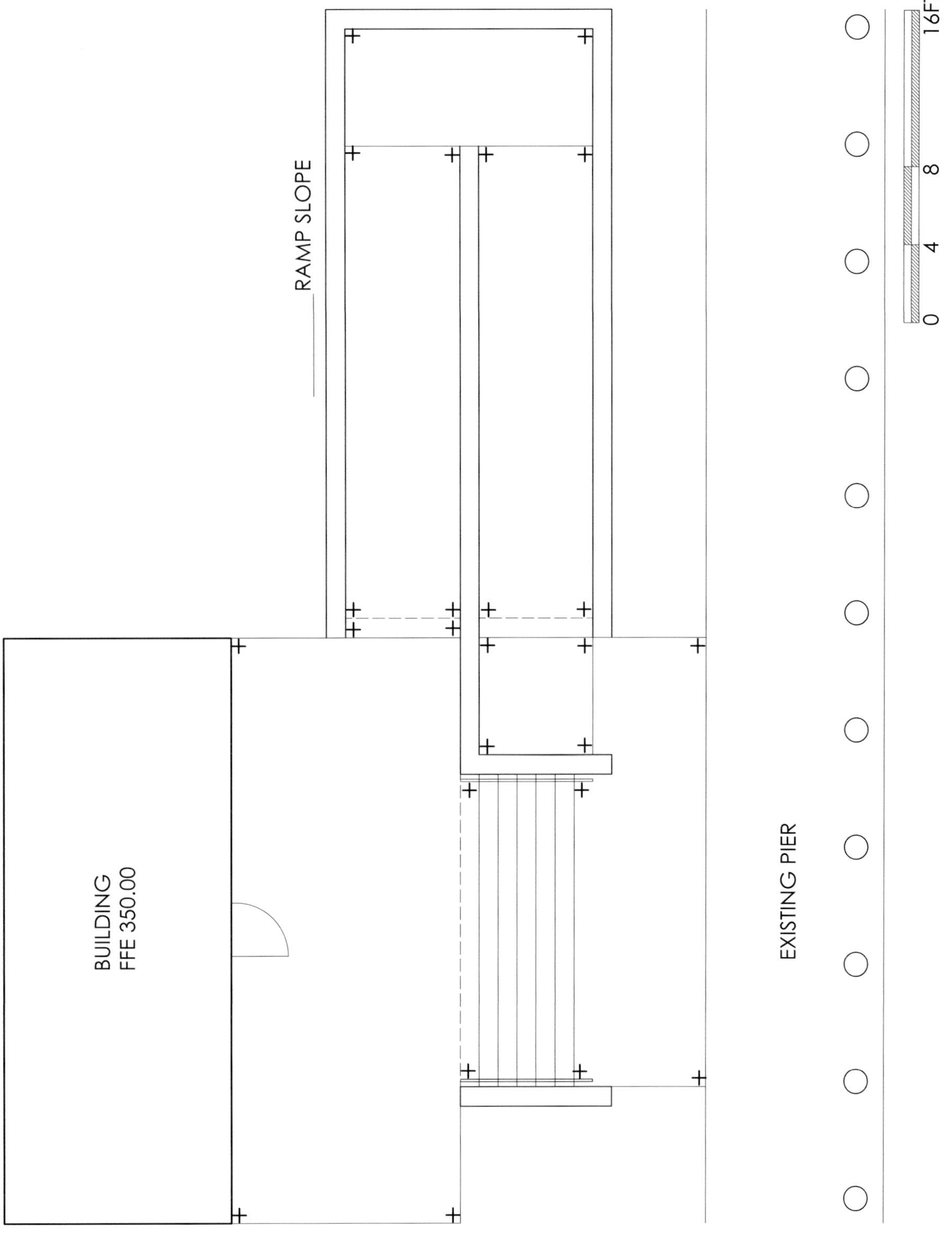
RAMP SLOPE
BUILDING
FFE 350.00
EXISTING PIER
0
4
8
16FT

Grading vignette 3 (answer on pages 214–216)

A building has been placed on the top of a berm. A circular concrete driveway has been provided. Grade the driveway, the adjacent sidewalk and the walk up to the building. Provide drainage around the building. Drainage from the sidewalk may flow into the driveway.

1. Road has a cross slope of 2%.
2. Sidewalk longitudinal slope matches the road.
3. Sidewalk cross slope is 2%.
4. Curbs are 6" above the road.
5. FFE is 1" higher than the sidewalk outside of the door.
6. Landscape outside of the building is 6" lower than the FFE.
7. Slope away from the outside of building at 2% for 5'.
8. Sidewalk slopes away from the door at 2%.
9. The center of the circular drive is a planted berm.
10. Label all proposed contours.
11. Show all spot elevations.
12. Minimum slope is 2%; maximum slope is 3:1.

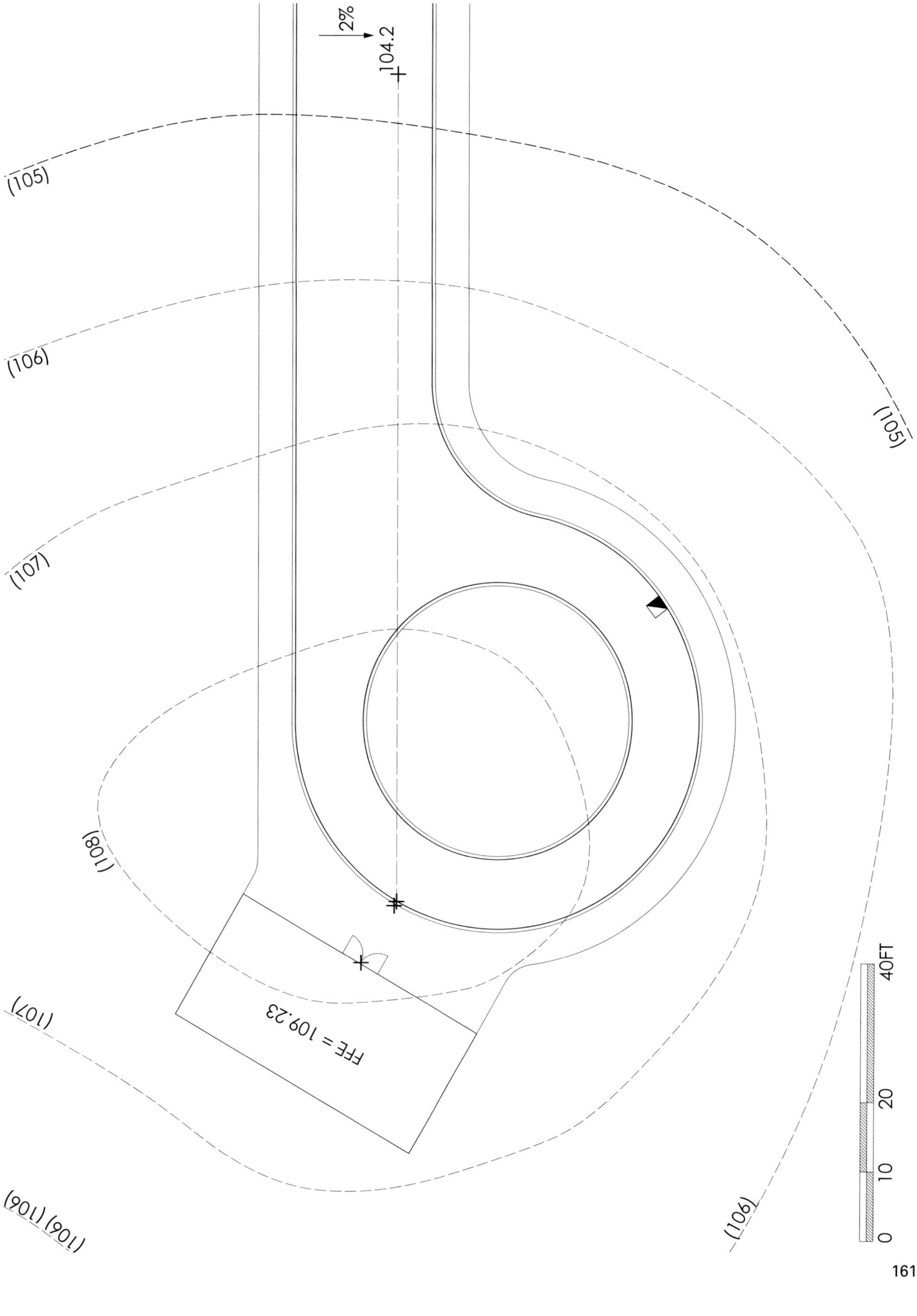
2%
104.2
(105)
(106)
(107)
(108)
FFE = 109.23
(105)
(107)
(106)
(106)
(106)
0
10
20
40FT

Grading vignette 4 (answer on pages 217–219)

Provide two flights of stairs and a 6' wide ramp to reach the platform.

1. Minimum slope is 2%; maximum slope is 3:1.
2. Risers = 6" (0.5'); treads = 12" (1').
3. Provide three evenly spaced, evenly sloped landings on the path.
4. Label TS and BS for each flight of stairs
5. Draw the handrails for the stairs.
6. Label slope for the landings.
7. Provide a 1' wide continuous cheek-wall on both sides of the stairs.
8. The cheek-walls shall align with the back edge of the top step and extend 1' below the bottom step.
9. Label TW and BW for the cheek-walls.
10. Top of cheek-wall (TW) is 6" higher than the elevation of the top step at the beginning of the wall and 6" higher than the elevation of the bottom step at the end of the wall.
11. Bottom of cheek-wall (BW) is 3" lower than the top of the cheek-wall on the landscape side.
12. C-shaped (90°) ramp shall start clear of the handrail and cheek-wall above the first flight of stairs.
13. End the ramp just clear of the handrails and cheek-wall below the bottom flight of stairs.
14. Grade the ramp so that each of the three ramp runs have a uniform slope.
15. Provide spot elevations for each ramp run and landing (x.xx).
16. Cross slope on the first and last ramp runs to match the stair landings.
17. Water can flow over the ramp.
18. Show proposed contours on the ramp and tie proposed contours back to the existing topography.

DISTANCE IN FEET	STAIRS	LANDINGS	TOTAL
VERTICAL			
HORIZONTAL			

TOTAL NUMBER OF STEPS ________

SLOPE OF LANDINGS ________

LENGTH OF A TYPICAL LANDING________

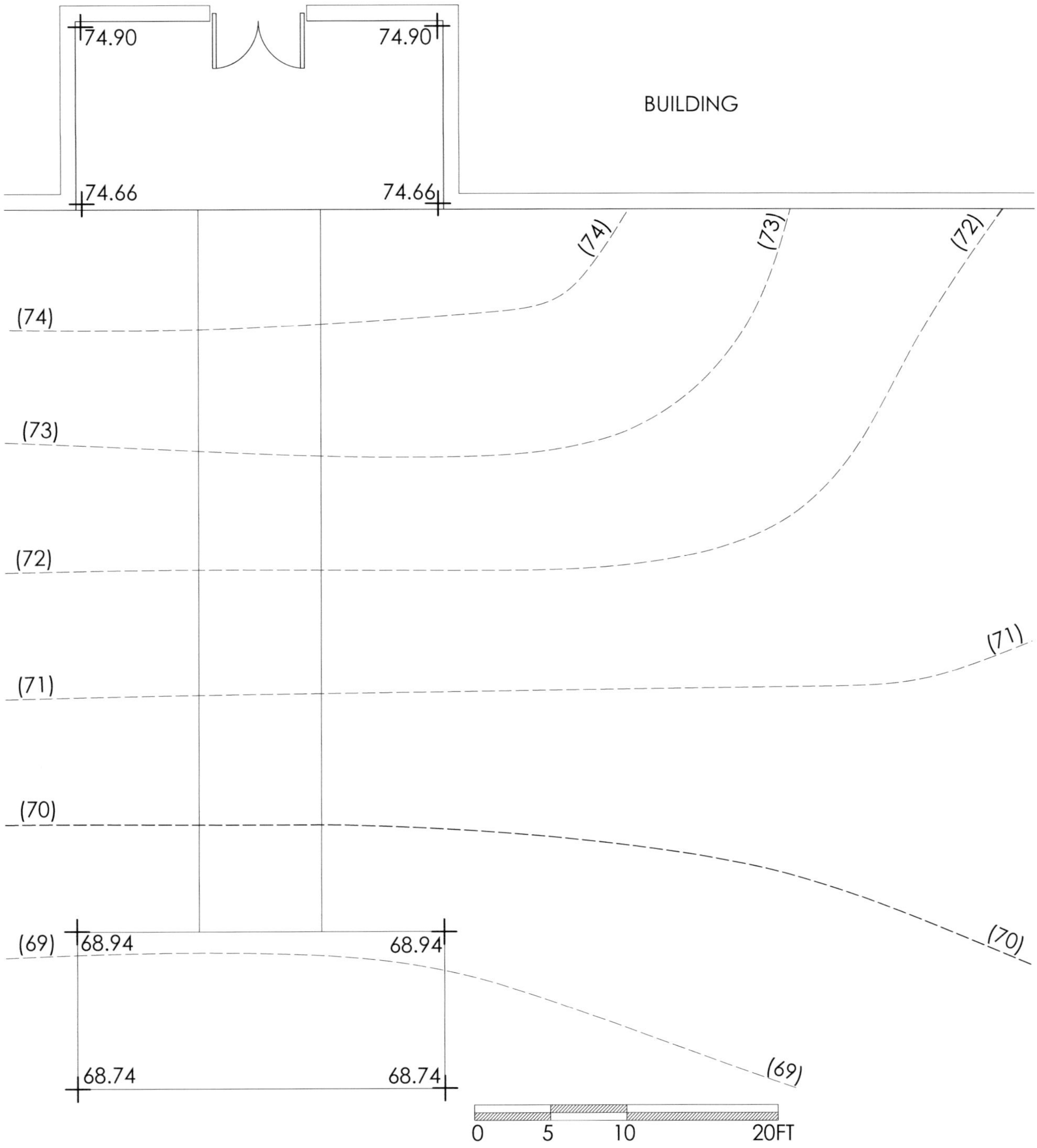

Grading vignette 5 (answer on pages 220–221)

Grade the proposed asphalt road, concrete gutters, sidewalks and retaining wall.

1. Determine a uniform gradient for the road.
2. Road has a crown of 1/4" per foot.
3. Sidewalk longitudinal slope matches the road.
4. Sidewalk cross slope is 2% and drains towards the gutters.
5. Label all proposed contours.
6. Show all spot elevations.
7. Retaining wall is 1' tall. Soil behind the wall is 2" lower than the TW.
8. Minimum slope is 2%, maximum slope is 3:1.

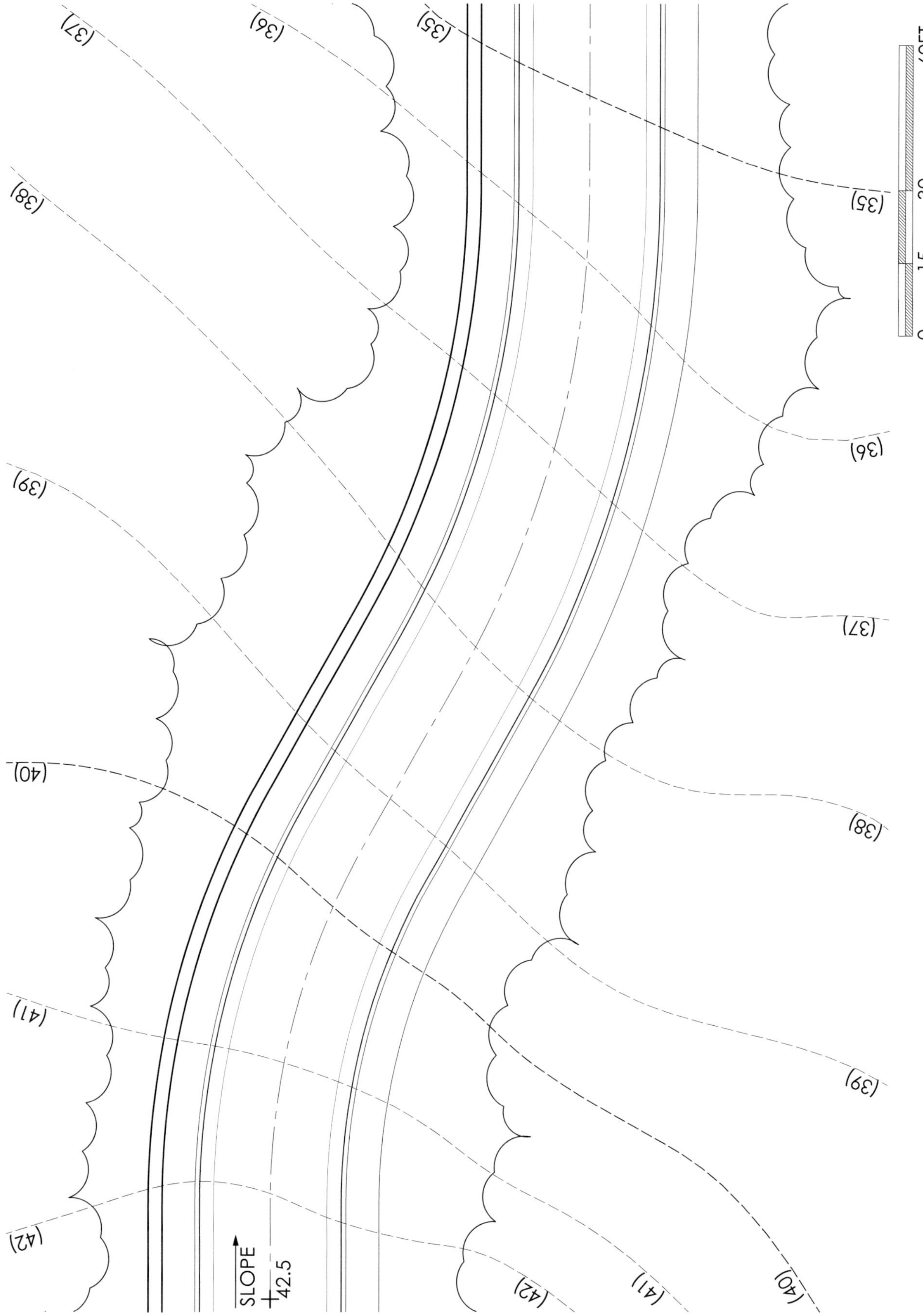
(35)
(36)
(37)
(38)
(39)
(40)
(41)
(42)
SLOPE
42.5
0
15
30
60FT

Grading vignette 6 (answer on pages 222–223)

Grade the four single family houses.

1 Provide a slope for the main walkway.
2 Main walkway has a cross slope of 2% draining to the east.
3 Paths to the houses are sloped uphill at 2%.
4 Porch is 6" above the walkway.
5 Porch slopes uphill at 2%.
6 FFE is 1" higher than porch.
7 Grade outside each house is 6" lower than FFE.
8 Provide a 5' buffer around each building sloped at 2% away from the building.
9 HPS for each building is located 5' from the edge of the buffer and slopes away from the buffer at 2%.
10 Connect swales to the drain inlets and provide rim elevations.
11 Minimum slope is 2%; maximum slope is 5:1.
12 Round all spot elevations to the nearest tenth (1/10) x.x'.

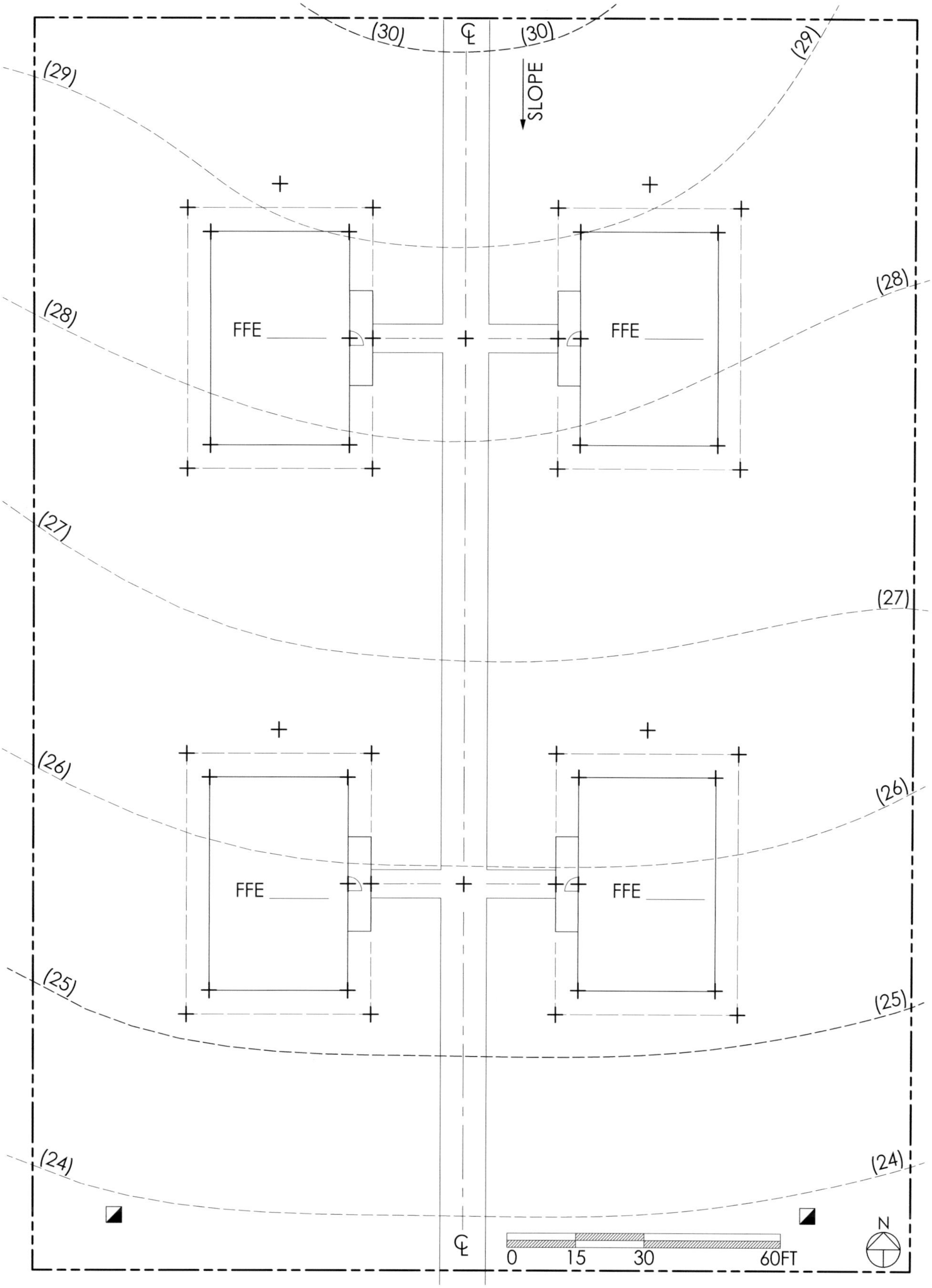

(30)
℄
(30)
(29)
(29)
SLOPE
(28)
(28)
FFE
FFE
(27)
(27)
(26)
(26)
FFE
FFE
(25)
(25)
(24)
(24)
℄
0
15
30
60FT
N

Grading vignette 7 (answer on pages 224–226)

Divert water around the building to the retention pond.

1. Provide FFE for the building.
2. Pathway to the building is level.
3. Porch is 6" higher than the pathway.
4. Porch slopes away from door at 2%.
5. FFE is 1" higher than porch.
6. Outside of the building is 6" lower than the FFE, ground slopes at 2% away from the building for 10'.
7. HPS is located 5' away from the 10' buffer and slopes at 2% downhill from the buffer.
8. Divert surface water through the culvert.
9. Culvert has 2' of soil cover over it.
10. Provide spot elevations for the culvert.
11. Pond is 2' deep.
12. Provide spot elevations for the dam/embankment.
13. Label all proposed contours.
14. Minimum slope is 2%; maximum slope is 3:1; pipe slope is 0.5% (0.005).

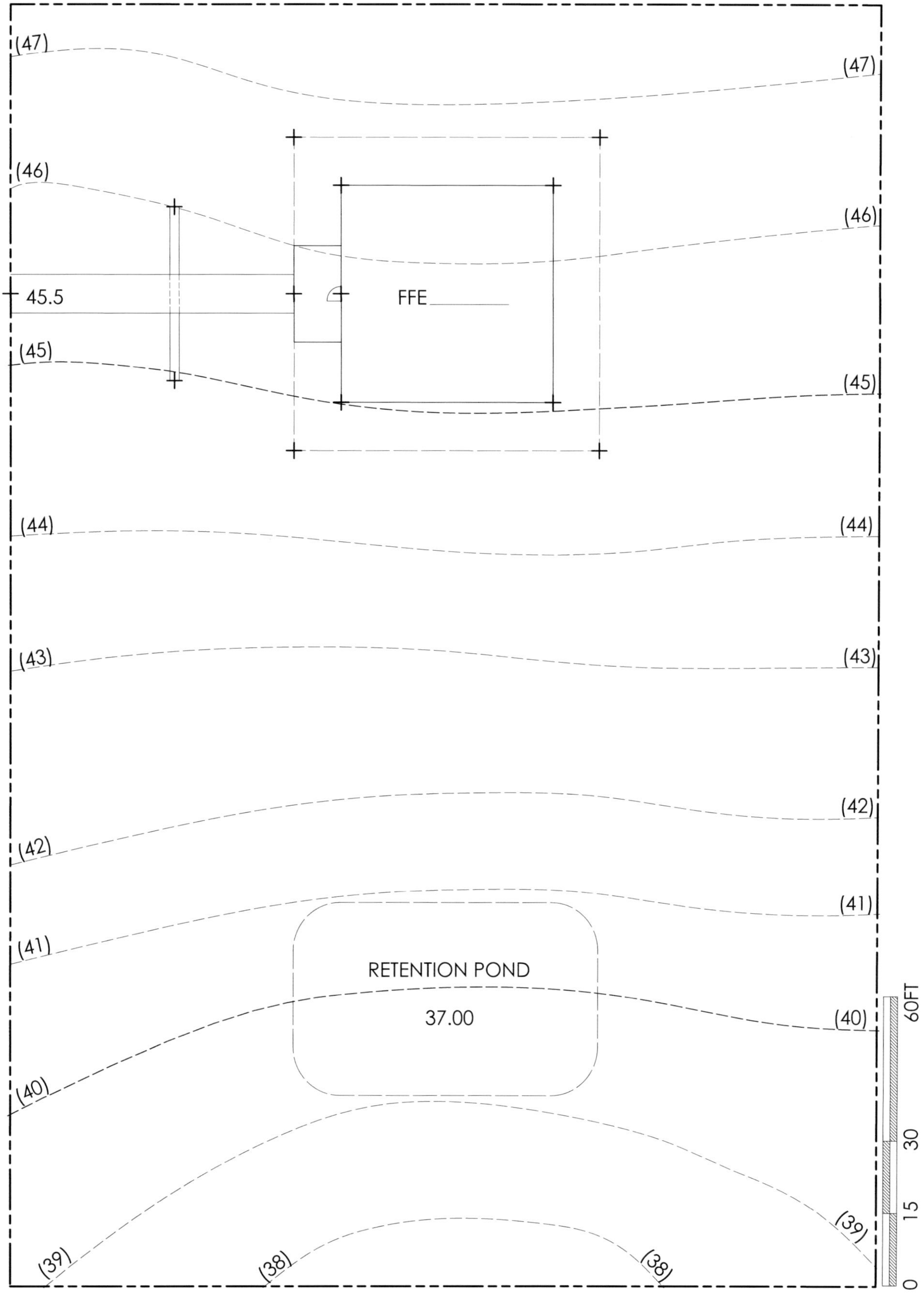
(47)
(47)
(46)
(46)
45.5
FFE
(45)
(45)
(44)
(44)
(43)
(43)
(42)
(42)
(41)
(41)
RETENTION POND
37.00
(40)
(40)
(39)
(38)
(38)
(39)
0
15
30
60FT

Grading vignette 8 (answer on pages 227–229)

Create a 24' wide asphalt road with a shoulder and divert water through the culvert to a retention pond.

1 Road has a crown of 2%.
2 Use the vertical curve to locate the whole number spot elevations on the road centerline.
3 Provide a 5' shoulder for both sides of the road with a 2% cross slope.
4 Slope of the culvert pipe is 0.5% (0.005).
5 Top of the culvert is 1'-4" lower than the LP of the road.
6 The bottom invert elevation of the pipe is 2' lower than the top of the pipe.
7 The minimum slope is 2% and the maximum slope is 4:1.
8 A swale for the road is not needed.

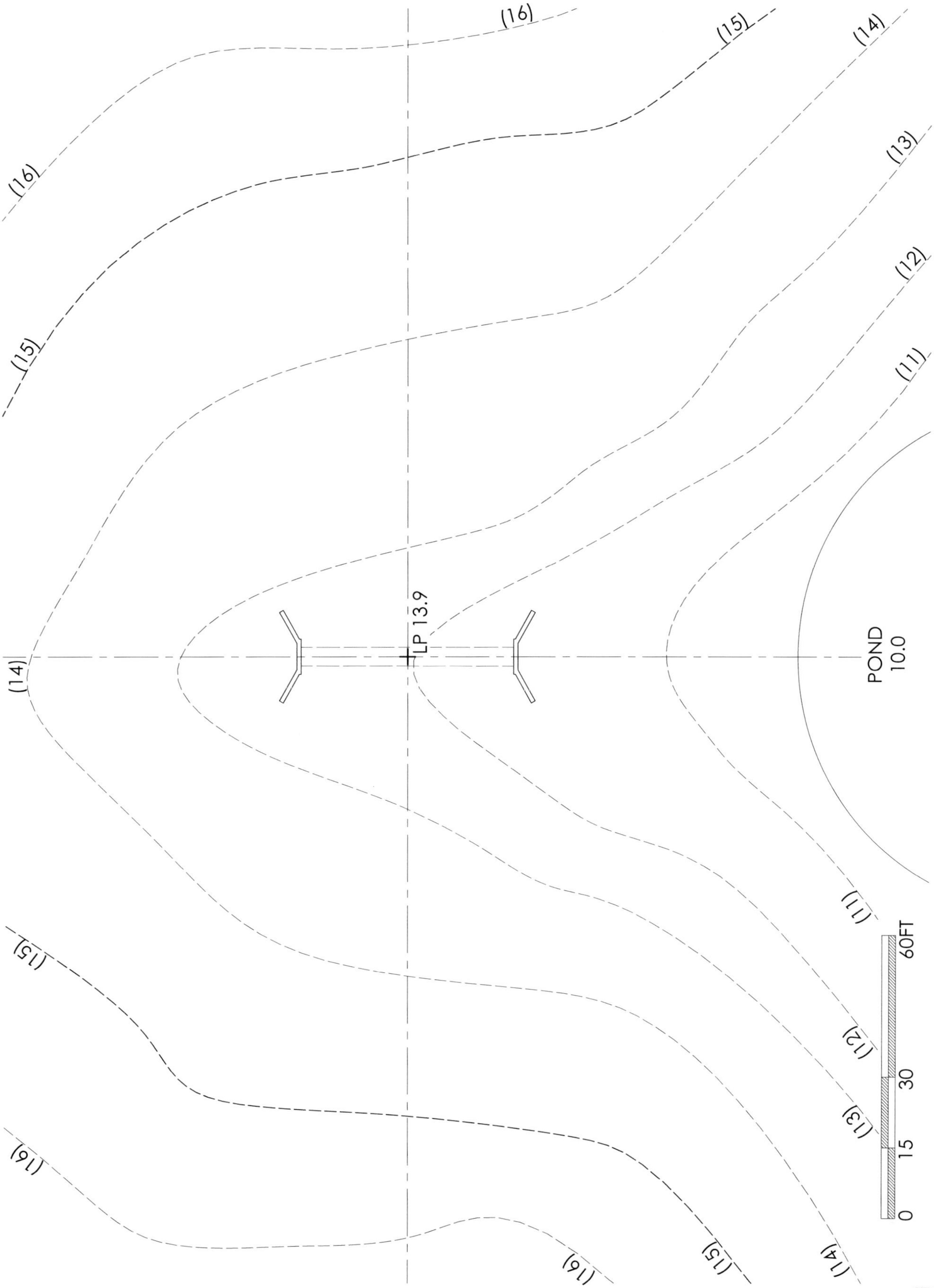
(16)
(15)
(14)
(13)
(12)
(11)
LP 13.9
POND
10.0
0
15
30
60FT

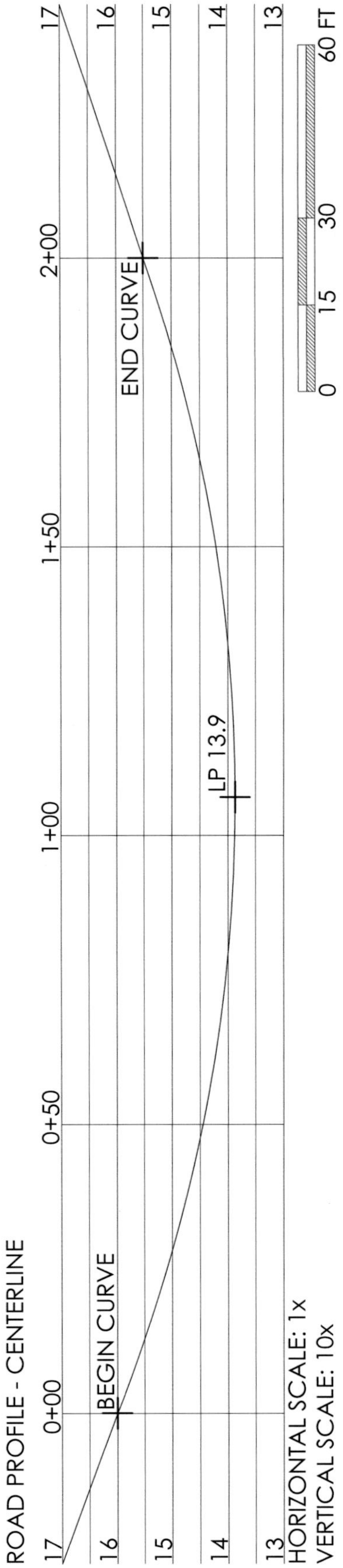
ROAD PROFILE - CENTERLINE
17
16
15
14
13
0+00
0+50
1+00
1+50
2+00
BEGIN CURVE
LP 13.9
END CURVE
HORIZONTAL SCALE: 1x
VERTICAL SCALE: 10x
0
15
30
60 FT

Grading vignette 9 (answer on pages 230–231)

Grade the house into the sloping landscape. Provide a horseshoe swale around the property and direct the water into the drain inlets.

1 Finish grade around the north, west and south sides of the house is 6" lower than the FFE.
2 A 5' buffer slopes away from the house at 2%.
3 HPS is located 5' away from the buffer and slopes downhill at 2%.
4 Front porch is 1" lower than FFE and slopes at 2% towards the front edge of the stairs.
5 Bottom of the stairs is 1.5' lower than the front edge of the porch.
6 Patio is flush to the FFE and slopes away from the house at 2%.
7 Lawn slopes away from the patio at 2%.
8 Water must not pool in the corner of the retaining wall.
9 Top of the retaining wall is 2' higher than the patio and lawn.
10 Minimum slope for the landscape is 4% and for the swales is 5%.
11 Maximum slope for the landscape is 4:1.
12 Cross slope on the path is 2%.
13 Water from the path must flow into the landscape and not off the property.
14 Do not grade within the tree drip line.
15 Swale must not encroach on the lawn.

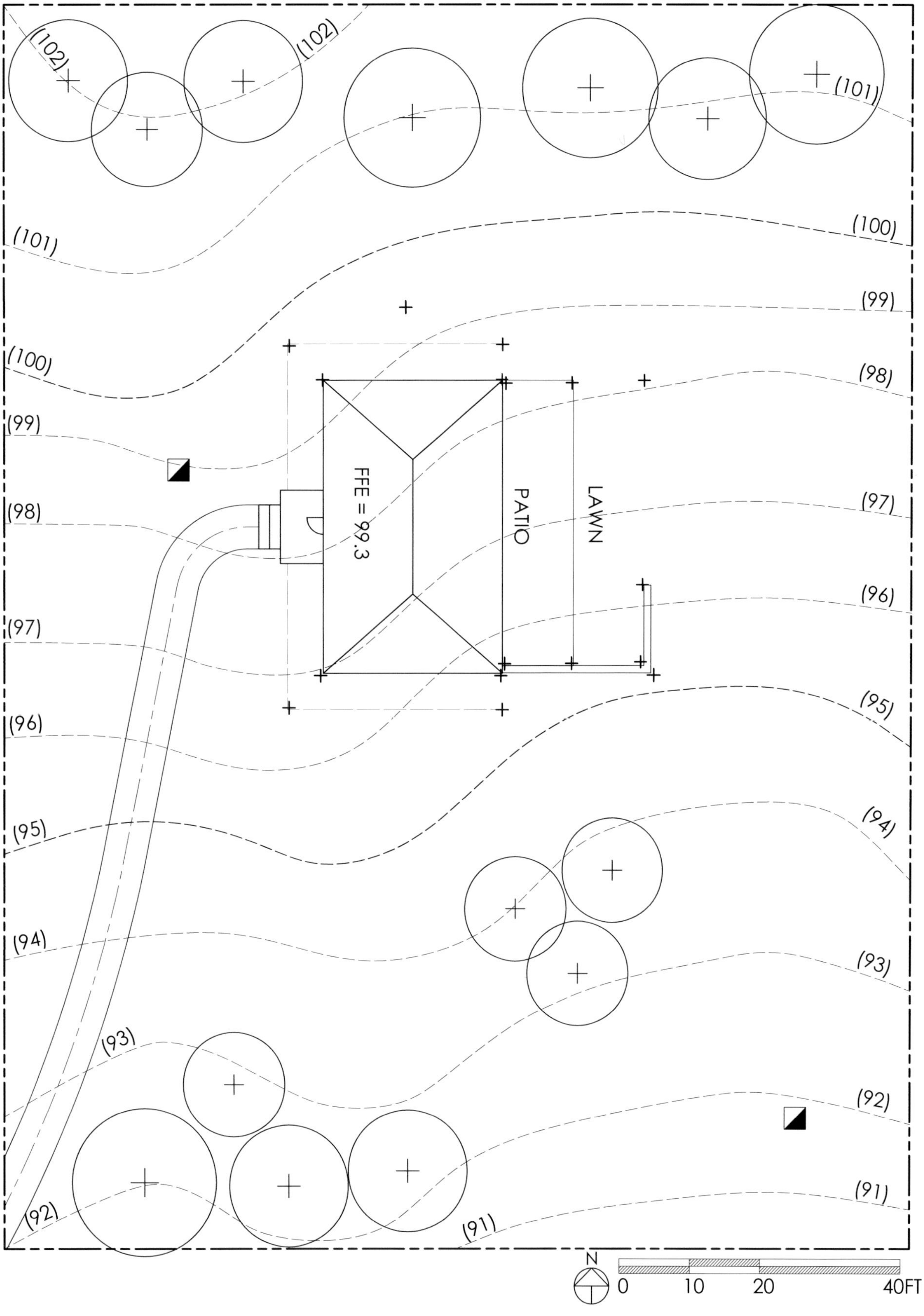
(102)
(102)
(101)
(101)
(100)
(100)
(99)
(99)
(98)
(98)
FFE = 99.3
PATIO
LAWN
(97)
(97)
(96)
(96)
(95)
(95)
(94)
(94)
(93)
(93)
(92)
(92)
(91)
(91)
N
0
10
20
40FT

Grading vignette 10 (answer on pages 232–233)

From a deck on a building overlooking a road, provide a sloping berm to block the view of traffic on the road.

1 Minimum slope is 2% and maximum slope is 5:1.
2 Label all proposed contours.
3 Assume an eye height of 5' for a person standing on the deck.
4 Assume an 8' truck is traveling on the road.

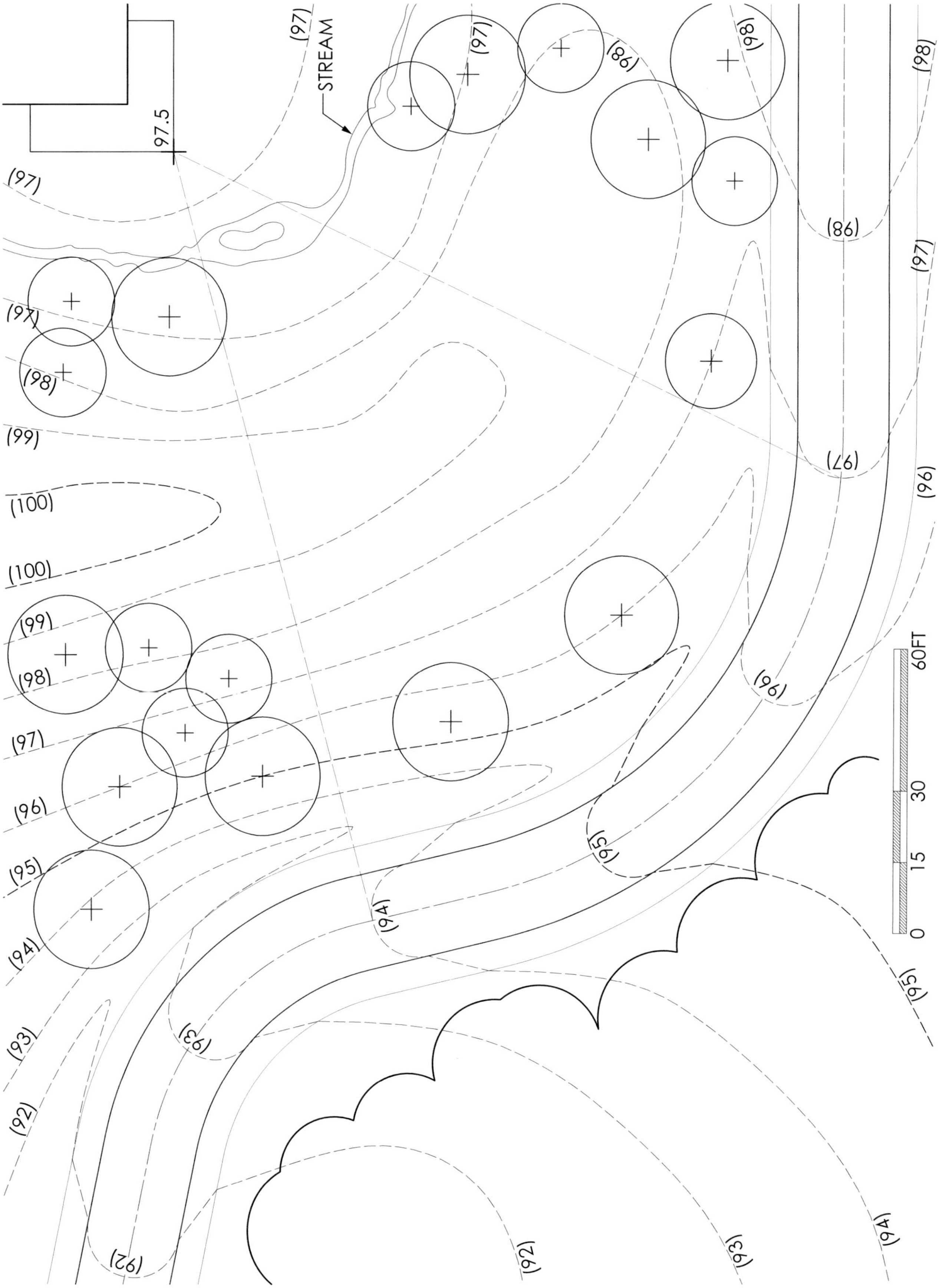
STREAM
97.5
(97)
(98)
(99)
(100)
(96)
(95)
(94)
(93)
(92)
0
15
30
60FT

Section Nine
Solutions to exercises and vignettes

Section One

1-1) Find the decimal foot equivalents for the following architectural measurements:

4'-6" = 4.5'	53'-9" = 53.75'
111'-2" =111.1$\overline{66}$'	40'-3" = 40.25'
75'- 0 1/4" = 75.0208'	32'-1" = 32.08$\overline{33}$'

1-2) Find the architectural measurements for the following decimal foot equivalents:

97.$\overline{33}$' = 97'-4"	28.91$\overline{66}$' = 28'-11"
7.75' = 7'- 9"	56.58$\overline{33}$' = 56'-7"
112.0104' = 112'-0 1/8"	250.0208' = 250'- 0 1/4"

1-3) Find the VERTICAL distance for each of the following:

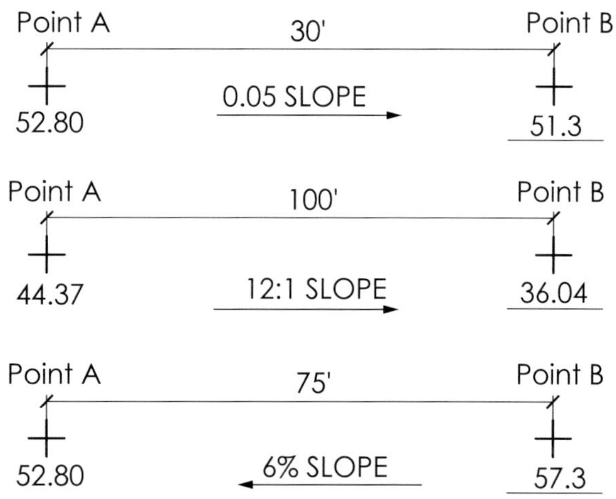

1-4) Find the SLOPE for each of the following:

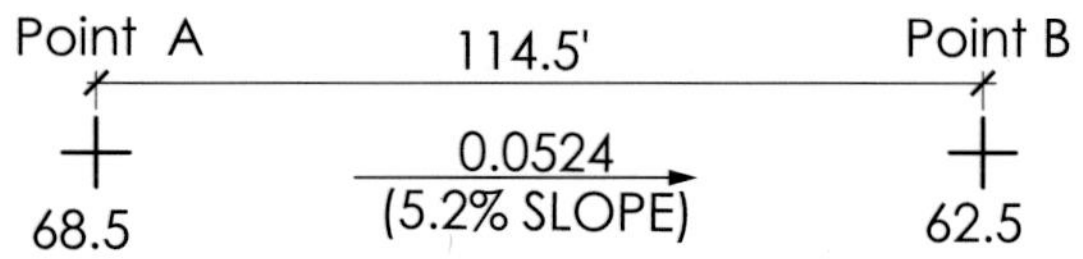

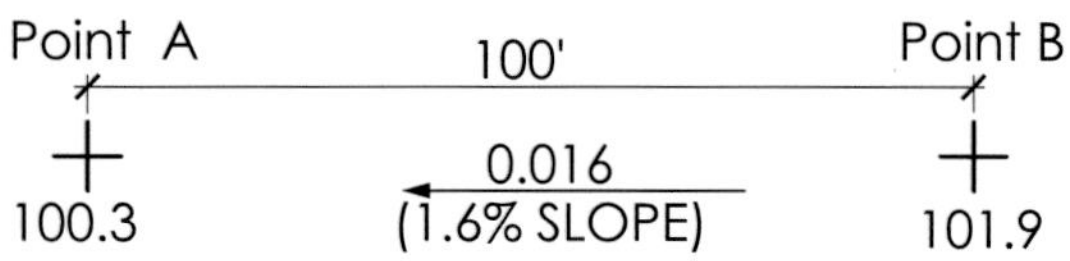

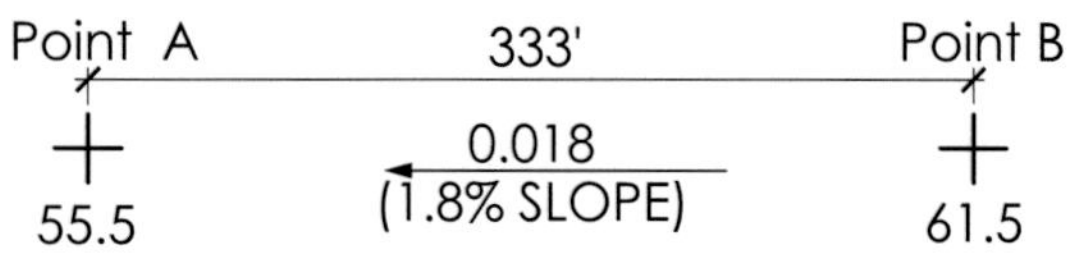

1-5) Find the HORIZONTAL distance for each of the following:

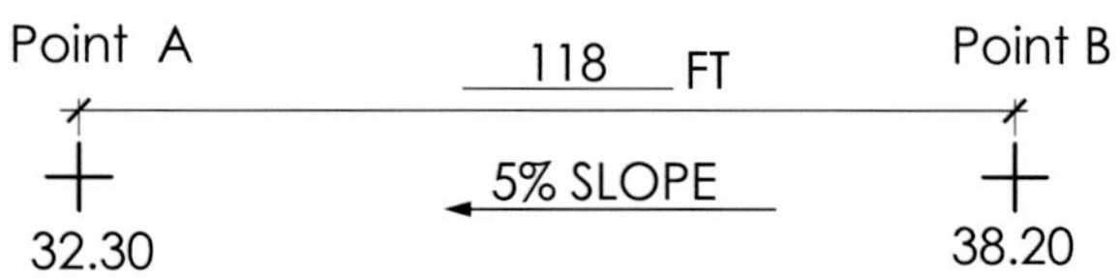

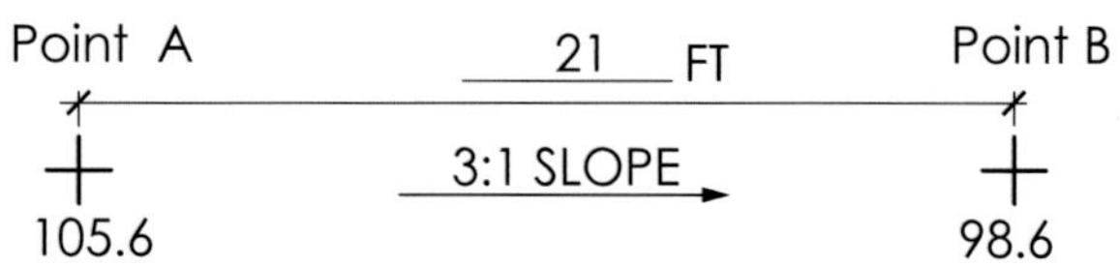

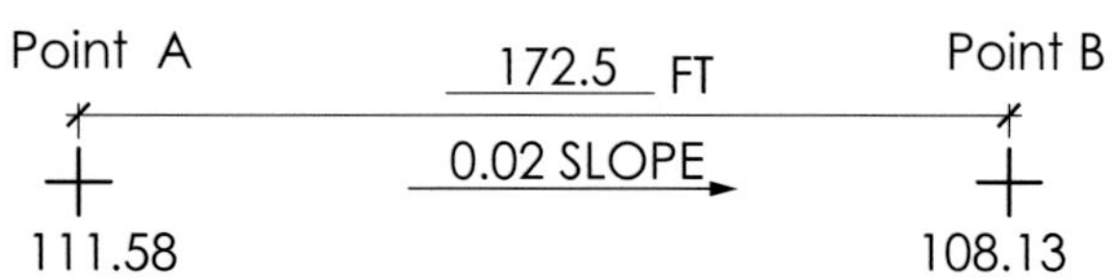

1-6) Complete the 10 x 10 grid to find the contour lines:

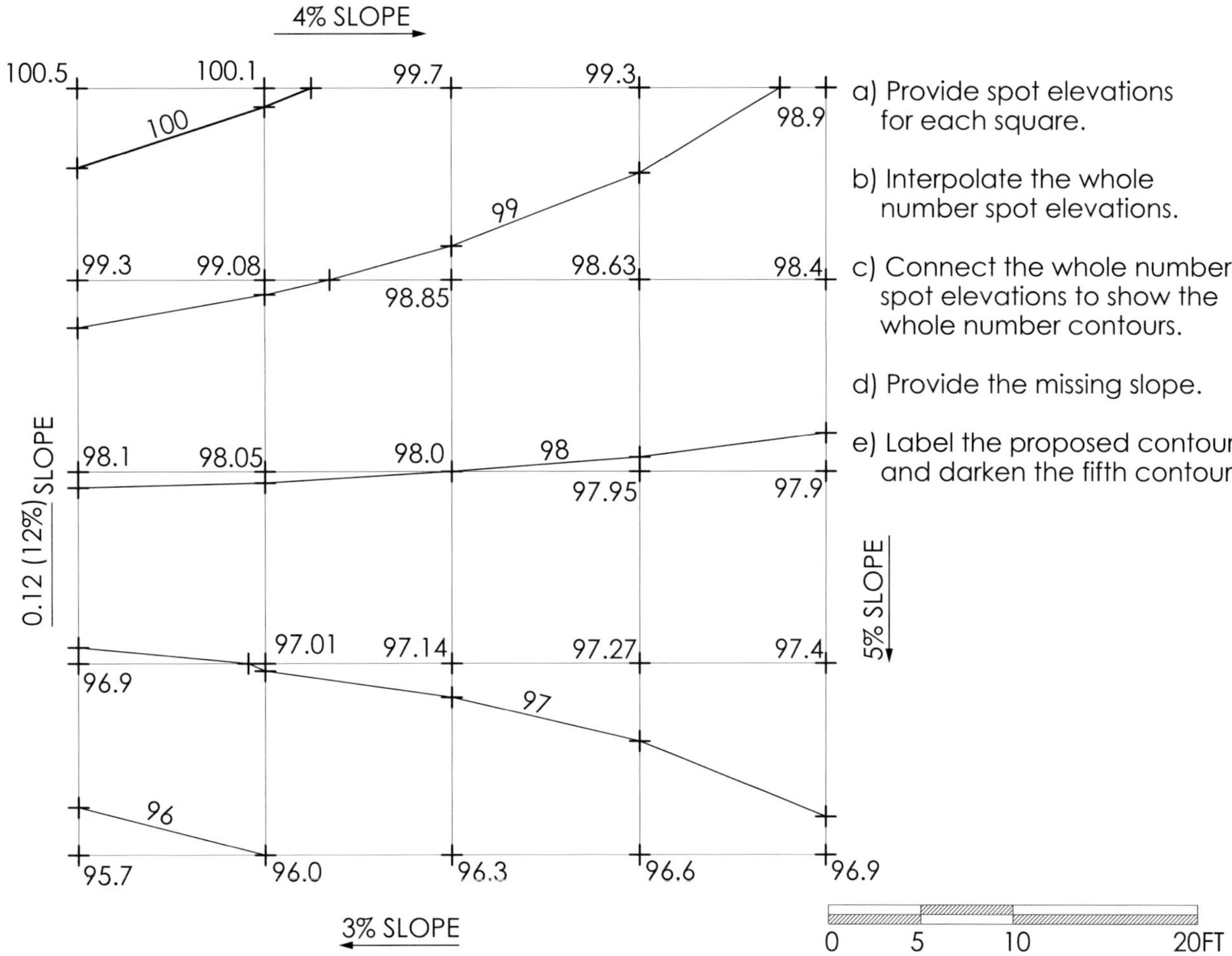

a) Provide spot elevations for each square.

b) Interpolate the whole number spot elevations.

c) Connect the whole number spot elevations to show the whole number contours.

d) Provide the missing slope.

e) Label the proposed contours and darken the fifth contour.

Section Two

2-1) Number and label the existing contour lines at 1' contour intervals to create one peak, two high points (HP) and three valleys.
a) Show center flowlines for the valleys.
b) Darken every fifth contour line; label existing contours on the uphill side.

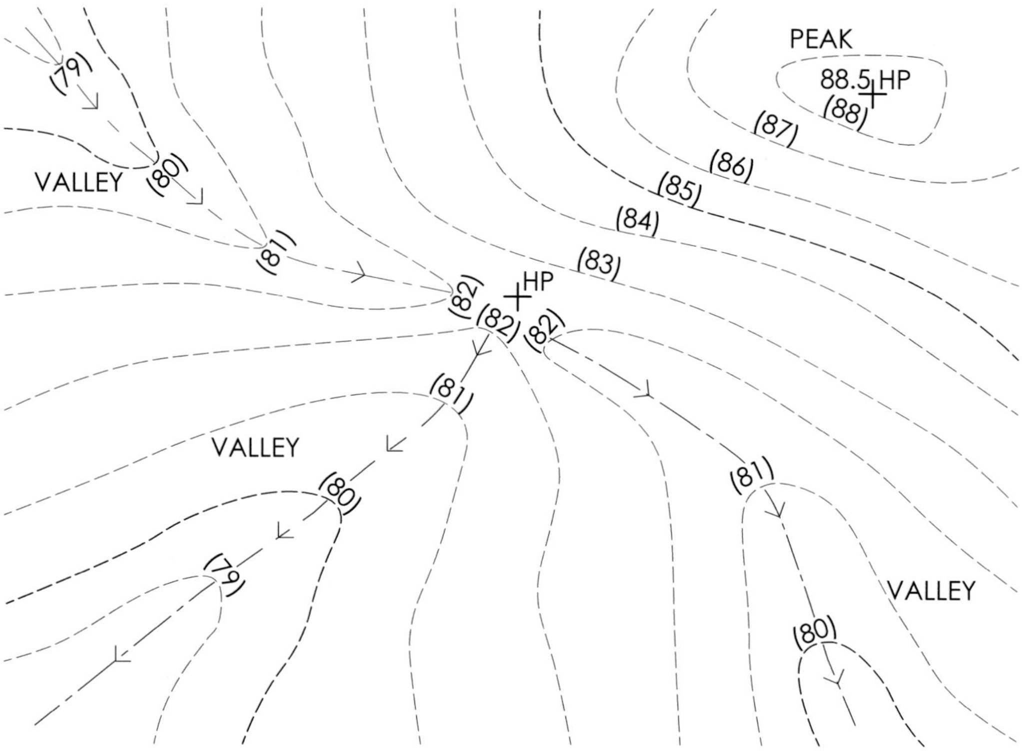

2-2) Number and label the existing contour lines at 1' contour intervals to create a depression, two ridges and a valley.

a) Show center flowlines for the valley and ridgelines for the ridges.

b) Darken every fifth contour line; label existing contours on the uphill side.

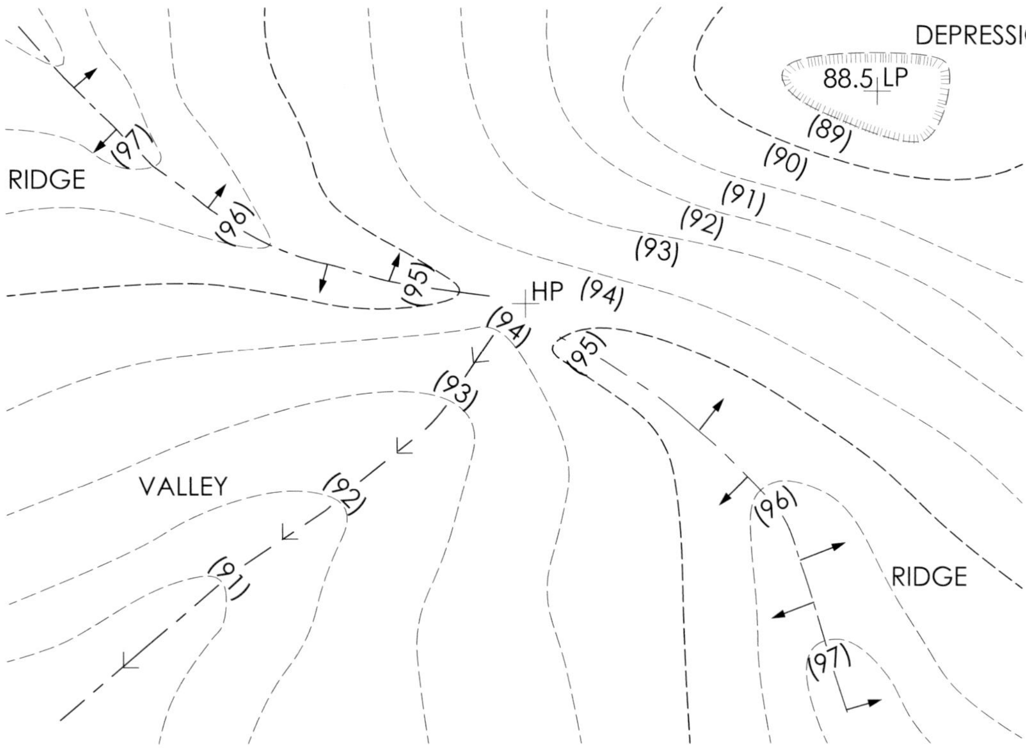

2-3) Draw and label the ridgelines for the three watersheds. Write HP and LP along the ridgelines. Locate four 1000 x 1500 SF planes on ridges with slopes between 2–3%. Draw flowlines for the valleys in the watersheds.

Section Three

3-1) Grade the two surfaces and show proposed contour lines. Minimum slope is 2%, maximum slope is 5:1; use sheet drainage, no swale is required. Draw two overlapping sections showing the existing and proposed grade.

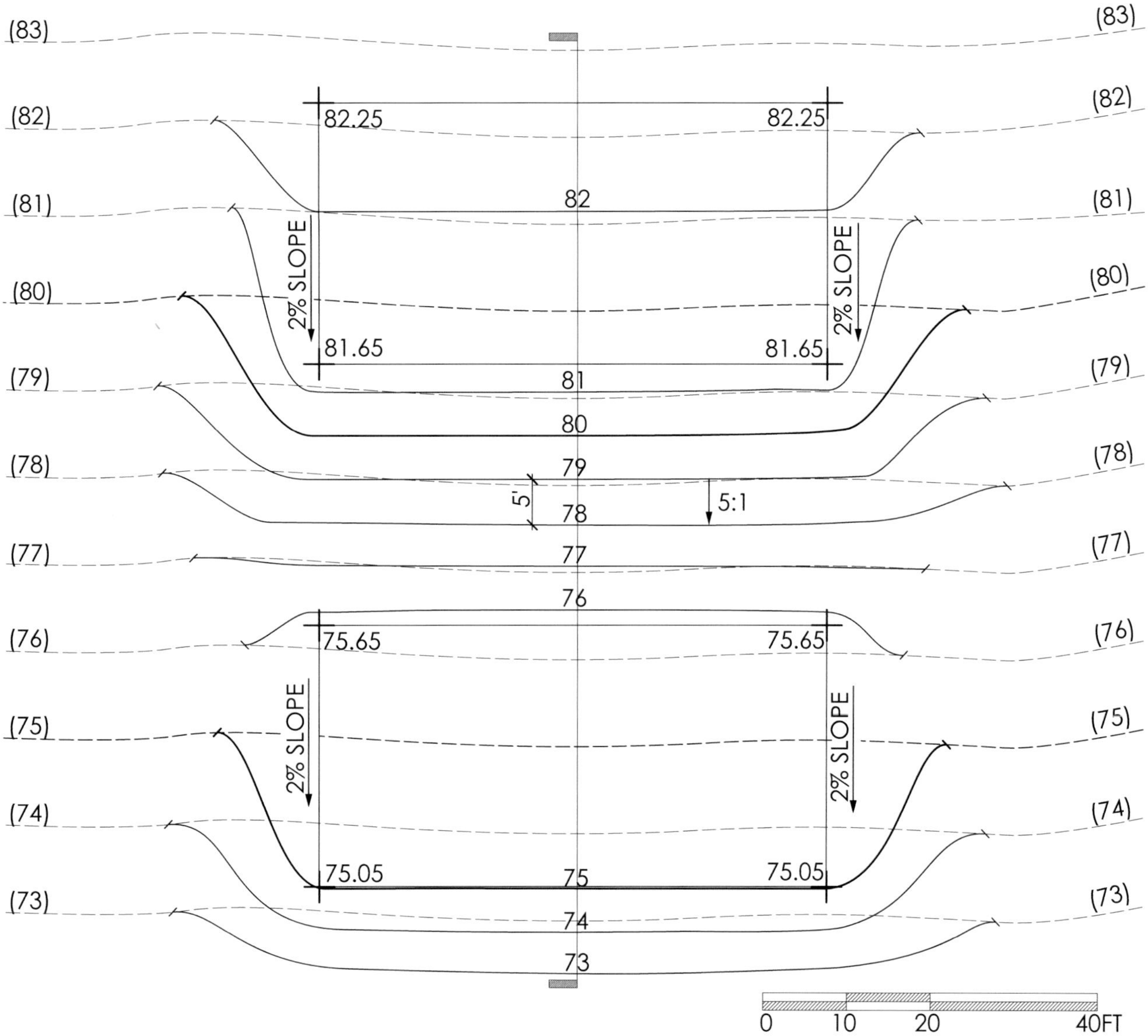

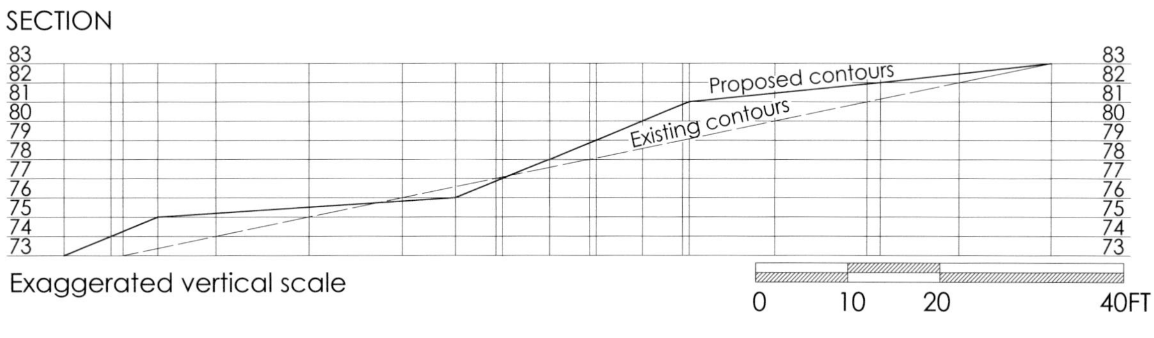

3-2) Draw a 18" swale.

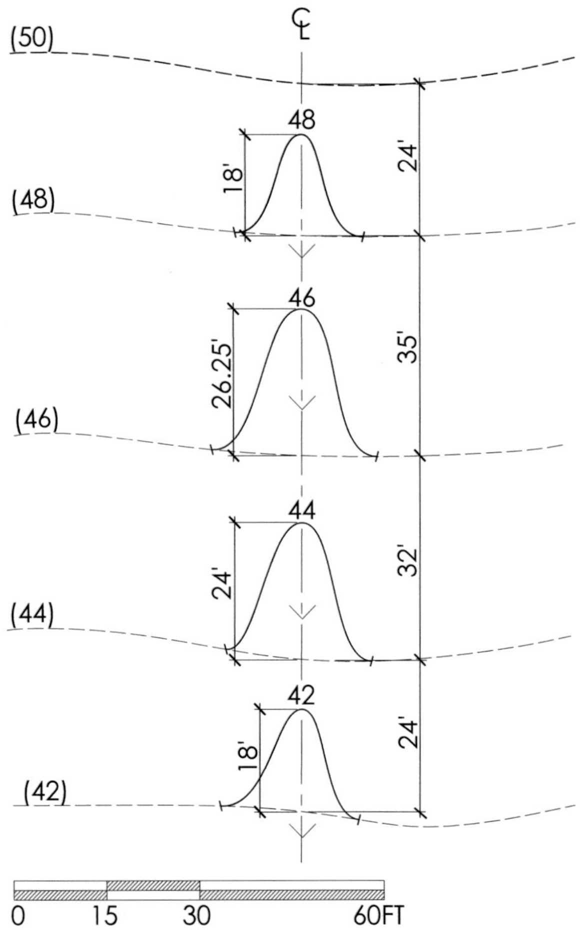

1) Measure distance between 2' contour lines.
2) Multiply by 0.75.
3) Repeat for each contour line.

3-3) Draw a swale with an 8% minimum slope and a 3:1 maximum slope.

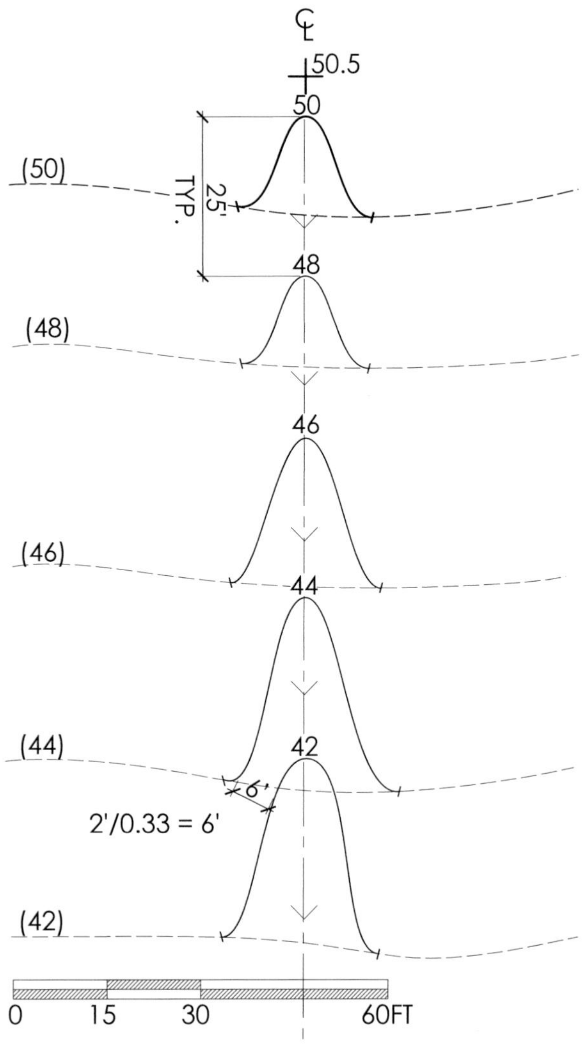

1) Interpolate to find the 50 contour.
2) Use slope formula to find slope interval.
   H=V/S 2'/0.08 = 25'
3) Measure uniform distance along centerline.
4) Draw swale.

3-4) Draw a 12" ridge.

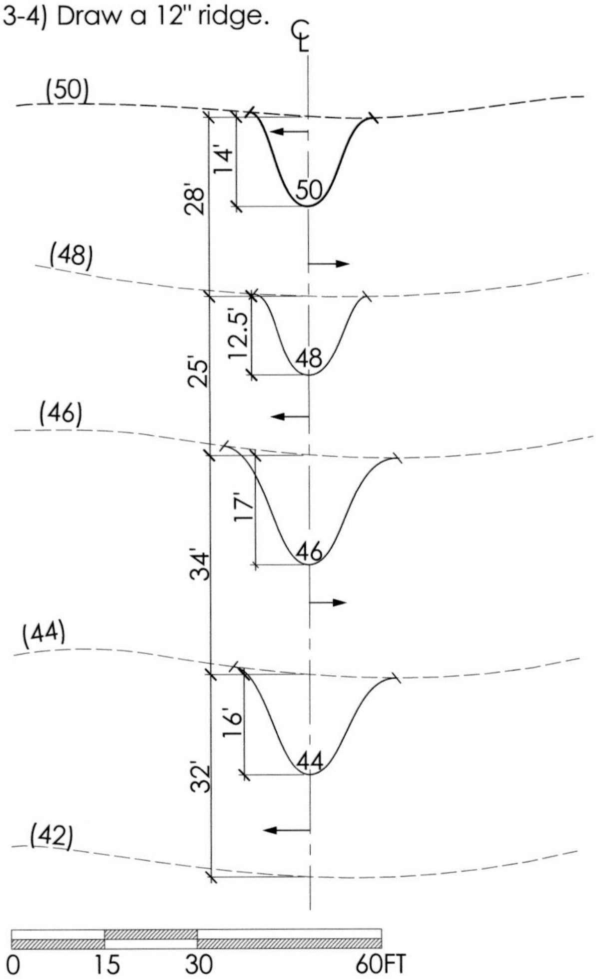

1) Measure distance between 2' contours.
2) Divide by 2.
3) Repeat for each contour line.

3-5) Draw a ridge with an 8% minimum slope and a 3:1 maximum slope.

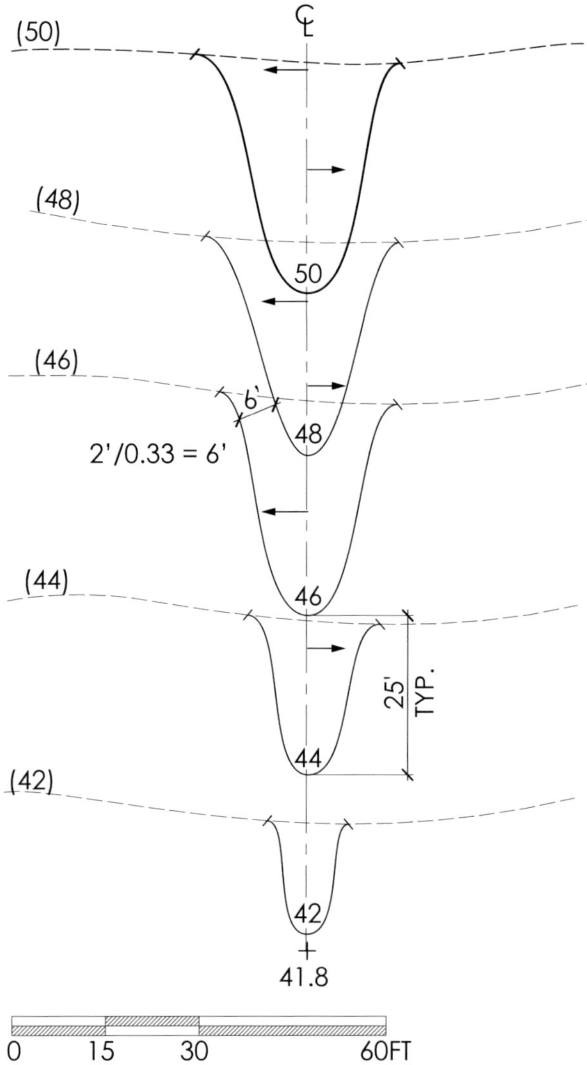

1) Interpolate to find 42 contour.
2) Use slope formula to find slope interval.
   H= V/S 2'/0.08 = 25'
3) Measure slope interval along centerline.
4) Draw ridge.

3-6) Draw a swale with a 3' deep retention pond.

a) Swale should have a minimum slope of 4% and a maximum slope of 5:1.
b) Bottom of the retention pond should be 323 and the sides should have a maximum slope of 5:1.
c) Label all parts of the pond and swale.

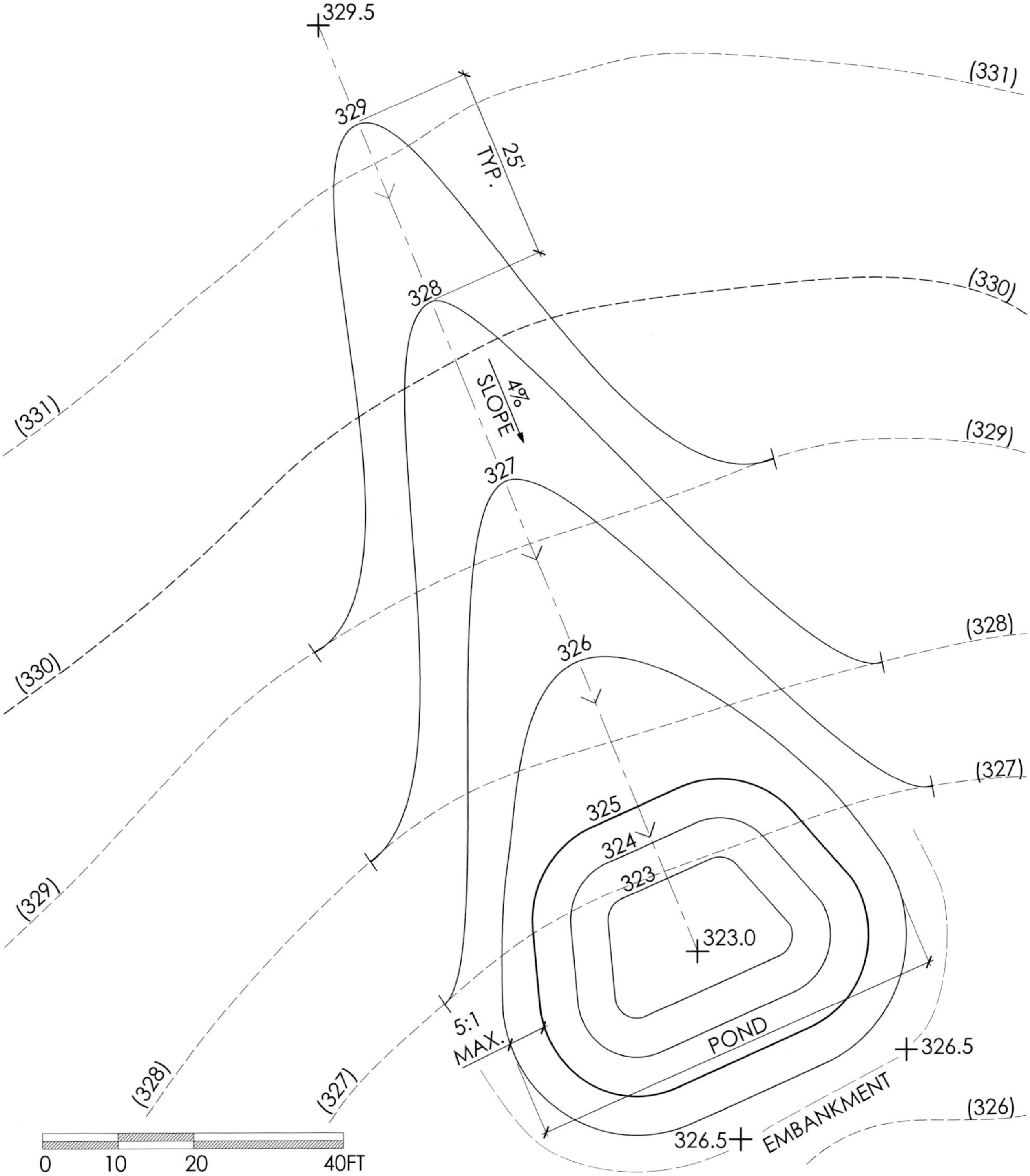

3-7) Draw a horseshoe swale around the concrete pad.

a) Provide a 10' buffer around the pad with a 2% slope.
b) Provide a HPS at 10' from edge of buffer at 2% slope.
c) Minimum slope is 2%; maximum slope is 10:1.

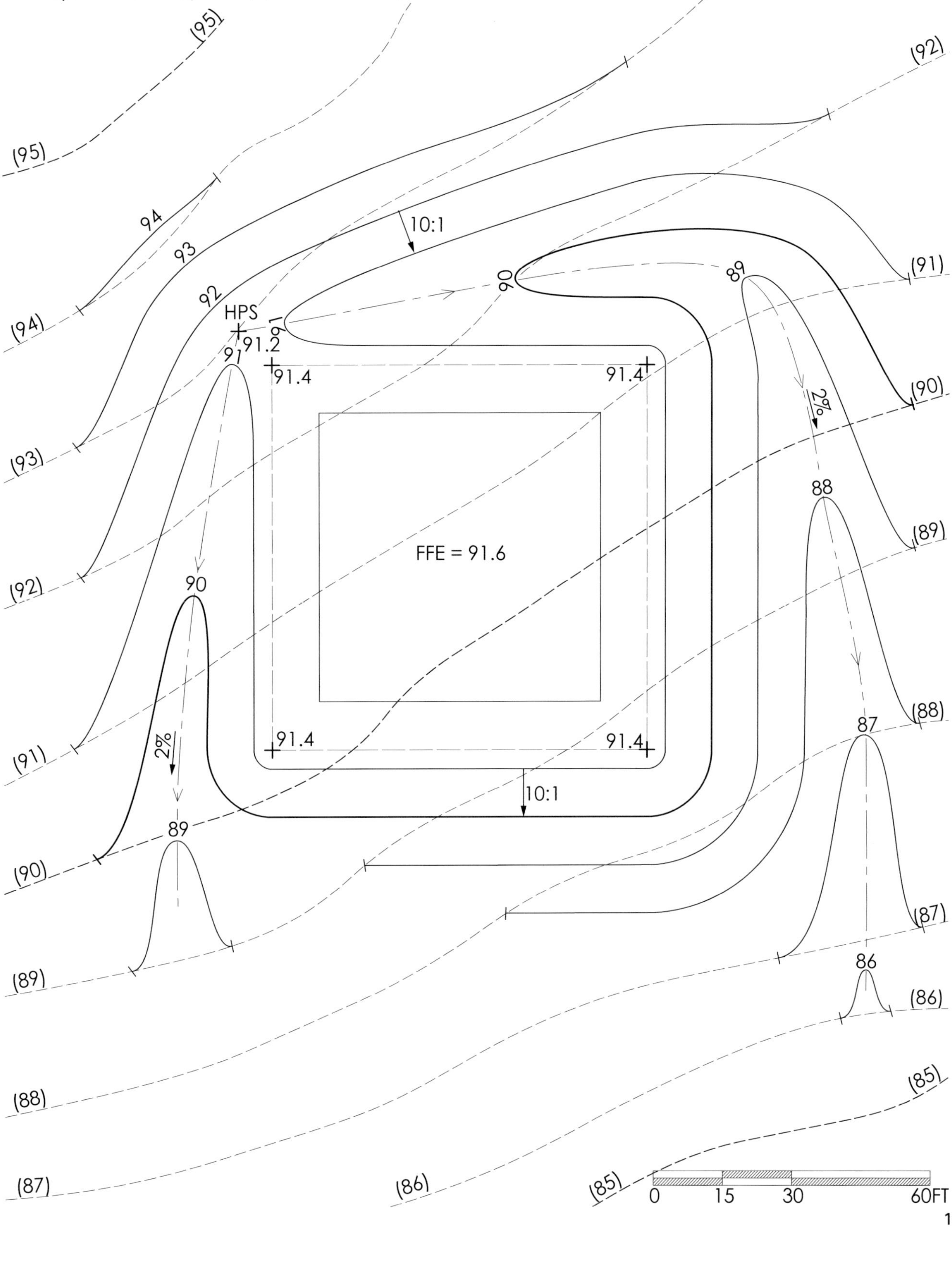

3-8) Draw a berm and swales.

a) Person standing at Point A has an eye height of 5'.
b) Minimum slope is 2%; maximum slope is 4:1.
c) Total building height is 10'.
d) Divert water to two drain inlets.
e) Label RIM EL for both drain inlets.

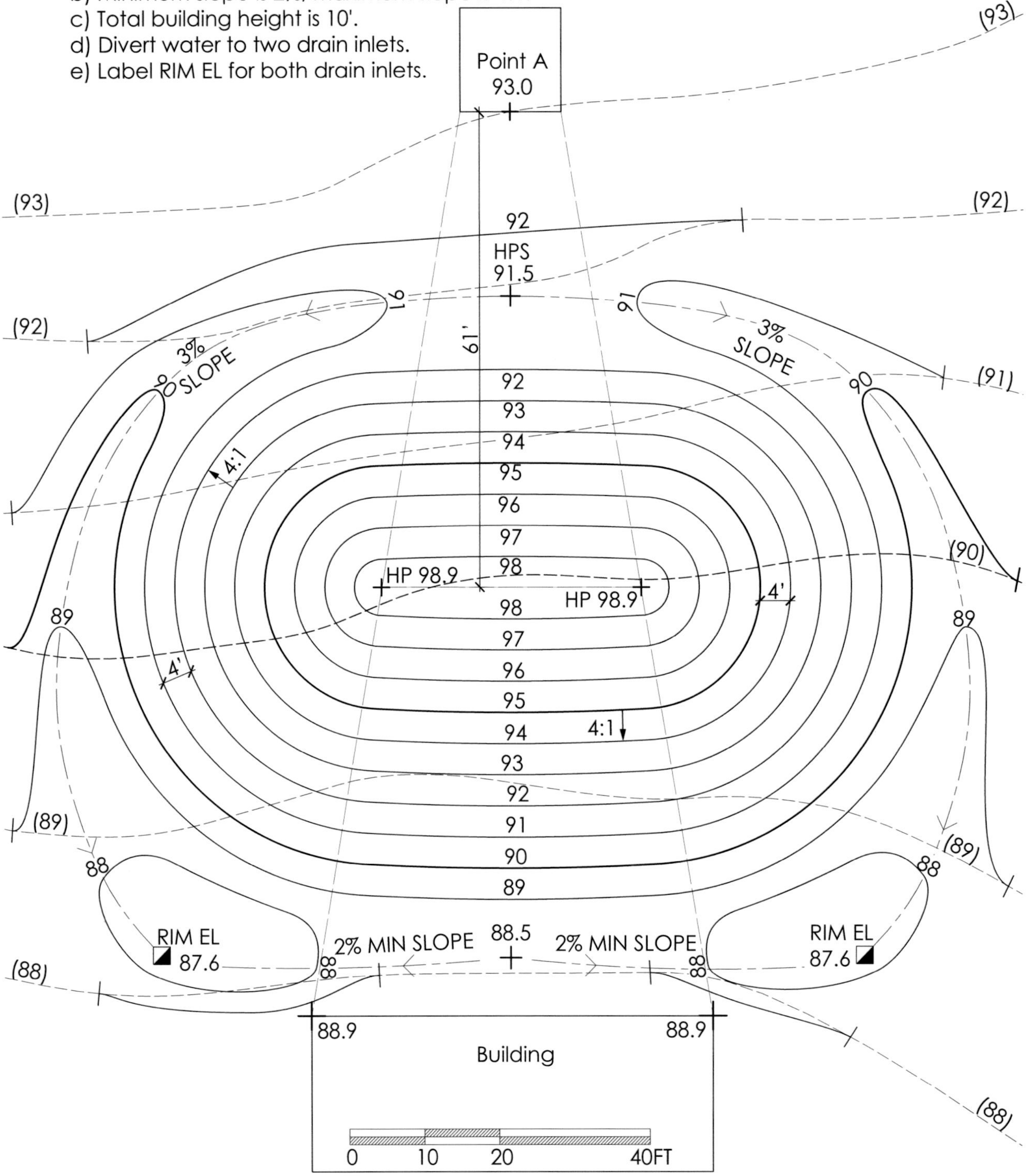

3-8) Line of sight Section

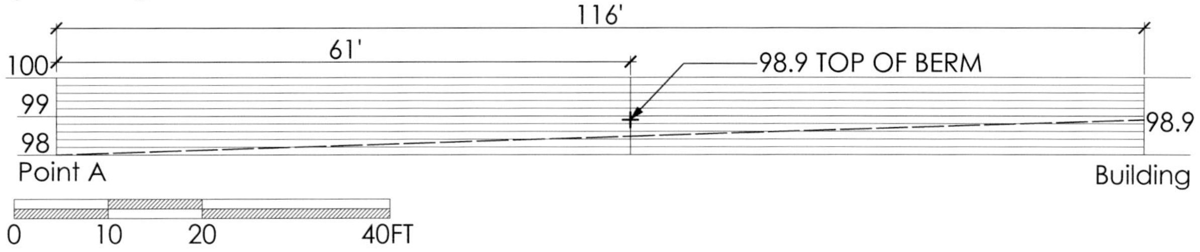

Section Four

4-1) Draw and grade a 6' wide path connecting Point A and Point B.

a) Path longitudinal slope is 3%; path cross slope is 2%; maximum slope is 4:1.
b) Calculate the total length of path needed.
c) Site path to avoid existing trees and connect to existing topography.

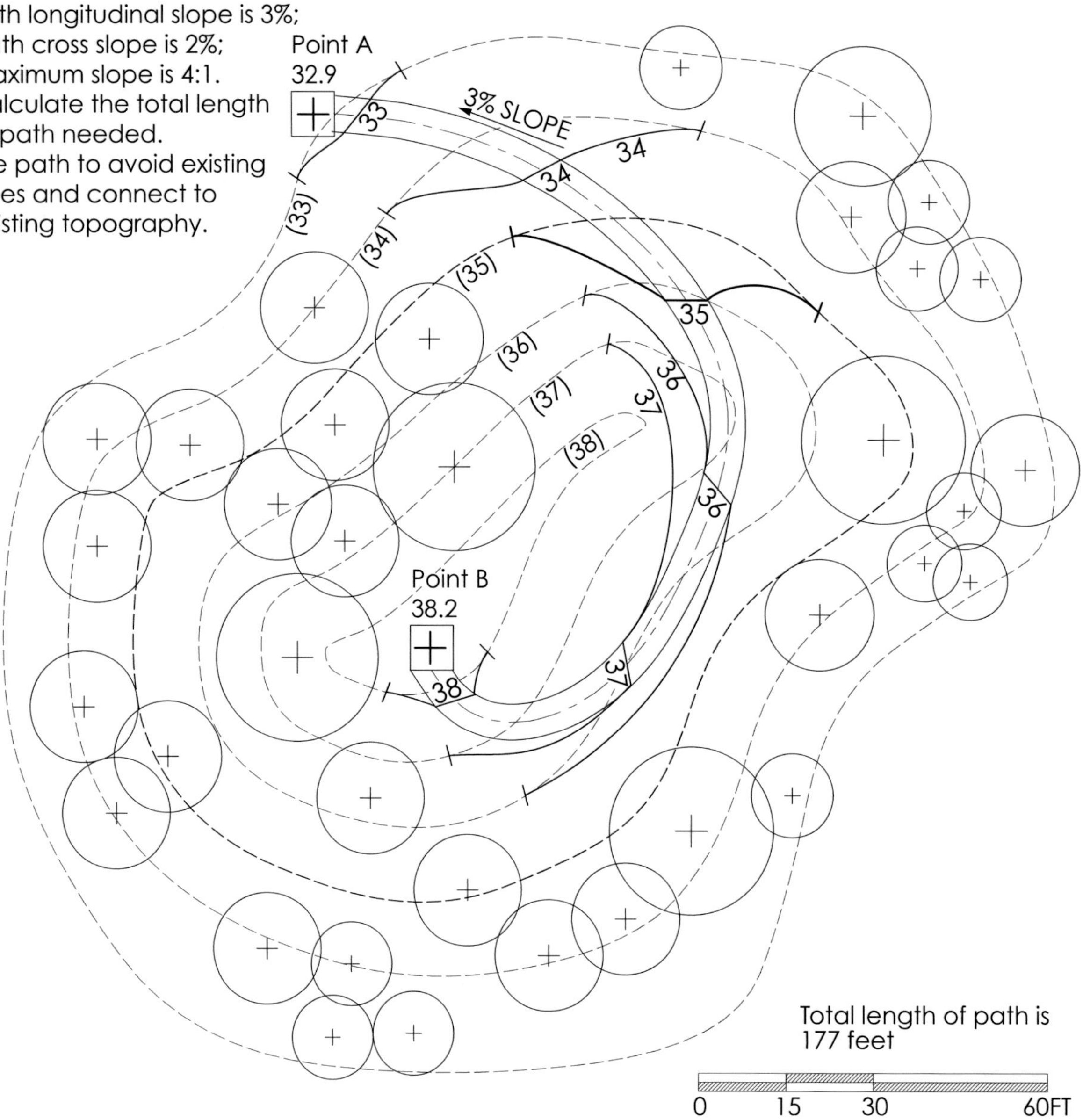

4-2) Grade the asphalt road, 6" curb and sidewalk.

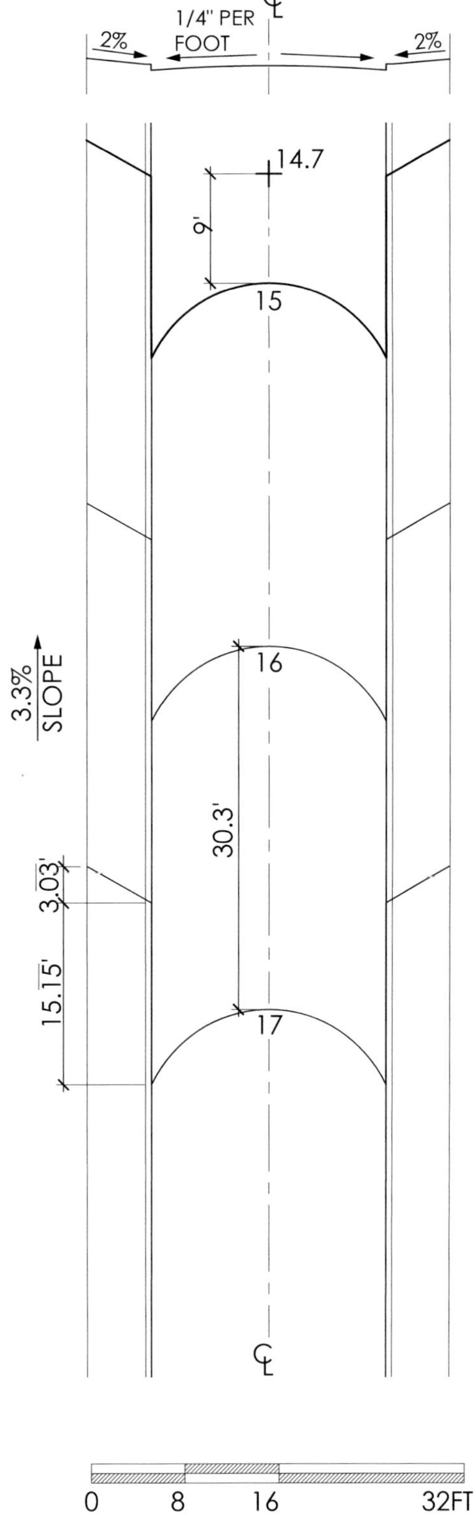

4-3) Grade the asphalt road, shoulders and swale.

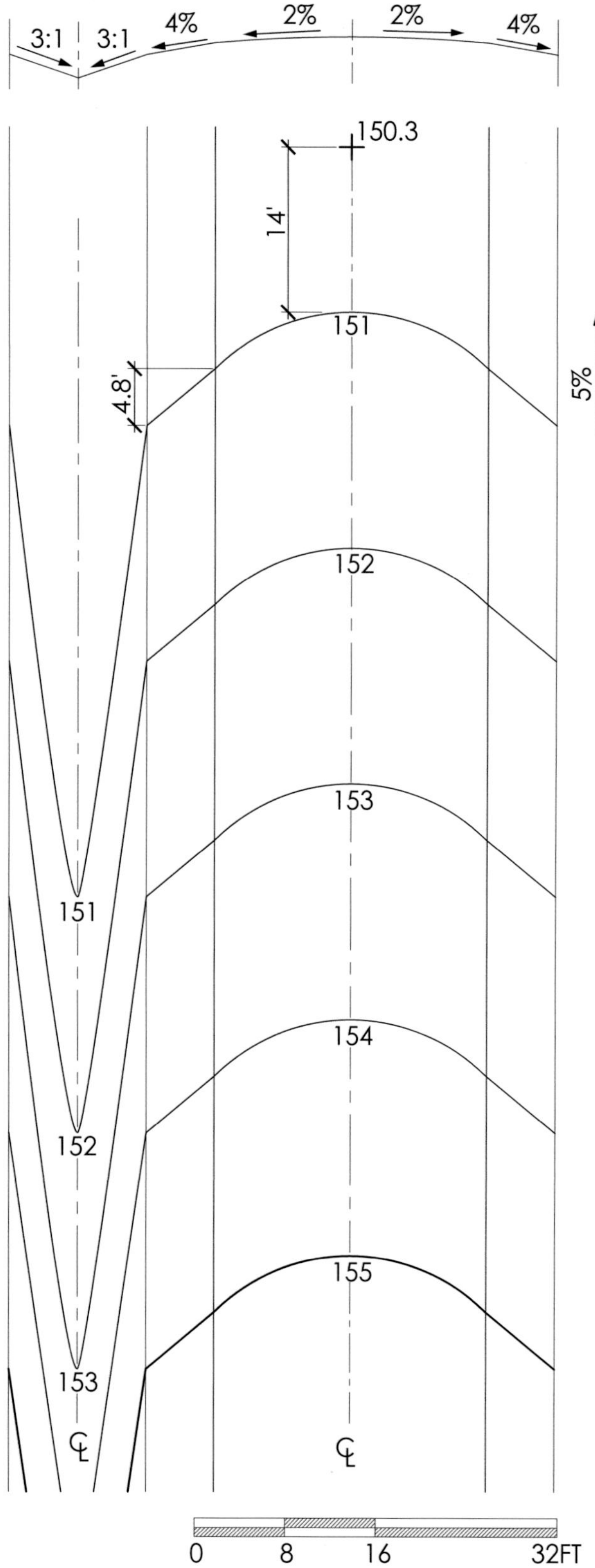

4-4) Grade the ramp with uniform slopes.

GENERAL:
A. Total change in vertical distance = 5.08'.
B. Provide critical spot elevations at 1/100th of a foot (x.xx').
C. Show required handrails.

RAMPS:
D. Label all parts of the ramp.
E. Label the slope on each leg of the ramp.
F. Provide a 2% cross slope for drainage where needed.

LANDINGS
G. Slope on landings not to exceed 1:48 (2%).

HANDRAIL EXTENSIONS:
H. Slope handrail extension areas at 2% slope downhill.

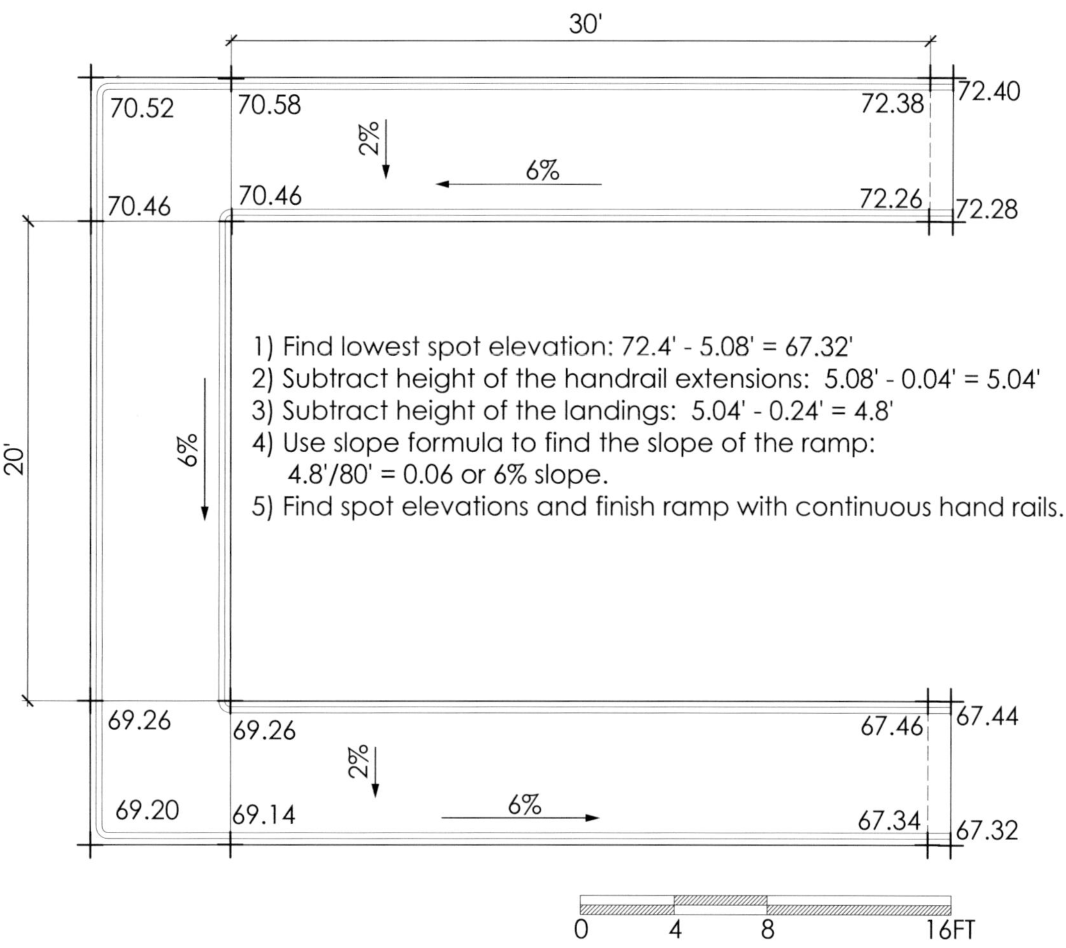

4-5) Grade stairs, cheek-walls and retaining walls. Provide all spot elevations and show proposed contours.

STAIRS:

A. Create two flights of stairs with five steps each separated by a 5' landing.
B. Riser height = 6", Tread length =12"
C. Slope on stair landing is 2%.
E. Show all parts of the stairs.
F. Label TS and BS for each set of stairs.

CHEEK-WALLS:

G. Cheek-walls are 1' wide and continuous.
H. Cheek-walls extend horizontally 1' above and 2' below the stairs.
I. Top of cheek-walls are 2.5' higher than first TS and last BS elevations.
J. BW at the landscape side of the cheek-wall is 3" lower than TW.

RETAINING WALLS:

K. 1' wide retaining wall extends perpendicular to the bottom of the cheek-walls for 14' on both sides.
L. Elevation of top of retaining wall at the corner is the same as the top of the adjacent cheek-wall.
M. Top of the retaining wall slopes at 2% for 14' on either side of the stairs.
N. BW at the back of the retaining wall is 3" lower than the TW.

PROPOSED CONTOURS:

O. Interpolate along cheek-wall to find proposed contours.
P. Tie back to existing contours.
Q. Maximum landscape slope is 3:1.
R. Label contours.

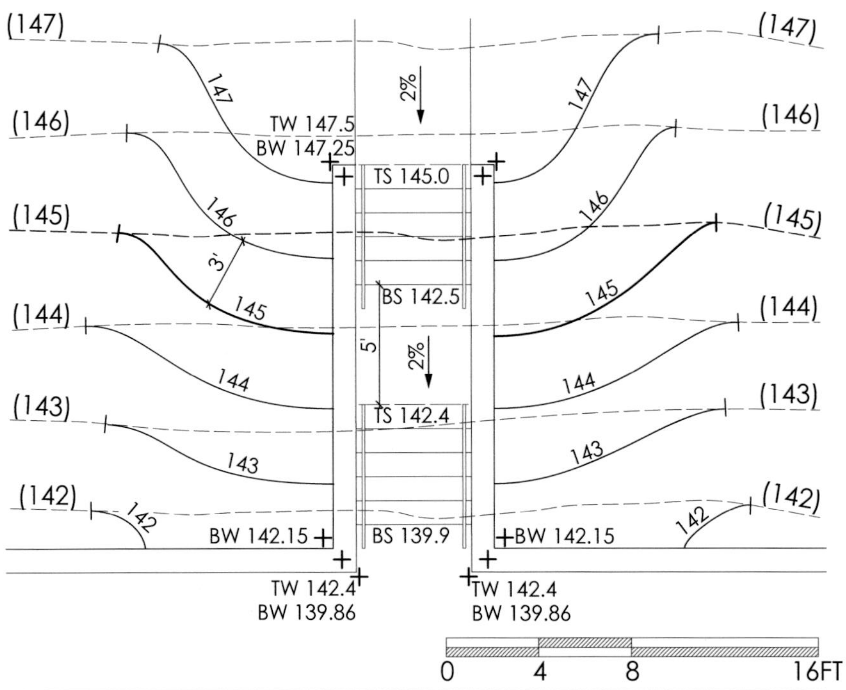

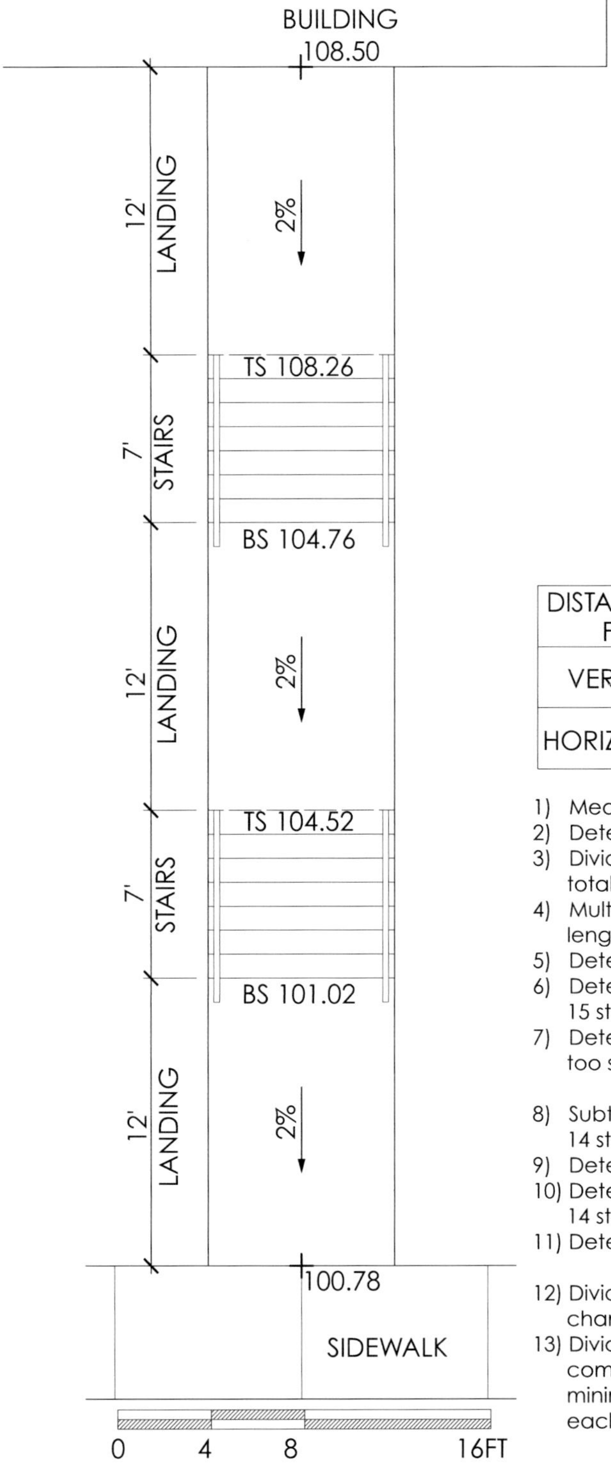

4-6) Create two flights of stairs and three equal landings from the building to the sidewalk.

a) Riser height = 6"; Tread length = 12"
b) Write the slope on the landings, show handrails as needed and provide dimensions for the stairs and landings.

DISTANCE IN FEET	STAIRS	LANDINGS	TOTAL
VERTICAL	7'	0.72'	7.72'
HORIZONTAL	14'	36'	50'

Total number of steps 14

Slope of landings 0.72'/36' = 0.02 or 2%

Length of a typical landing 12'

DISTANCE IN FEET	STAIR 1	STAIR 2	TOTAL
VERTICAL	3.5'	3.5'	7'
HORIZONTAL	7'	7'	14'

DISTANCE IN FEET	LANDING 1	LANDING 2	LANDING 3	TOTAL
VERTICAL	0.24'	0.24'	0.24'	0.72'
HORIZONTAL	12'	12'	12'	36'

1) Measure the total horizontal distance: 50'
2) Determine the vertical change in elevation: 108.5' - 100.78' = 7.72'
3) Divide the vertical change by the riser height (0.5') to determine total number of steps: 7.72'/0.5' = 15.44 or 15 steps.
4) Multiply number of steps by the tread length to determine total length of stairs: 15' x 1' = 15'
5) Determine landing length: 50' - 15' = 35'
6) Determine vertical change in elevation for the landing: 15 steps x 0.5' = 7.5'; 7.72' - 7.5' = 0.22'
7) Determine the slope of the landing: 0.22'/35' = 0.0063'. Landing is too shallow.

8) Subtract one step from the overall stair height and stair length. Use 14 steps with a total stair height of 7.0' and a total stair length of 14'.
9) Determine landing length: 50' - 14' = 36'
10) Determine vertical change in elevation for the landing: 14 steps x 0.5' = 7.0'; 7.72' - 7.0' = 0.72'
11) Determine the slope of the landing: 0.72'/36' = 0.02 or 2%.

12) Divide the landing by 3 to create 12' long landings with vertical change of 0.24' each.
13) Divide the steps into two flights of stairs. You can use any combination, but note that a flight of stairs should consist of at minimum three steps. This solution uses two stairs with seven steps each. (Other options are 3 & 11, 4 & 10, 5 & 9, and 6 & 8.)

4-7) Grade the stairs and the ramp with uniform slopes. Show all parts of the stairs and ramps.

STAIRS:

A. Riser height = 6"; Tread length = 12"
B. Overall vertical elevation for the stairs = 5'
C. Provide a 5' clear landing at the bottom and top of stairs only.
D. Stair landings slope at 2%.
E. Draw and label all parts of the stairs.

RAMPS:

F. Calculate overall vertical elevation for the ramps.
G. Slope handrail extension at 2% downhill and subtract elevation from total vertical distance.
H. Calculate minimum length of ramp needed.
I. Determine the nearest whole number higher than the minimum length of ramp.
J. Calculate and label slope for each leg of the ramp.
K. Provide a 2% cross slope for the ramp where needed.
L. Label all spot elevations on ramp.

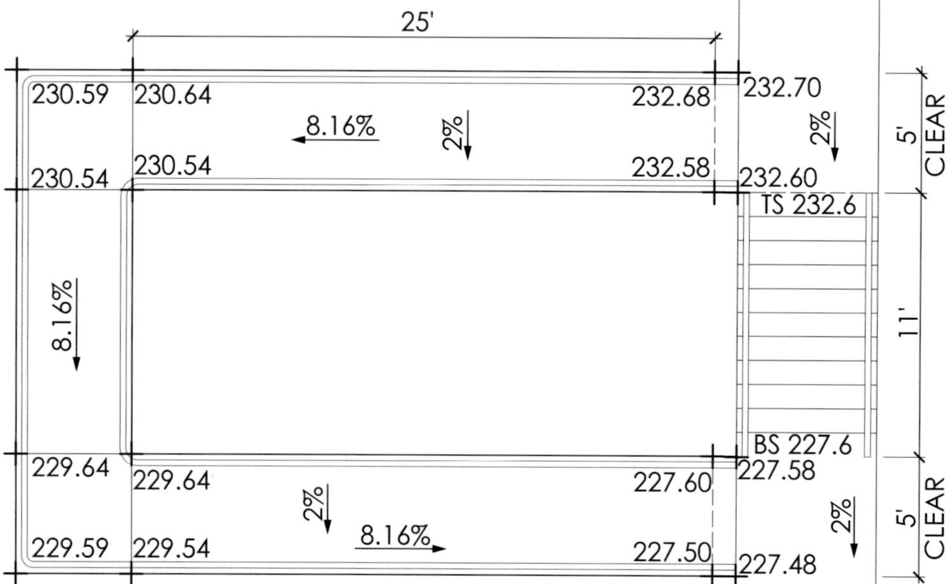

1) Calculate number of steps needed. 5'/ 0.5' = 10 steps.
2) Draw stairs including hand-rails. (1' at top of stairs + 1 tread length at the bottom)
3) Mark off 5' clear for a landing at the top and the bottom of the stairs and calculate spot elevations for the landings.
4) Determine total height for ramp: 232.70' - 227.48' = 5.22' total height.
5) Subtract the height of the two handrail extensions: 5.22' - 0.04' = 5.18'
6) Subtract the height of two 5' ramp landings. 5.18' - 0.2' = 4.98'.
7) Divide 4.98' by the maximum slope to find the minimum length of ramp needed. $4.98'/0.08\overline{33} = 59.76'$
8) Subtract the length of the stairs + handrails (11') from 59.76' to find the length needed for the two ramp runs perpendicular to the stairs: 59.76' - 11' = 48.76'
9) Divide by 2 for each ramp run: 48.76'/2 = 24.38'.
10) Round up to the nearest whole number: 25'
11) Find the slope of the ramp: 11'+ 25'+ 25' = 61'; 4.98'/61' = 0.081639 or 8.16%.
12) Draw the ramp with spot elevations, slopes labeled and handrails.

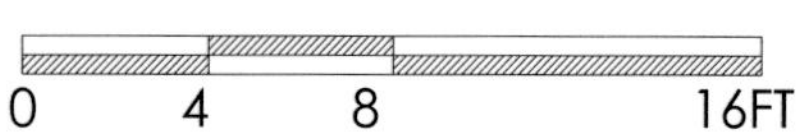

4-8) Create an ADA-compliant ramp with intermediate landings and determine the elevation at the top of the overlook deck. Topography slopes downhill from north to south. The sidewalk acts as the landing at the bottom of the ramp, no additional landings are needed. Proposed contours are not needed.

a. Grade the ramp using an 8% longitudinal slope.
b. Show all the necessary spot elevations along the ramp runs and landings.
c. Grade landings so that they drain; maximum gradient on the landings is 2%.
d. Cross slope on the middle ramp section is 2% and drains to the south.
e. Remember to add in handrail extension areas; slope on the handrail extensions is 2% downhill.
f. Show required handrails.
g. Show elevation of Overlook Deck.
h. Show elevations to the nearest 1/100th (x.xx').

1. Handrail extensions: 1' long x 0.02 = 0.02'
2. Landings: 6' long x 0.02 = 0.12'; or 6' long x 0.01 = 0.06'
3. Ramp run 1: 23' x 0.08 = 1.84'
4. Ramp run 2: 20' x 0.08 = 1.60'
5. Ramp run 3: 15' x 0.08 = 1.20'

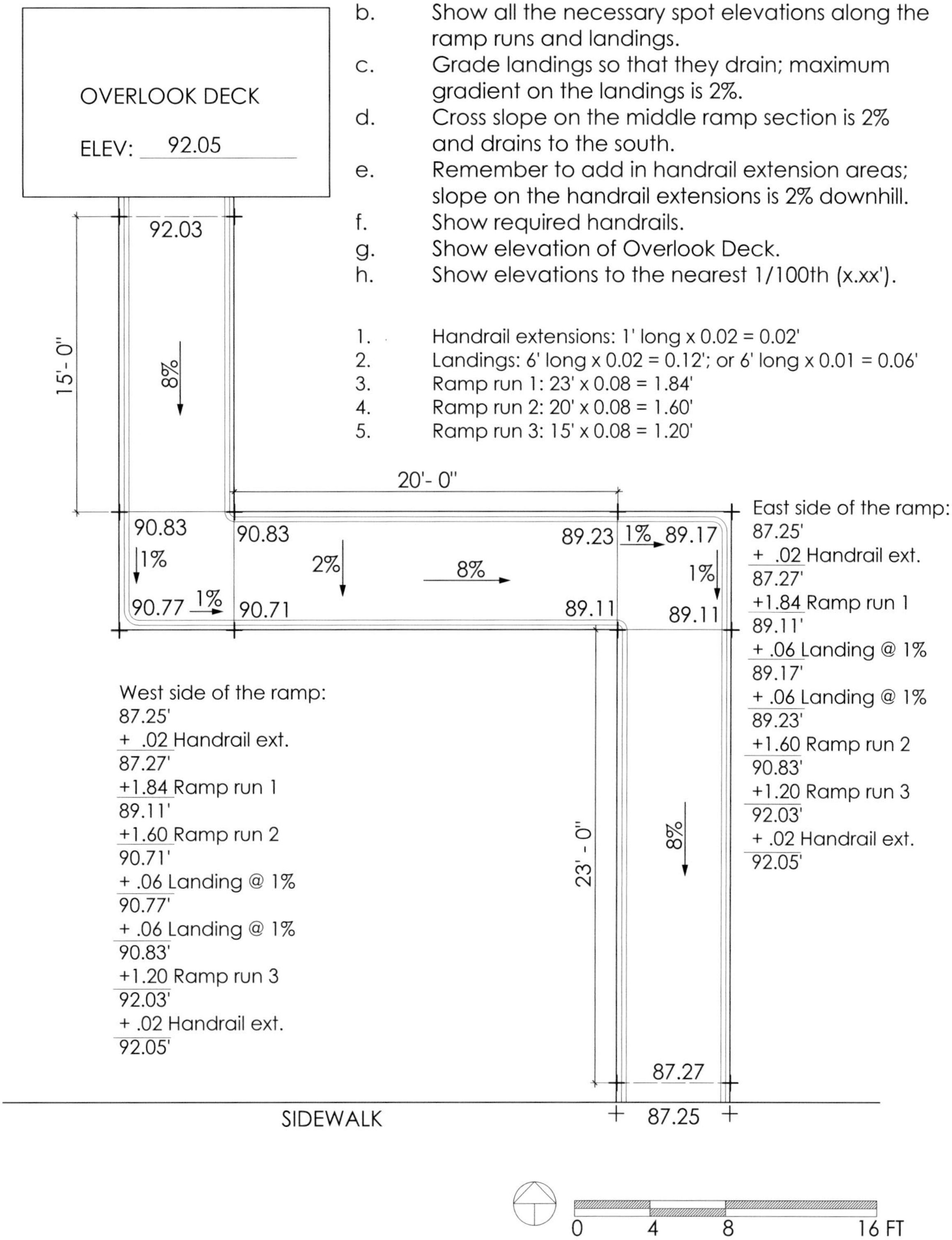

Section Five

5-1) Grade the asphalt road and parking lot.

a) Start at the spot elevation provided and grade the road with the gradients provided.
b) Runoff from the road must not drain into the parking lot.
c) All hardscape surface runoff from the parking lot must drain into the inlet provided.
d) Minimum slope is 2%, maximum slope is 4:1and curb height is 6".
e) Grade the landscape island in the middle into a mini-ridge.
f) Show TC/BC for each spot elevation in the parking lot.

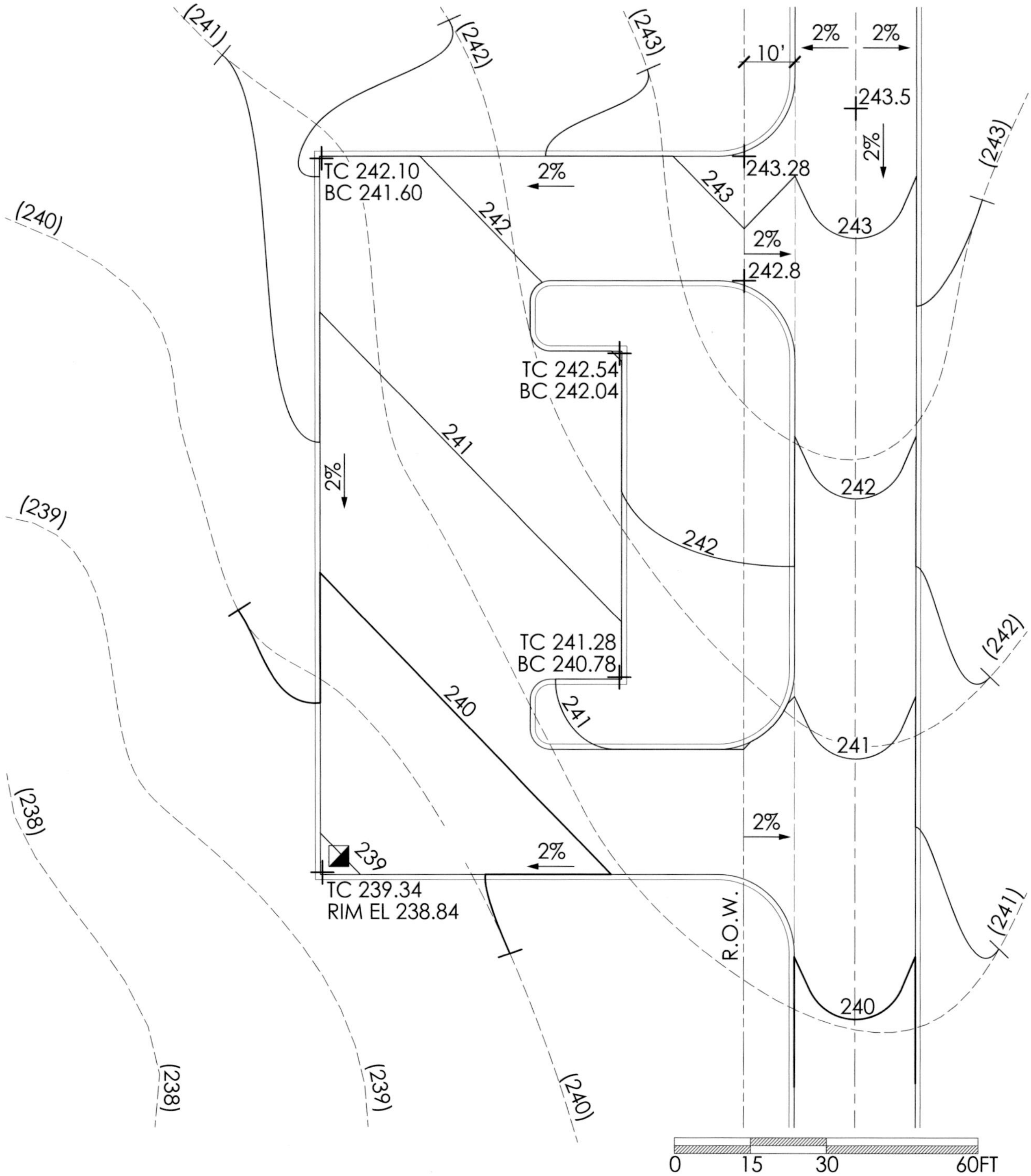

5-2) Grade the asphalt road, shoulders, swales and culvert pipe.

a) Crown of the road is 2%. Shoulder slopes at 2% away from the road. Swale along the side of the road is 6" deep; maximum slope is 5:1.

b) The pipe culvert slope is 0.5% (0.005). Calculate a slope for the swale.

c) Bottom of the pipe culvert is 2' lower than the spot elevation for the road.

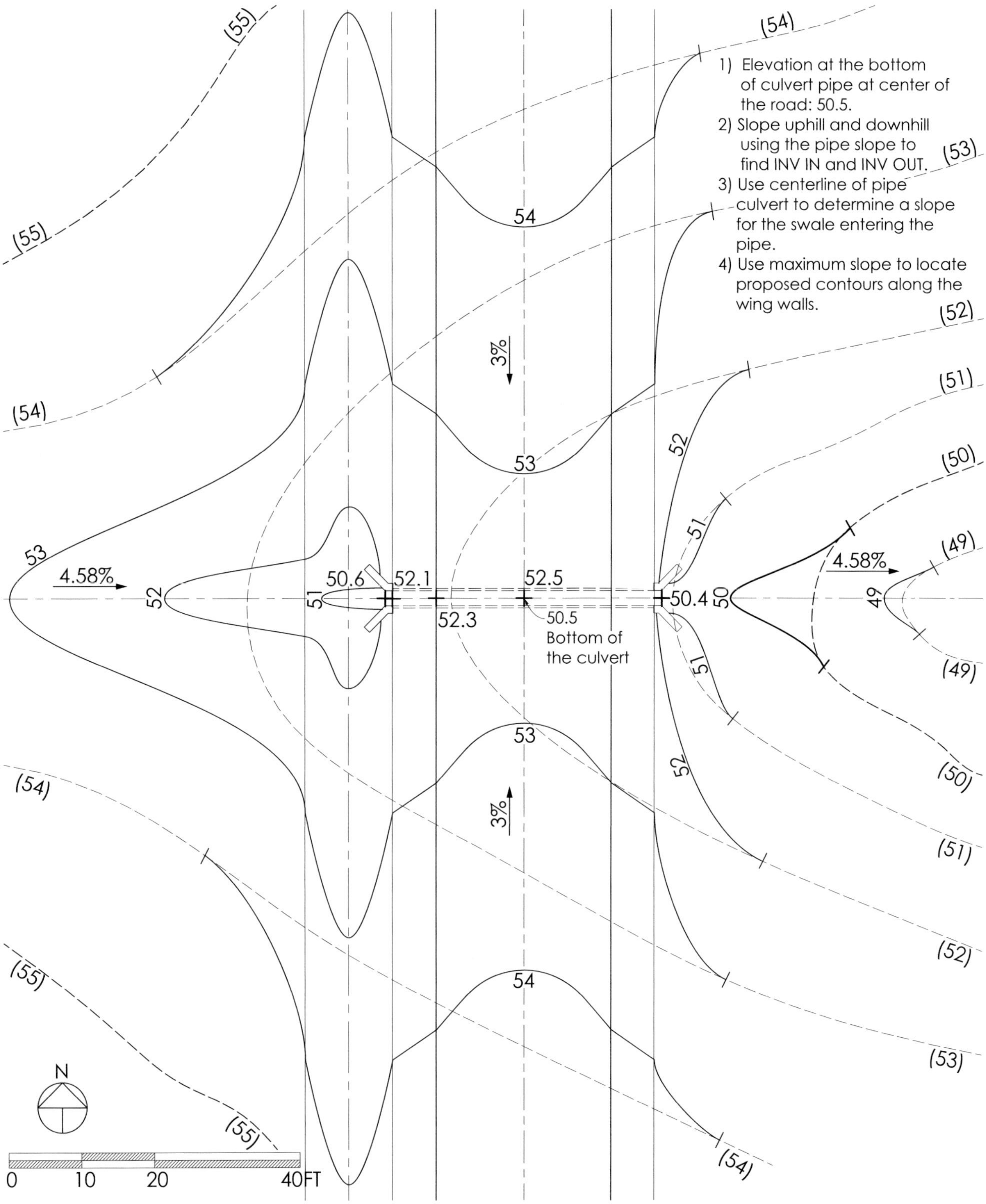

5-3) Draw a sloping berm.
   a) Measure the distance from Point A to Points B and C and create sections to find the height of the berm.
   b) Choose a distance on each line of the viewshed that corresponds to the topography.
   c) Label the spot elevations and slope for the berm and draw proposed contours; maximum slope is 4:1; minimum slope is 2%.
   d) No swales are needed; water may flow onto existing topography.
   e) Eye height at Point A is 5' and vehicle height on the road is 6'.

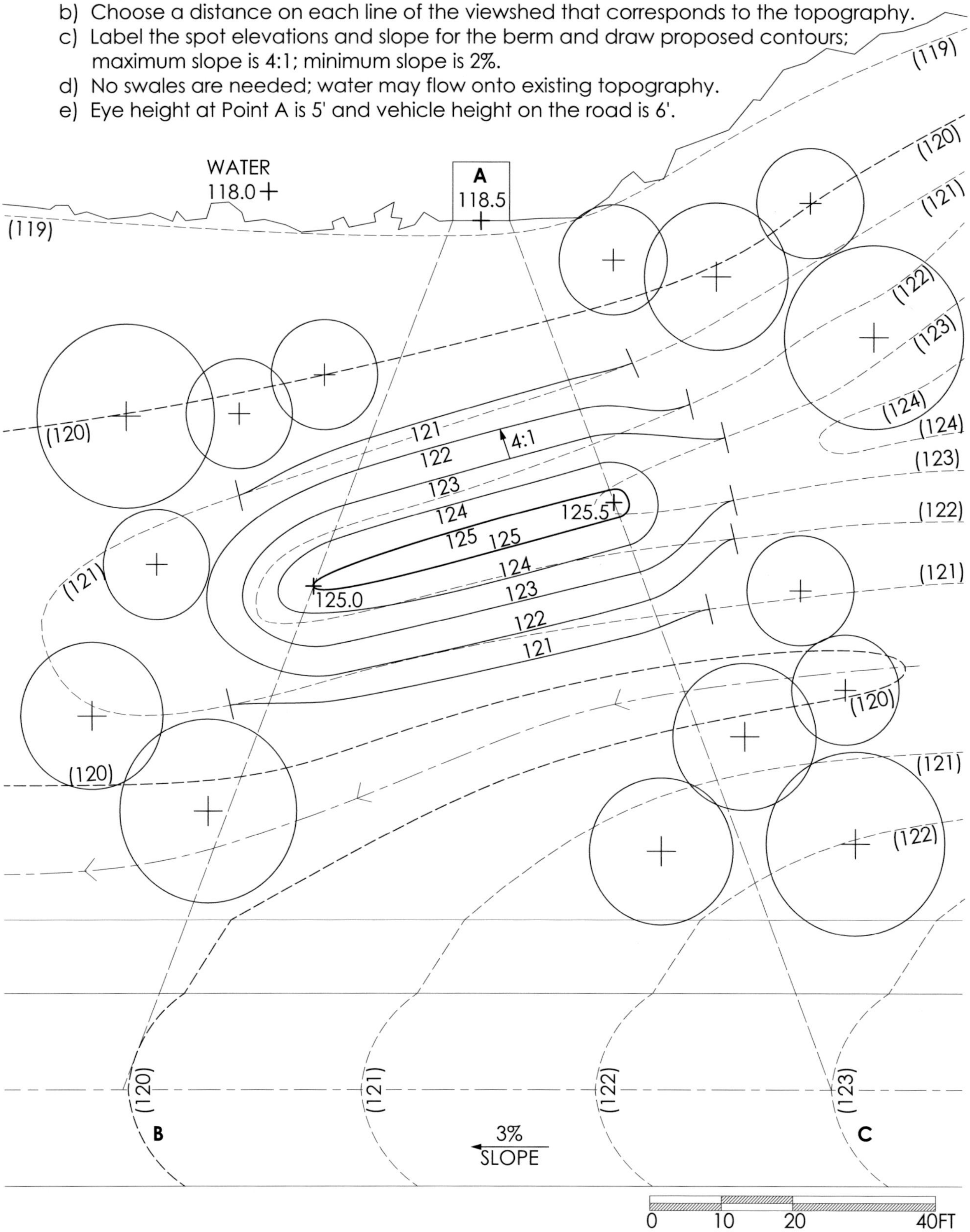

5-3) Line of Sight sections

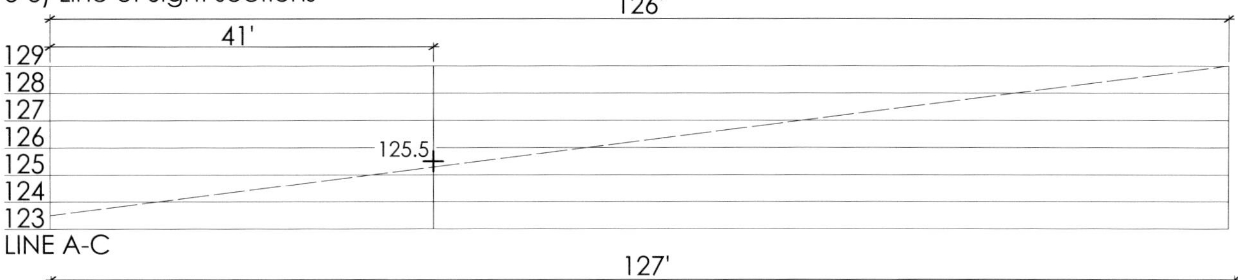

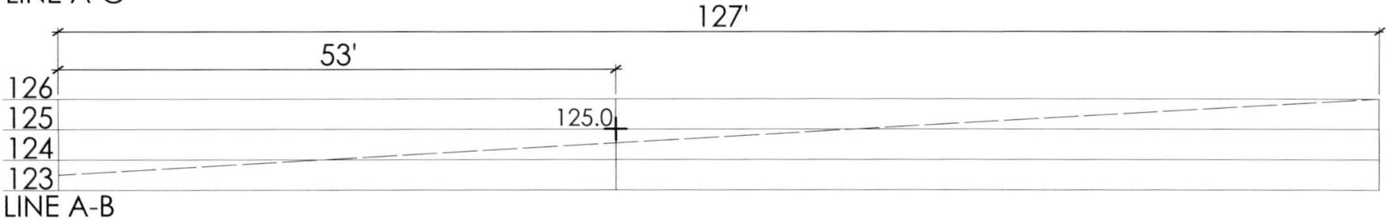

Section Six

6-1) Find the spot elevations along the length of the pipes and determine the depth of soil covering the pipes.

a) Calculate the missing slope.

b) Maintain a minimum depth of 2' of cover over the pipes.

c) Provide INV OUT at the catch basin and at the end of the culvert; and TW elevation.

d) Freeboard is 6" above the stream elevation.

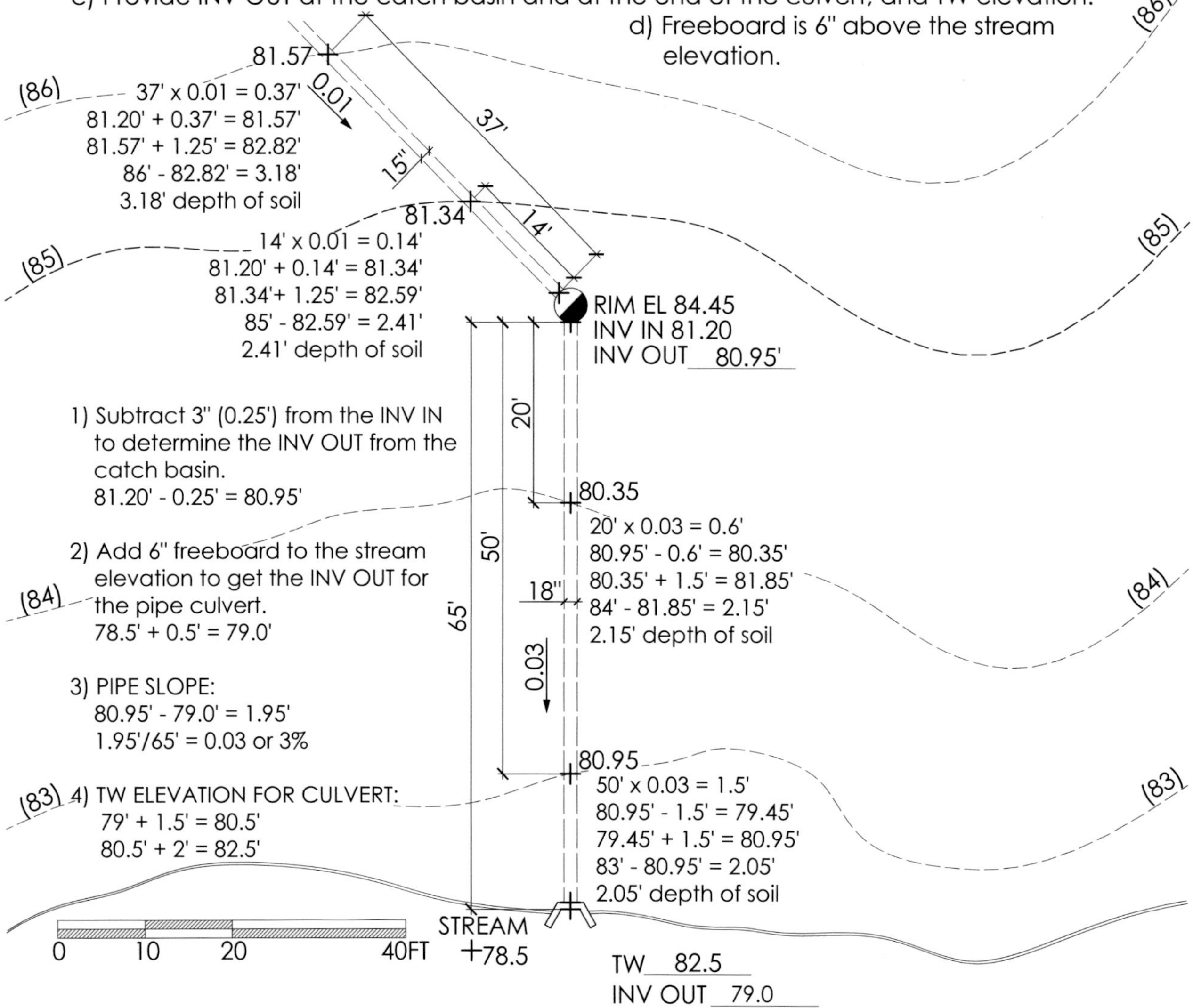

6-2) Find the FFE and create drainage around the building pad.

a) Longitudinal slope on the eastern swale is 3%; longitudinal slope on the western swale is 4.5%. Maximum slope is 4:1.
b) Slope away from building pad at 2% for the first 5'.
c) HPS is 5' from the edge of 5' clear area and slopes at 2%.
d) Provide RIM ELs for the drain inlets.

1) Find the highest and lowest spot elevation on the building pad by interpolation using the slope of the topography.
Highest spot elevation:
1'/14' = 0.071428 (7.14%)
11' x 0.071428 = 0.7857 or 0.79'
43' - 0.79' = 42.21'
Lowest spot elevation:
1'/18' = 0.055555 (5.55%)
14' x 0.055 = 0.777 or 0.78'
38' + 0.78' = 38.78'

2) Add highest and lowest spot elevation and divide by 2 to find the FFE.
42.21' + 38.78' = 80.99'
80.99'/2 = 40.49 or 40.5'

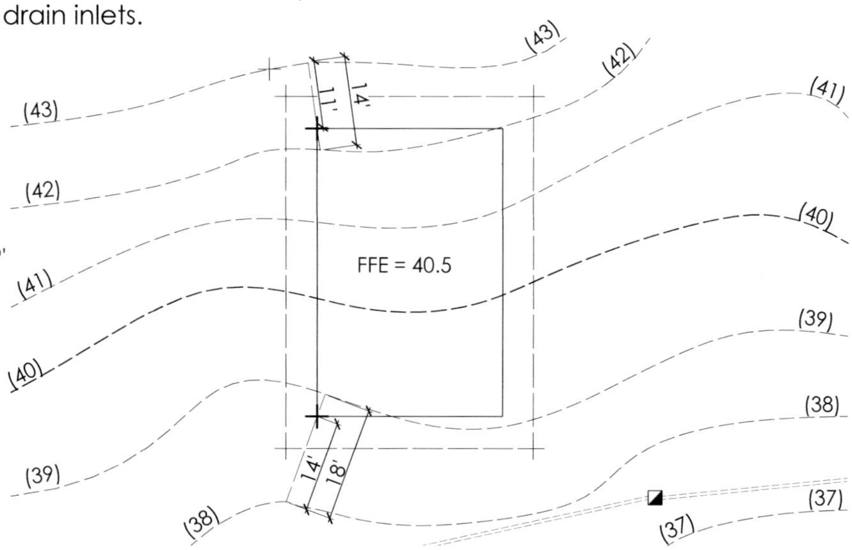

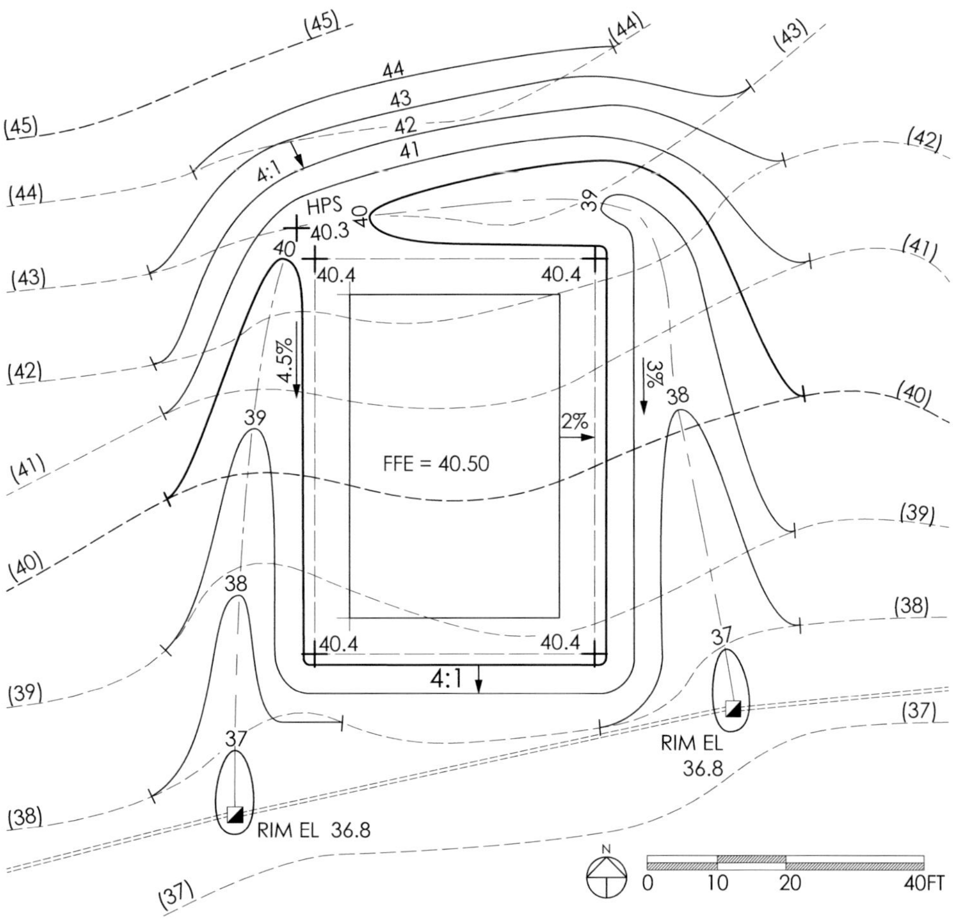

6-3) Create a swale.

a) Maintain a depth of 2' over the existing pipe.
b) Longitudinal slope of the swale is 3.5%.
c) Maximum slope is 5:1.
d) Provide slope of existing pipe.
e) Provide depth of cover where swale crosses the pipe.

1) Determine the pipe slope.
188.9' - 187.9' = 1'
1'/ 140.0' = 0.007

2) Mark slope interval for the swale along the centerline.
1'/ 0.035 = 28.57'

3) Create swale with maximum side slopes of 5:1.

4) Interpolate to find spot elevation along the centerline that intersects with the pipe.
23' x 0.035 = 0.805'
193' - 0.805' = 192.2'

5) Interpolate to find the spot elevation at bottom of the pipe.
70' x 0.007 = 0.49'
188.9' - 0.49' = 188.41'

6) Add 1.5' to find the spot elevation at the top of the pipe.
188.41' + 1.5' = 189.91'

7) Subtract the top of pipe elevation from the elevation of the swale.
192.2' - 189.91' = 2.29'

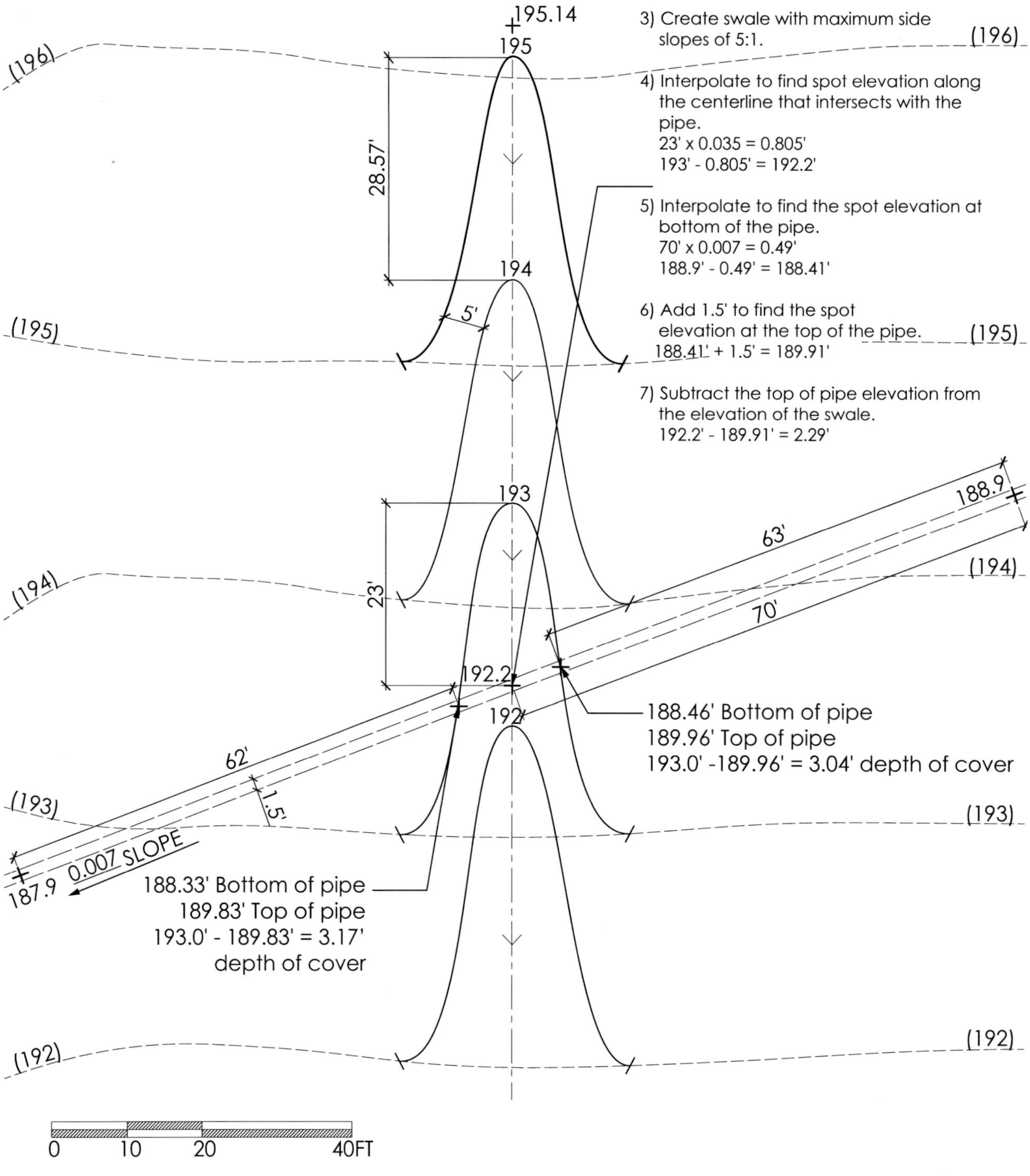

Grading vignette 1 – solution

A gazebo has been placed at the top of a hill. Provide an accessible concrete path from the gazebo to the existing sidewalk.

Step 1

Interpolate to find the spot elevation at the bottom of the 6' ramp at the gazebo. This is the starting point for the path.

Step 2

Identify the slope of the existing road and the existing sidewalk. Interpolate along the existing sidewalk using the longitudinal slope of the road to find a spot elevation for the start of the path. (I chose 65.8'; any point is fine.)

Step 3

Subtract the spot elevation at the existing sidewalk from the spot elevation at the end of the gazebo ramp and divide the change in elevation by 3% and 4% to determine the minimum and maximum length of path needed, 73.3'−65.8'=7.5'; 7.5'/0.03=250' maximum length 7.5'/0.04=187.5' minimum length.

Step 4

Draw a path between the minimum and maximum length, 217'. Determine the slope of the path, 7.5'/217'=0.03456 or 3.46%. Using the slope of the path, evenly space the proposed contours along the path.

Step 5

Add the 2% cross slope to the proposed contour lines across the path so that water drains off the path. Note that the cross slope in my solution changes direction between proposed contours 71 and 70.

Step 6

Tie the proposed contours back to the existing contours, make sure to allow a minimum of 4' between proposed contours. Label the proposed contours and the slopes on the path and topography.

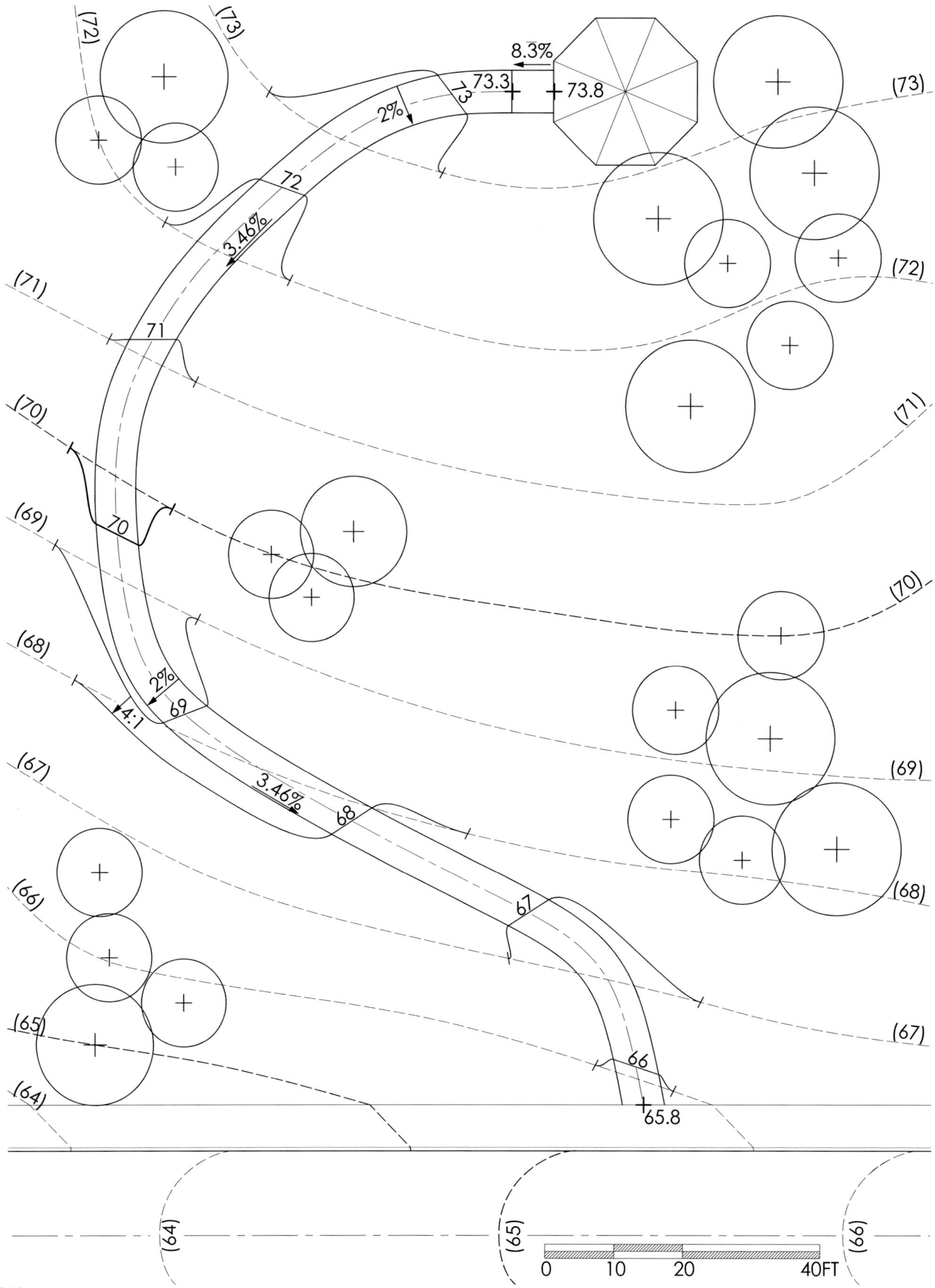
(72)
(73)
8.3%
73.3
73.8
73
2%
72
3.46%
71
70
2%
4:1
69
3.46%
68
67
66
65.8
(73)
(72)
(71)
(70)
(69)
(68)
(67)
(66)
(65)
(64)
(64)
(65)
(66)
0
10
20
40FT

Grading vignette 2 – solution

Grade a ramp and stairs for the proposed office building.

Step 1

Convert 1" to its decimal foot equivalent, 1"/12" = 0.0833'. Subtract 0.833' from the FFE, 350' − 0.0833' = 349.9167 or 349.92', rounded to the nearest hundredth.

Step 2

Measure the length of the courtyard and interpolate to find the spot elevations at the corners of the courtyard and at the beginning of the ramp, 12' × 0.02 = 0.24'; 349.92' − 0.24' = 349.68' and 6' × 0.02 = 0.12'; 349.92–0.12' = 349.80'.

Step 3

Label the top of the stairs (TS). Remember that the top of the stairs is the elevation at the back edge of the first tread which is the same elevation as the lower edge of the courtyard. Subtract 3' to get the bottom of stairs (BS) elevation, 349.68' − 3' = 346.68'.

Step 4

From the BS elevation, measure the length of the landing and interpolate to find the spot elevation at the edge of the landing where it meets the existing pier, 7' × 0.02 = 0.14'; 346.68' − 0.14' = 346.54'. Because it is level at the bottom edge both corners have the same elevation. Measure and interpolate to determine the spot elevation at the ramp landing. Remember to add the elevation, 6' × 0.02 = 0.12'; 346.54' + 0.12' = 346.66'.

Step 5

You now have the highest point, 349.80', and the lowest point, 346.66', for the ramp. Subtract the highest elevation from the lowest elevation to get the overall vertical change, 349.80' − 346.66' = 3.14'.

Step 6

Subtract the elevation required for the two handrail extensions, 1' × 0.02 = 0.02'; 0.02' × 2 = 0.04'; 3.14' − 0.04' = 3.10'. Calculate the vertical change for the middle landing and subtract it from the elevation, 13' × 0.02 = 0.26'; 3.10' − 0.26' = 2.84'. This is the vertical change for the ramp runs.

Step 7

Measure the length of the ramp runs and determine the uniform slope for the ramp, 2.84'/48' = 0.059166 or 5.9166%. Do not round the slope.

Step 8

Draw the continuous handrails for the ramp. Label the handrail extensions and the ramp runs with spot elevations. For the middle landing, calculate the slope for the corners. Measure the horizontal distance of the widths and length for the landing, 6' + 13' + 6' = 25'.

Find the slope using the vertical change for the middle landing, 0.26'/25' = 0.0104 or 1.04%. Label the corners of the middle landing, 6' × 0.0104 = 0.0624'; 348.36' − 0.06' = 348.30'; 13' × 0.0104 = 0.1352'; 348.30–0.14' = 348.16'; 348.16' − 0.06' = 348.10'.

Step 9

For the landing at the bottom of the ramp, measure the horizontal distances, 6' + 6' = 12', and divide by the vertical change, 0.12', to find the slope at the corners of the landing, 0.12'/12' = 0.01 or 1.0%. Calculate the spot elevations for the corners of the bottom landing. Label slopes where required.

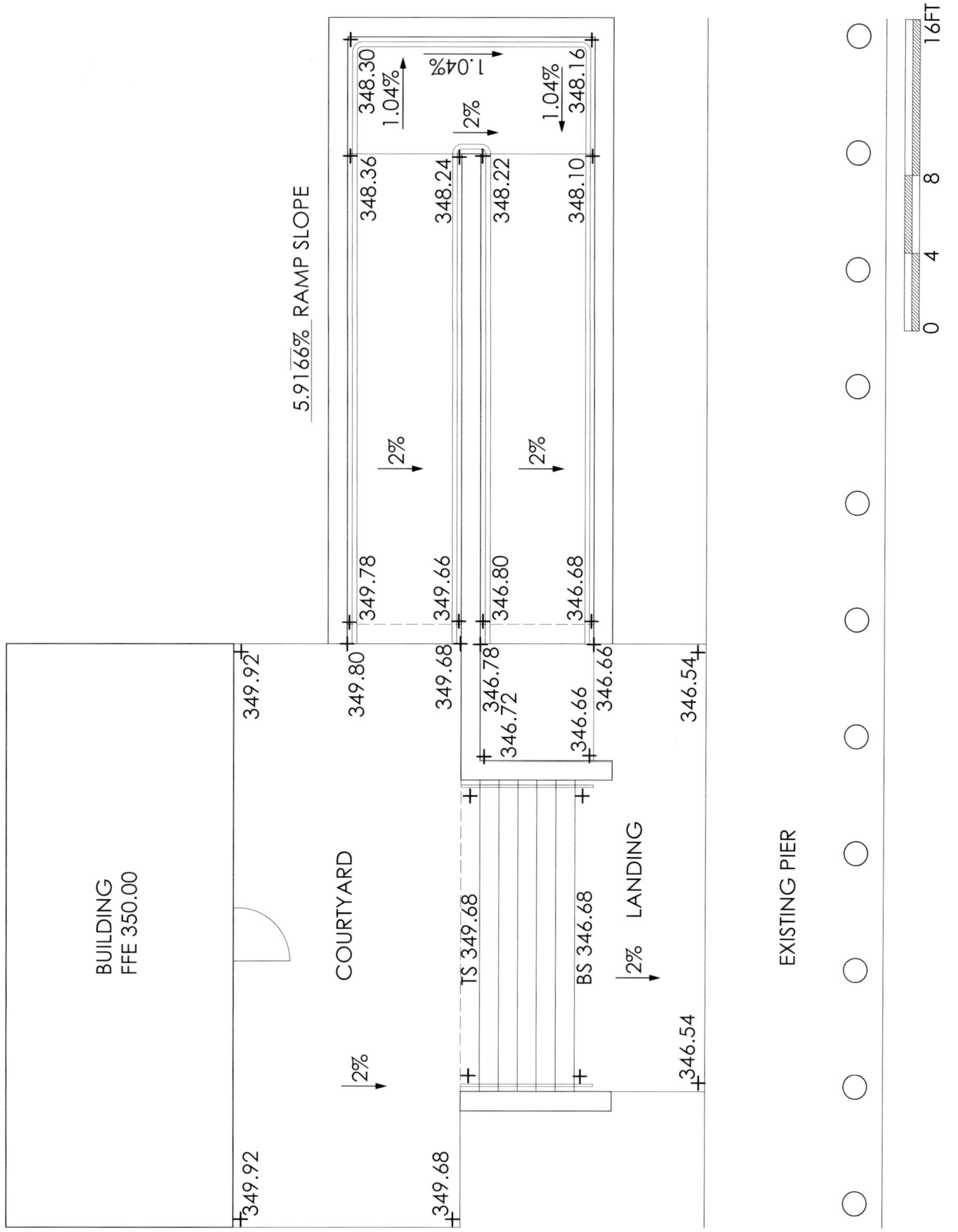
BUILDING
FFE 350.00
COURTYARD
LANDING
EXISTING PIER
5.9166% RAMP SLOPE
TS 349.68
BS 346.68
2%
1.04%
348.30
348.16
348.36
348.24
348.22
348.10
349.78
349.66
346.80
346.68
349.92
349.80
349.68
346.78
346.72
346.66
346.54
0
4
8
16FT

Grading vignette 3 – solution

A building has been placed on the top of a berm. A circular concrete driveway has been provided. Grade the driveway, the adjacent sidewalk and the walk up to the building. Provide drainage around the building. Drainage from the sidewalk may flow into the driveway.

Step 1

Calculate all the necessary spot elevations as noted in the problem statement, 109.23'−0.833' = 109.15'; 109.15'−0.2' = 108.95', at the top of the curb; 108.95'−0.5' = 108.45' at the bottom of the curb.

Step 2

Measure the horizontal distance between the given spot elevation and the spot elevation calculated at the bottom of the curb, 119'. Find the slope for the driveway (108.45' − 104.2' = 4.25'; 4.25'/119' = 3.57% OR 3.6%).

Step 3

Starting at the spot elevation at the end of the driveway, grade the driveway using the calculated slope. Remember that the driveway is not crowned but is pitched (cross slope) to one side. Use the cross slope to find the elevations at the top and bottom of the curb and for the sidewalks. Use the road gradient slope interval to quickly mark off the whole number spot elevations for the road and sidewalk.

Step 4

Connect the proposed contours across the road. Calculate the slope between the proposed contours using the Pythagorean theorem, see inset. Use this slope to locate the proposed contours around the circular part of the driveway. $a^2+b^2=c^2$; $3.6^2+2^2=16.96$. The square root of 16.96 = 4.12. Use 4.12% slope for the proposed contours on the road, 1'/0.0412 = 24.27'. Each of the proposed road contours should be approximately 24' apart horizontally.

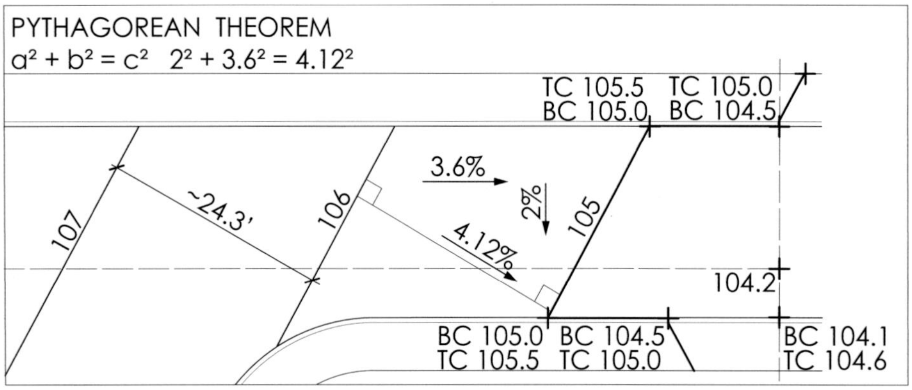

Step 5

Grade the circular berm using 4.12% slope. Proposed contours should point downhill to ensure that it is a berm.

Step 6

Create the proposed contour lines around the building and tie proposed contour lines back into existing contours. Note that the 109 contour runs along the side of the sidewalk and building. Label proposed contours and include all three slopes for the driveway.

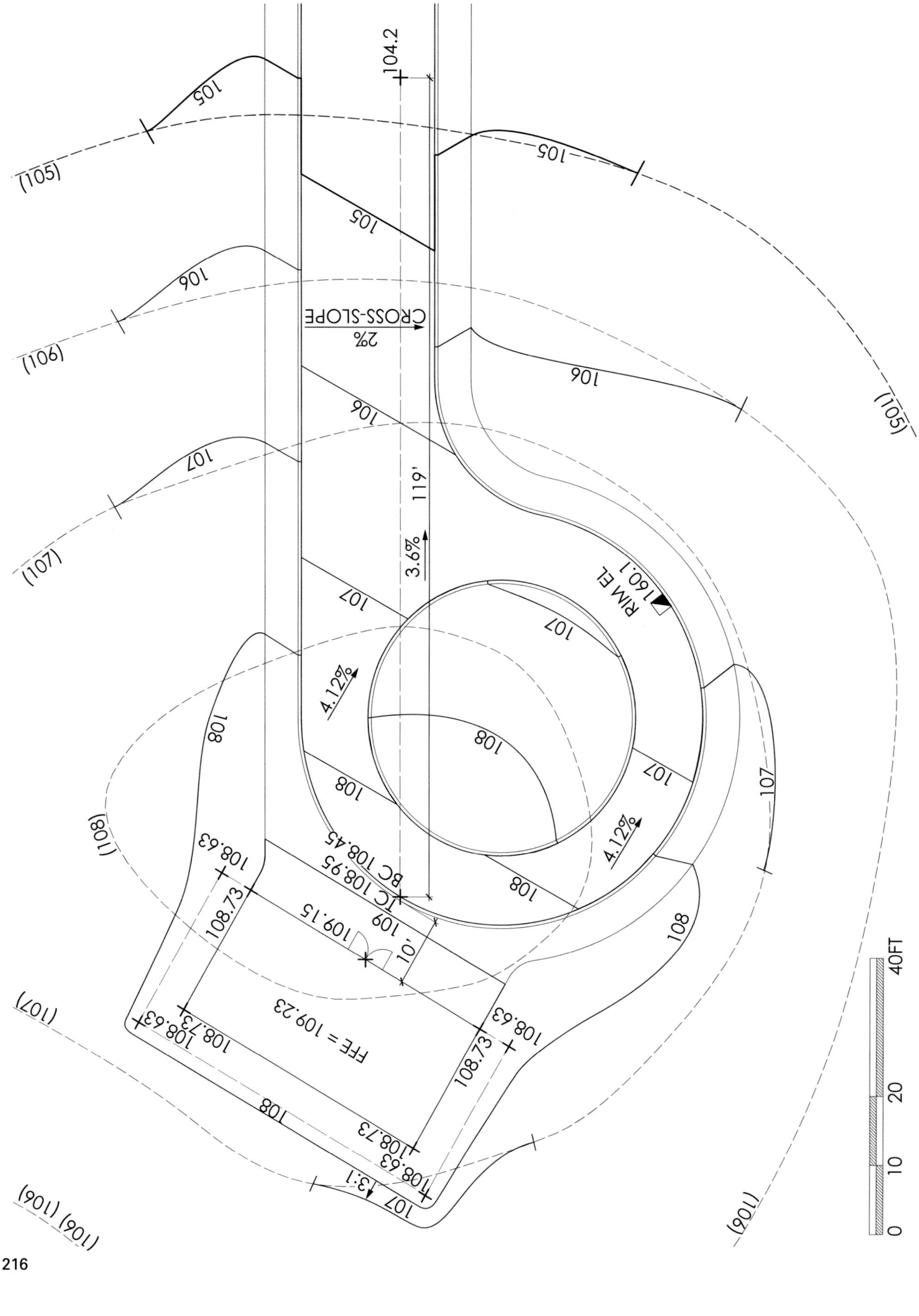
104.2
105
(105)
(105)
106
(106)
107
(107)
(107)
2%
CROSS-SLOPE
119'
3.6%
RIM EL
4.12%
4.12%
108
(108)
TC 108.95
BC 108.45
109.15
109
10'
108.63
108.73
FFE = 109.23
3:1
(106)
(106)
(106)
0
10
20
40FT

Grading vignette 4 – solution

Provide two flights of stairs and a 6' wide ramp to reach the platform.

Step 1

Start with the path, measure the horizontal distance from the building courtyard to the concrete pad at the bottom of the path, 46'. Calculate the vertical change, 74.66' − 68.94' = 5.72'. Calculate the slope for the path, 5.72'/46' = 12.43%. The path requires stairs.

Determine the number of steps needed, 5.72'/0.5' = 11.44 or 11 steps. Multiply the numbers of steps by the tread length, 11 steps × 1' = 11' to determine the length for the stairs.

Determine the length of the landings, 46' − 11' = 35'. Determine the height of the stairs, 11 steps × 0.5' = 5.5'. Subtract the height of the stairs from the total height to determine the height of the landing, 5.72' − 5.5' = 0.22'. Determine the slope for the landing with 11 risers, 0.22'/35' = 0.006 or 0.6%. The slope is too shallow.

Subtract 1 step from the stairs, 11 steps − 1 = 10 steps. Determine the total height of the stairs, 10 steps × 0.5' = 5'. Determine the total length of the stairs, 10 steps × 1' = 10'. Determine the length of the landings, 46' − 10' = 36'. Calculate the new landing height, 5.72' − 5' = 0.72'. Calculate the new landing slope, 0.72'/36' = 0.02 or 2%.

Divide the stairs into two flights of stairs, five steps each with three evenly spaced landings, 36'/3 = 12'. Draw the stairs, remember to add the railings and TS/BS spot elevations. Show the slope on the landings.

Step 2

Draw the continuous cheek-walls. Start at the top of the first flight of stairs and align it with the back edge, the dashed line. Extend the cheek-wall to the end of the bottom handrail for the second flight of stairs. The cheek-wall should be the length of the two flights of stairs, the middle landing length plus the length of the handrail at the bottom of the stairs, 5' + 12' + 5' + 1' = 23'. This will be the length of the middle ramp run for the 90° ramp.

Step 3

The ramp starts 6' above the cheek-wall and extends 6' below the cheek-wall. Interpolate where the ramp meets the path to determine the highest, 74.54' and lowest, 69.04' ramp elevations. Calculate the overall vertical elevation for the ramp, 74.54' − 69.04' = 5.5'.

Subtract the vertical change for the handrail extensions, 1' × 0.02 = 0.02'; 0.02' × 2 = 0.04', and the vertical change for two landings, 6' × 0.02 = 0.12'; 0.12' × 2 = 0.24', to find the vertical change for the ramp runs, 5.5' − 0.04' = 5.46'; 5.46' − 0.24' = 5.22'.

Calculate the minimum and maximum possible lengths for the ramp. 5.22'/0.0833 = 62.66' or 63' minimum and 5.22'/0.05 = 104.40' or 104' maximum length. You can choose any length of ramp within this range as space allows. The length of the middle ramp run must be 23 to match the length of the two stairs and the middle landing. The longest ramp will be 83', 30' + 23' + 30' = 83' and the shortest ramp will be 63', 20' + 23' + 20' = 63'. Any ramp longer than 83' will require additional landings.

I chose to make each ramp run the same length. Multiply the length of the middle ramp run by 3 to determine the length of ramp needed, 23' × 3 = 69'. This length falls between the minimum and maximum ramp lengths. Calculate the slope for the ramp, 5.22'/69' = 0.07565. Do not round the slope.

Draw the ramp with handrail extension areas, two intermediate landings and continuous handrails. Label all spot elevations for the ramp. Do not round any numbers to the nearest hundredth on the landings until you are complete. Interpolate along the ramp and the cheek-walls to find the proposed contours. Draw smooth continuous lines and label the proposed contours. Remember the proposed 69 contour line at the bottom of the path.

DISTANCE IN FEET	STAIRS	LANDINGS	TOTAL
VERTICAL	5'	0.72'	5.72'
HORIZONTAL	10'	36'	46'

TOTAL NUMBER OF RISERS 10

SLOPE OF LANDINGS 2%

LENGTH OF A TYPICAL LANDING 12'

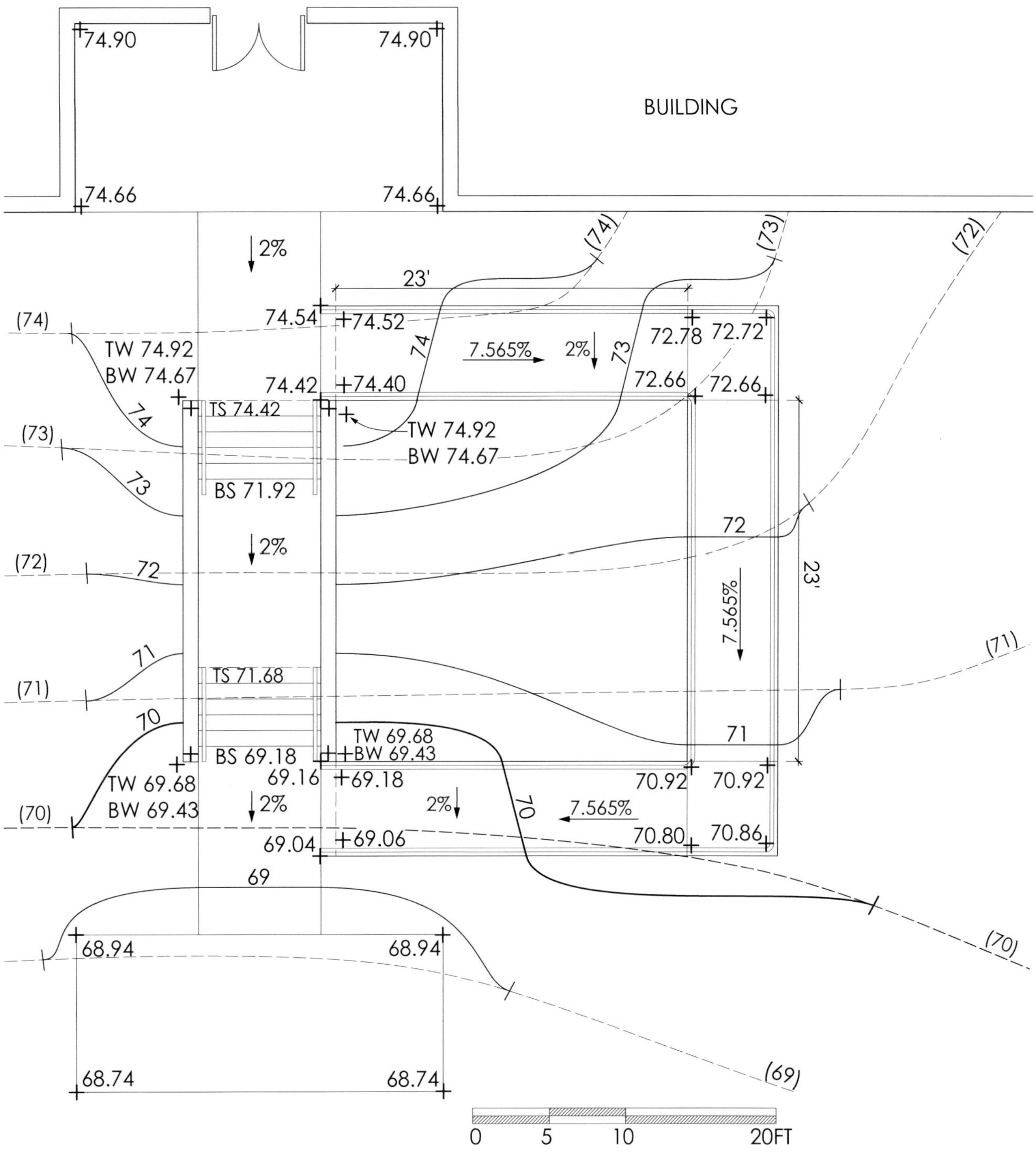

Grading vignette 5 – solution

Grade the proposed asphalt road, concrete gutters, sidewalks and retaining wall.

Step 1

Using the centerline of the road determine the change in elevation from the given spot elevation to the lowest existing contour line, 42.5' – 35' = 7.5'.

Determine the length of road from the spot elevation to contour 35 using the centerline of the road, 262'.

Find the slope of the road, 7.5'/262' = 0.02862 or 2.86%.

Step 2

Mark off uniform points along the centerline of the road based on the gradient of the road. Label them with the whole number spot elevations. Calculate the crown of the road and draw in the proposed contours for an asphalt road.

Step 3

Interpolate to find the 3" deep gutter on both sides of the road. Interpolate to find the proposed contour lines for the 6" curb, and the sidewalk.

Step 4

Label the TW and BWs for each contour line along the retaining wall. Measure the distance between each spot elevation along the back of the wall and determine a slope between each.

Interpolate to find the whole number spot elevation at the back of the wall.

Step 5

Tie the proposed contour lines into the existing contour lines along the sidewalk and behind the wall. Avoid the drip line of the trees.

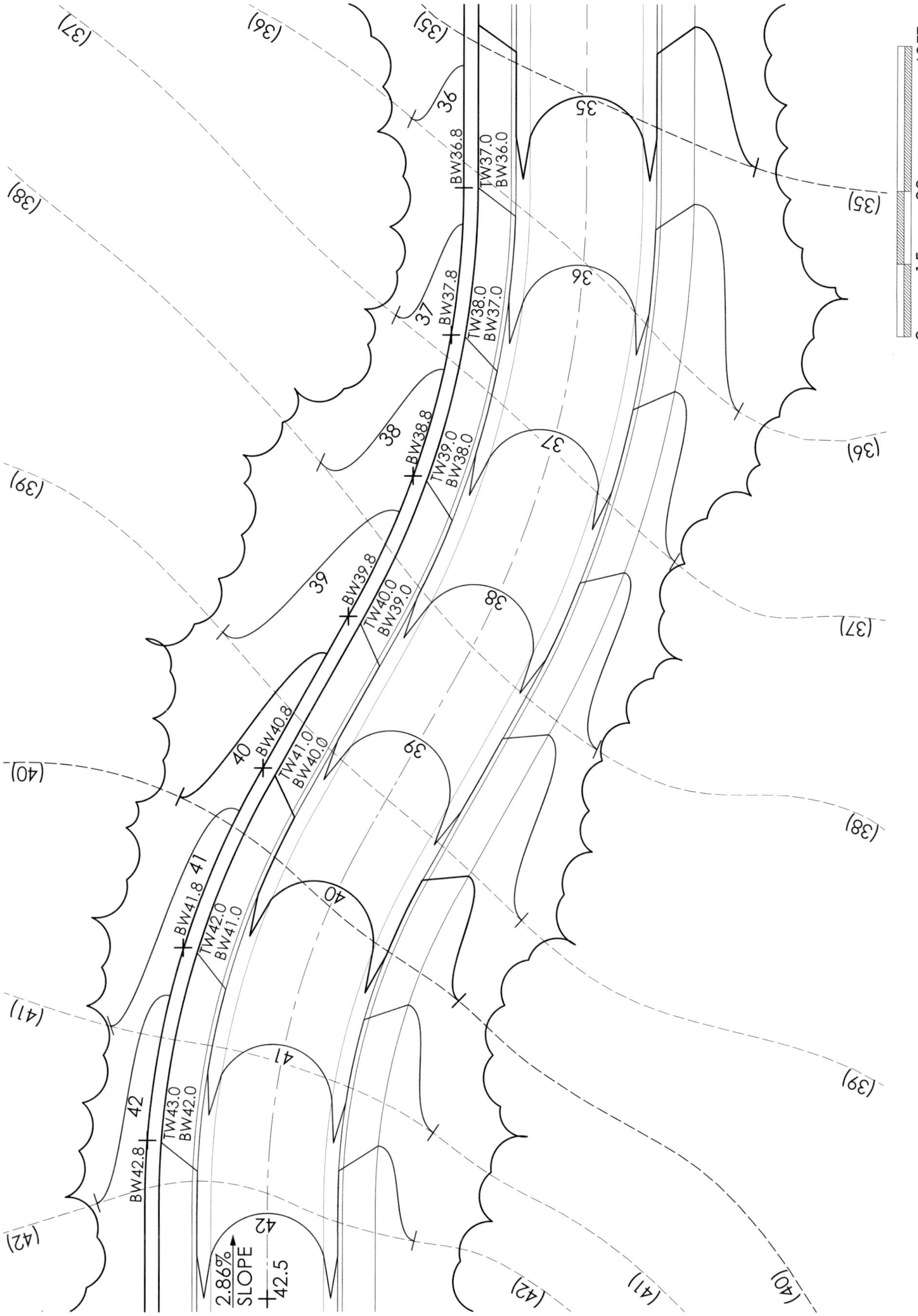
(37)
(36)
(35)
36
BW36.8
TW37.0
BW36.0
35
(35)
60FT
30
15
0
(38)
37
BW37.8
TW38.0
BW37.0
36
(36)
38
BW38.8
TW39.0
BW38.0
37
(39)
39
BW39.8
TW40.0
BW39.0
38
(37)
40
BW40.8
TW41.0
BW40.0
39
(40)
(38)
41
BW41.8
TW42.0
BW41.0
40
(41)
41
(39)
42
TW43.0
BW42.0
BW42.8
(42)
42
2.86%
SLOPE
42.5
(42)
(41)
(40)

Grading vignette 6 – solution

Grade the four single-family houses.

Step 1

Find the slope for the path based on the total slope of the existing topography. Measure the distance from the (30) contour line to the (24) contour line along the centerline of the path, 245'. Determine the change in elevation, 30' − 24' = 6'. Calculate the slope, 6'/245' = 0.0244897 or 2.45%.

Step 2

Locate the proposed whole number contour lines along the length of the path. Use interpolation to find the spot elevations for the intersections along the path.

Step 3

Provide the necessary spot elevations for all four buildings and find the FFE for each structure.

Step 4

To prevent the water from flowing back towards the buildings, create swales that drain to the drain inlets. Use the longitudinal slope for the path as the longitudinal slope for the swales to determine the RIM ELs for both drain inlets. Note that you can use other slopes for the swales, however using the same slope as for the path will minimize earthwork.

Step 5

Label the swales with the longitudinal slope, maximum slope, flowlines, and tie the proposed contours back to the existing contours.

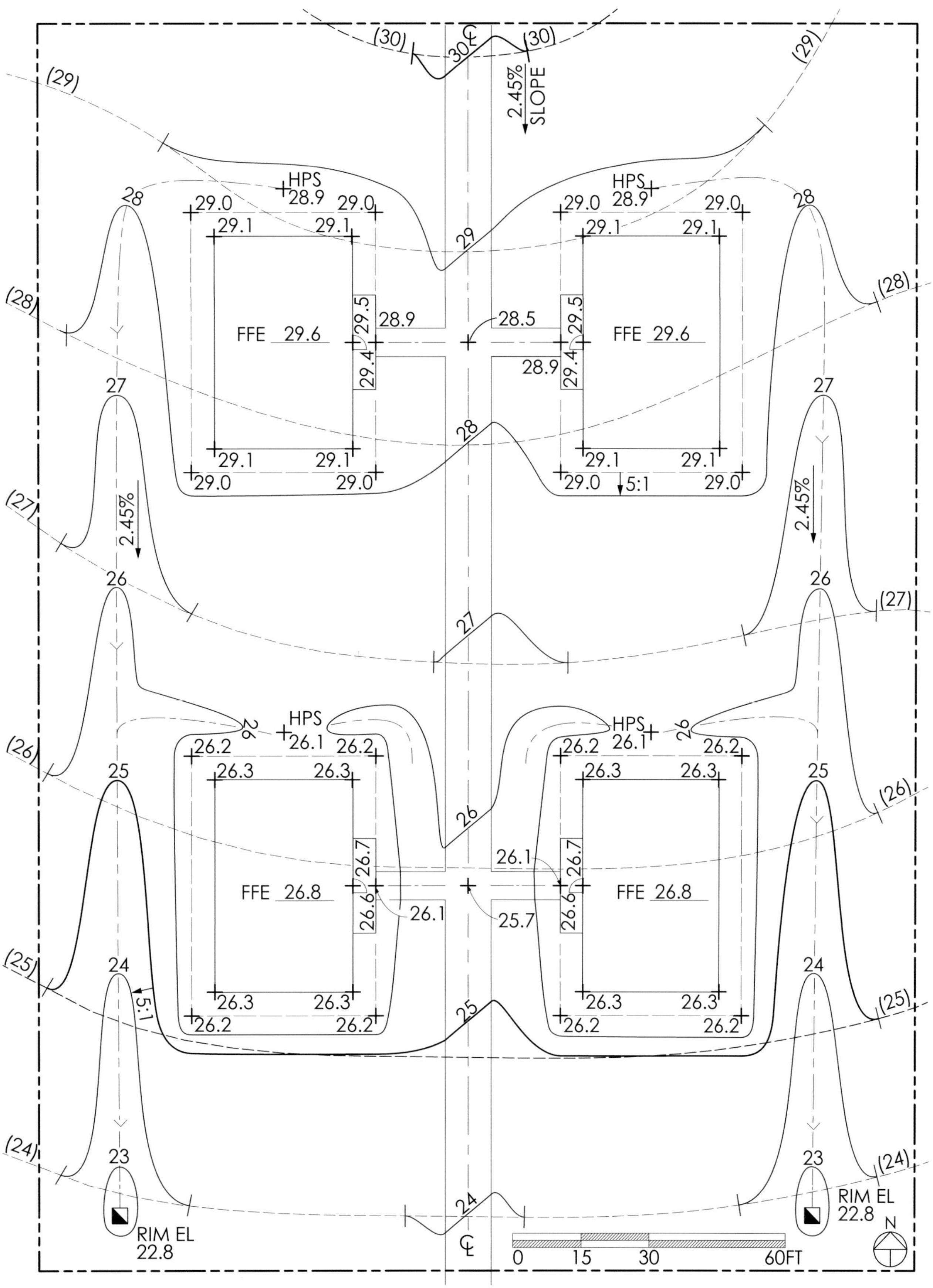
(30)
30
(30)
2.45% SLOPE
(29)
(29)
HPS 28.9
HPS 28.9
29.0
29.1
28
FFE 29.6
FFE 29.6
29.5
29.4
28.9
28.5
(28)
(28)
27
2.45%
5:1
(27)
(27)
26
HPS 26.1
26.2
26.3
FFE 26.8
FFE 26.8
26.7
26.6
26.1
25.7
(26)
(26)
25
(25)
(25)
24
(24)
(24)
23
RIM EL 22.8
RIM EL 22.8
0
15
30
60FT
N

Grading vignette 7 – solution

Divert water around the building to the retention pond.

Step 1

Draw a solid line around the dashed line for the retention pond. This is the bottom elevation for the pond. Add two additional closed contours lines at the maximum slope to complete the 2' of required depth for the pond.

Step 2

Using the provided spot elevation, determine the FFE for the building. Add 6" to the spot elevation at the edge of the porch, 45.5' + 0.5' = 46.0', and slope up towards the building at 2%, 8' × 0.02 = 0.16' or 0.2'; 46.0' + 0.2' = 46.2'. Add 1" to the find the FFE (46.2' + 0.0833' = 46.3').

Step 3

Find the spot elevations at the corners of the house, 46.3' − 0.5' = 45.8', and those sloping away from the house at 2% for 10'. 10' × 0.02 = 0.2'; 45.8' − 0.2' = 45.6'.

Step 4

Find the highest point in the topography, slope away from the building an additional 5' at 2% to find the HPS 45.5'. Label the HPS. Draw in two flowlines that connect with the retention pond. One line flows through the culvert and the other flows around the building.

Step 5

Determine the slope for each flowline from the HPS to the edge of the retention pond. Use either the calculated slope or the nearest whole number slope equivalent (e.g. 3.27% → 3.25% and 3.56% → 3.5% or 4%). This will allow you to easily measure whole number spot elevations along the flowlines using the slope interval.

Do not count the length of the pipe culvert in the calculation of the slope for the swale. Measure the distance to the start of the pipe culvert and calculate the INV IN elevation using the longitudinal slope of the swale. Use the pipe slope to determine the INV OUT elevation. Make sure the depth of the pipe meets the problem parameters.

Step 6

Create the swales around the building and connect them to the retention pond.

Step 7

Create an embankment above the two enclosed contour lines of the pond and add a spot elevation 6" higher than the highest contour of the retention pond. Add proposed contour lines below the embankment using the maximum slope and tie them back into the existing contours.

Step 8

At the top of the swale, add proposed contour lines and tie back into the existing contours using the maximum slope.

Step 9

Double-check everything and label everything with slopes, flowlines, embankment spot elevations and any dimensions that add clarity.

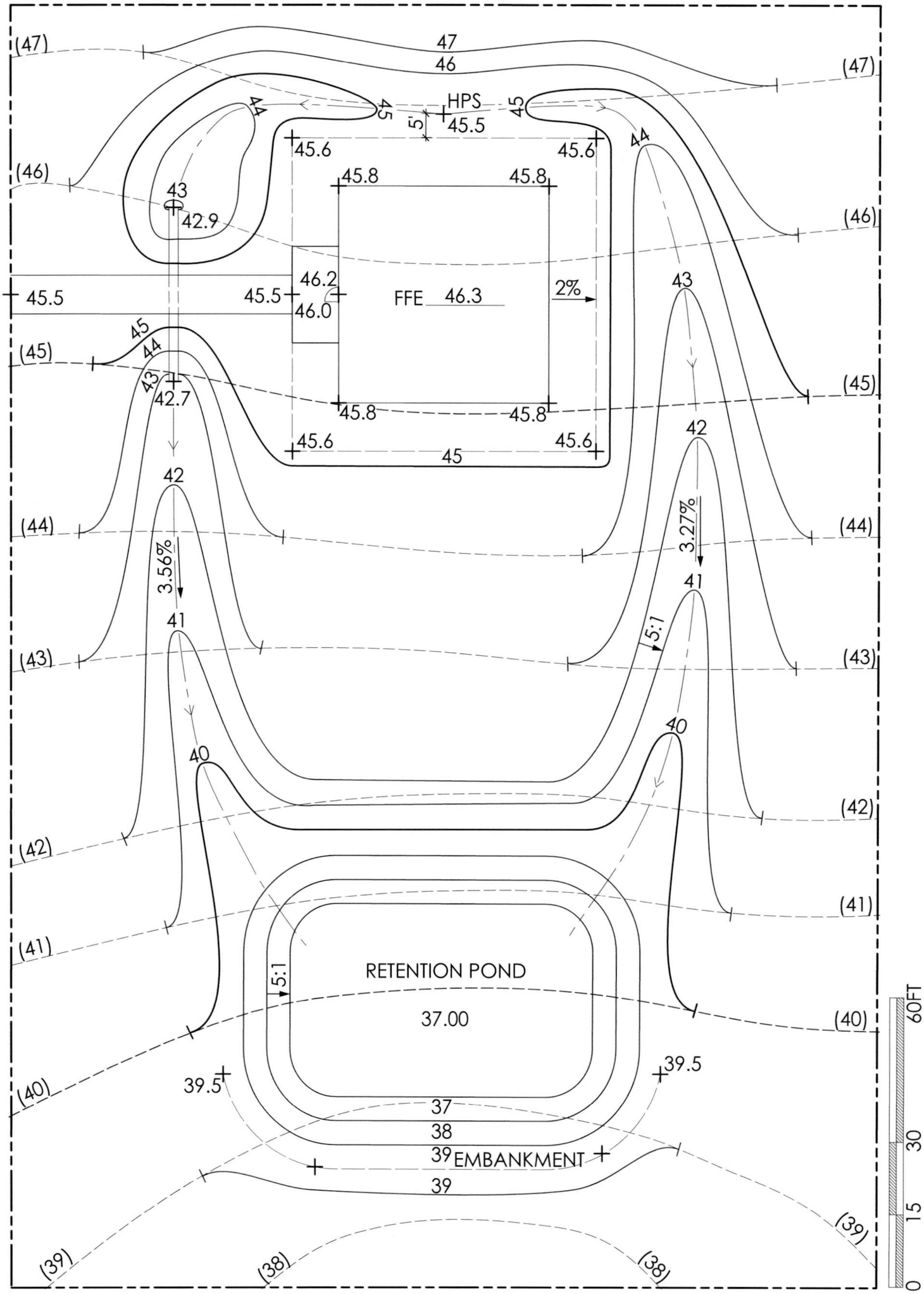
HPS
45.5
5'
45.6
45.6
45.8
45.8
43
42.9
44
45
46.2
46.0
45.5
45.5
FFE 46.3
2%
42.7
45.8
45.8
45.6
45.6
45
42
3.56%
41
40
43
42
3.27%
41
5:1
40
RETENTION POND
37.00
5:1
39.5
39.5
37
38
39 EMBANKMENT
39
(47)
(46)
(45)
(44)
(43)
(42)
(41)
(40)
(39)
(38)
0
15
30
60FT

Grading vignette 8 – solution

Create a 24' wide asphalt road with a shoulder and divert water through the culvert to a retention pond.

Step 1

Draw the proposed road contours using the vertical curve provided. Measure the distance to each whole number on the curve from the LP 13.9. Transfer this distance to the road centerline.

Use the road crown to find the elevation at the edge of the road. Interpolate between spot elevations at the edge of the road to find the location of the whole number contours along the edge of the road and along the shoulders.

Step 2

Determine the invert elevation of the culvert pipe at the intersection of the two centerlines. Top of culvert is 1'-4" lower than the low point of the road, 13.9' – 1.33' = 12.57'. Subtract the diameter of the pipe from the top of pipe elevation, 12.57' – 2' = 10.57'. The bottom center of the culvert is 10.57'.

Find the invert elevations at the beginning and end of the pipe. Since the pipe is 48' long, multiply half of the length by the pipe slope and add it to the invert elevation at the center of the pipe to find the INV IN at the entrance to the pipe, 24' × 0.005 = 0.12'; 10.57' + 0.12' = 10.69'. Repeat the same procedure for the INV OUT, but subtract the vertical change, 10.57' – 0.12' = 10.45'.

Step 3

Interpolate from the crown of the road across the road to find the spot elevations at the edge of the road and the shoulder, 12' × 0.2 = 0.24'; 13.9' – 0.24' = 13.66'; 5' × 0.02 = 0.1'; 13.66' – 0.1' = 13.56'. Use the maximum slope at the edge of the shoulder to find the spot elevation at the top of the culvert, 6' × 0.25 = 1.5'; 13.56' – 1.5' = 12.06'. Proposed contour 13 will cover the culvert pipe.

Step 4

Interpolate along the wing walls of the culvert at 4:1 to determine where the side slopes of the proposed 11 and 12 contours will connect to the wall.

Calculate the longitudinal slope for the swale entering the culvert between the existing (14) contour and the INV IN elevation for the pipe, 14' – 10.69' = 3.31'/58' = 0.05706 or 5.7%. Locate the whole number spot elevations and draw the proposed swale contours.

Step 5

Connect the proposed contours 11 and 12 back to the existing contours on the exiting side of the culvert using the maximum slope along the wing walls. Use a slope between 2% and 5.7% for the longitudinal slope for the swale exiting the culvert. Note that the minimum slope for the landscape is 2% so the maximum horizontal distance of the proposed 10 contour from the pond should not exceed 50'.

Step 6

Tie the proposed contours back to the existing contours. Maintain at least a 4:1 slope.

Step 7

Double-check everything. Label all the proposed contours, the longitudinal slope, the flowlines and the embankment. Include any dimensions that add clarity.

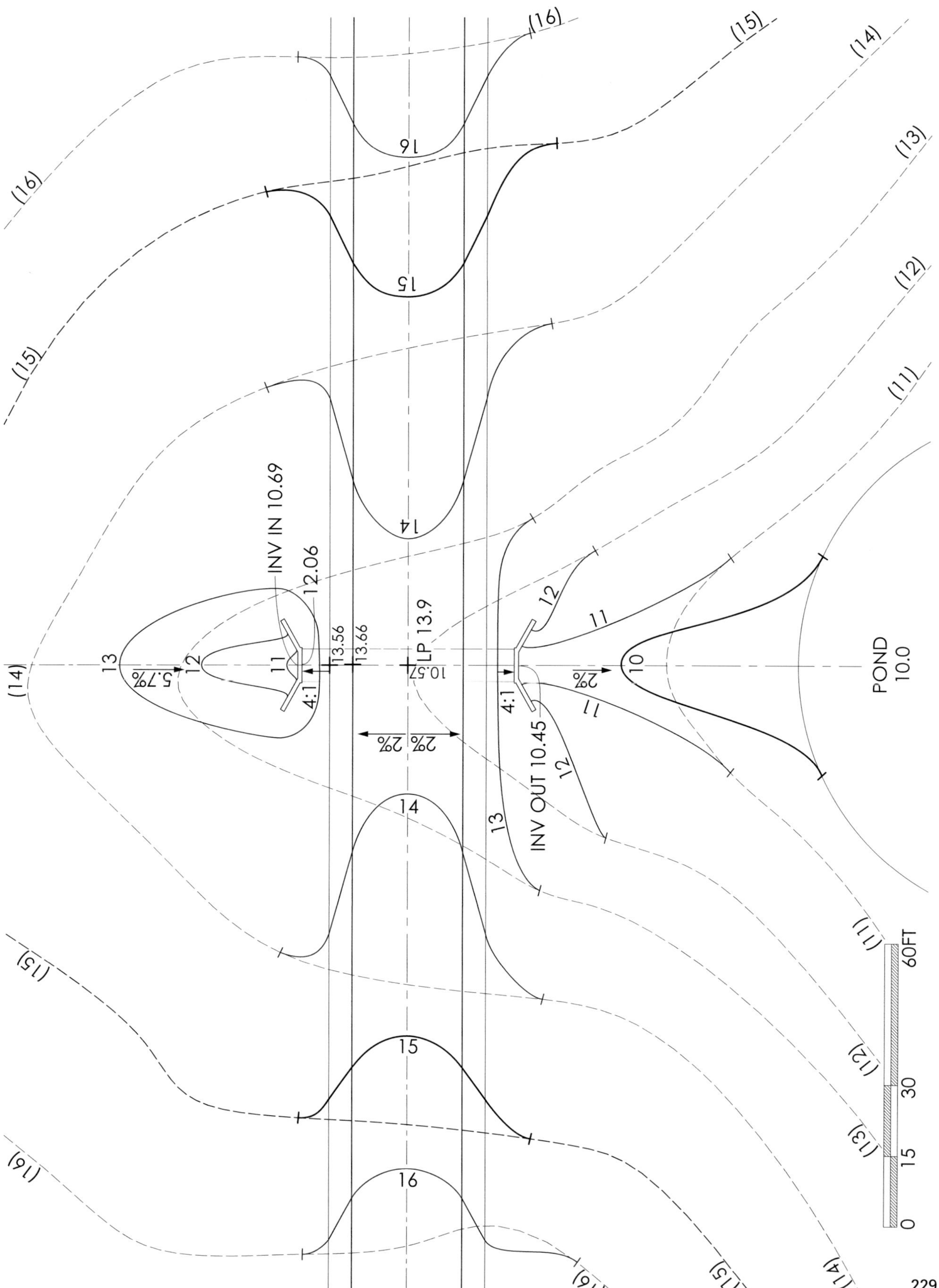

(16)
(15)
(14)
(13)
(12)
(11)
16
15
14
13
12
11
10
INV IN 10.69
12.06
13.56
13.66
LP 13.9
10.57
5.7%
4:1
2% 2%
2%
INV OUT 10.45
POND
10.0
0
15
30
60FT

Grading vignette 9 – solution

Grade the house into the sloping landscape. Provide a horseshoe swale around the property and direct the water into the drain inlets.

Step 1

Calculate the spot elevations for the front door, porch and at the bottom of the stairs. Place the spot elevations around the house on three sides and interpolate to find the elevations at the corners of the sloping buffer.

Step 2

Calculate spot elevations for the patio and the lawn. Slope away from the inside corner of the retaining wall at 2% to ensure that water does not pool in the corner.

Step 3

Using the maximum slope and elevation at the end of the retaining wall, calculate the spot elevation at the bottom of the wall on the east side of the wall, $12' \times 0.25 = 3'$; $98.7' - 3.0' = 95.7'$.

Step 4

Calculate a slope for the south side of the retaining wall using the elevation at the edge of the house and the elevation at the bottom corner of the wall, $98.8' - 95.7' = 3.1'$; $3.1'/21' = 0.14761$ or 14.76%. Locate the 98, 97, and 96 whole number spot elevations.

Step 5

Locate the HPS and calculate the elevation. Draw flowlines from the HPS to the drain inlets. Using 5% slope for the swales calculate the rim elevations for the drain inlets. Interpolate to find the location of the 98 contour and use the slope interval for 5% slope to mark off the whole number spot elevations along the flowlines.

Step 6

Calculate the slope for the path between elevations 97.6' and the existing (93) contour, $97.6' - 93' = 4.6'$; $4.6'/86' = 0.053488$ or 5.34%. Use the length of deflection equation to determine the horizontal distance for the cross slope, $(3' \times 0.02)/0.0534 = 1.1'$. Draw the path with the cross slope angled towards the open lawn, so water drains on to the property.

Step 7

Create the proposed contours using the maximum slope for side slopes and 5% for the swales. Connect the proposed contours to the path and retaining wall as required.

Create proposed contours above the HPS to tie the 99, 100 and 101 contours back to the existing topography.

Remember to locate the closed 99 contour that runs through the lawn and along the side of the patio and house.

Step 8

Check that all the proposed contours are connected back to the correct existing contour line. Label all contours, TW/BW for the retaining wall, flowlines and slopes for the swales, and slopes for the path.

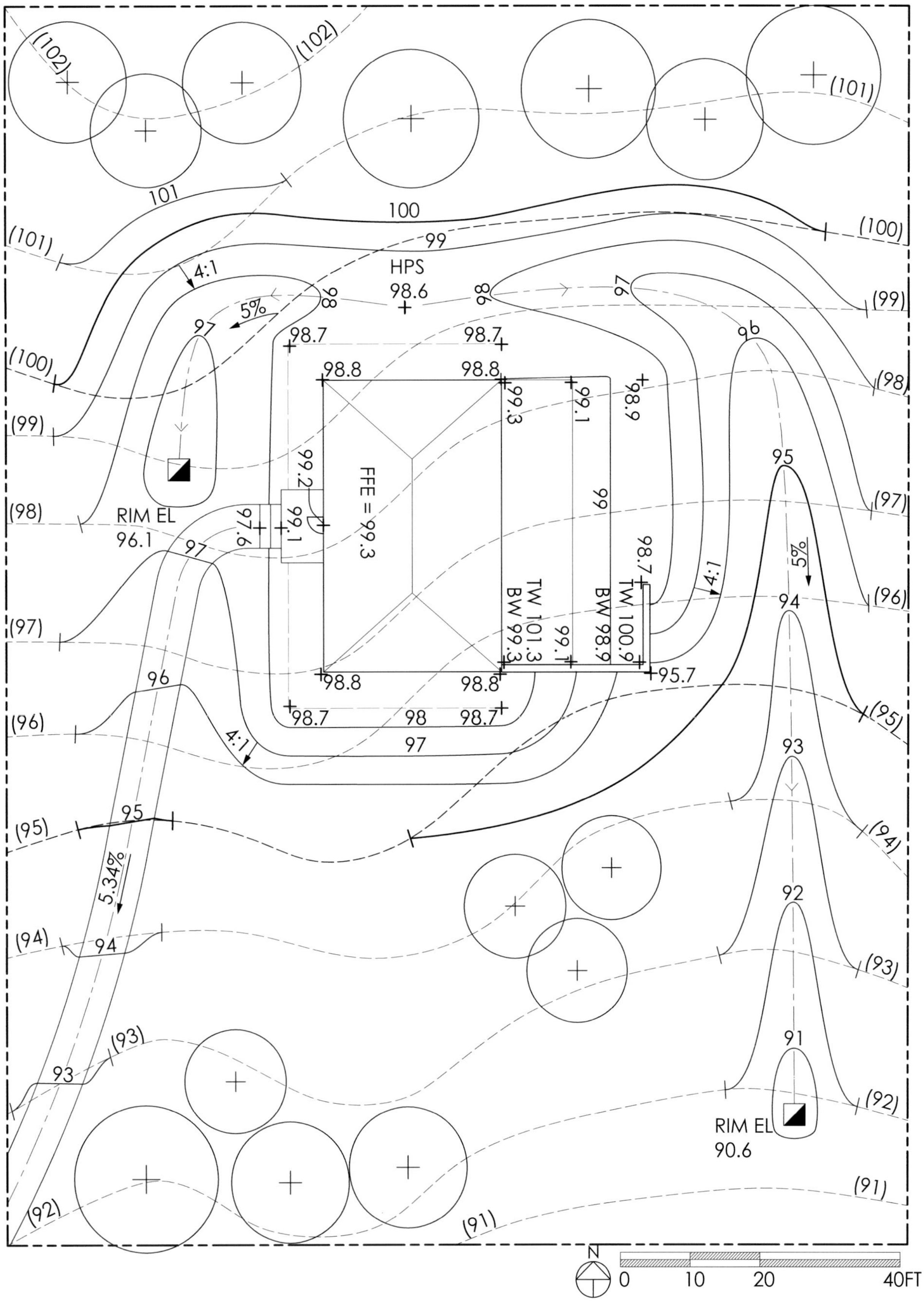
(102)
(102)
(101)
101
100
(101)
(100)
99
HPS
98.6
4:1
5%
97
98
98
97
96
(100)
(99)
(98)
(97)
(96)
(95)
(94)
(93)
(92)
(91)
98.7
98.7
98.8
98.8
99.3
99.1
98.9
99.2
FFE = 99.3
99
97.6
99.1
RIM EL
96.1
98.7
TW 100.9
BW 98.9
TW 101.3
BW 99.3
99.1
4:1
95
5%
94
95.7
98.8
98.8
98.7
98
98.7
97
96
4:1
95
5.34%
94
93
92
91
RIM EL
90.6
93
N
0
10
20
40FT

Grading vignette 10 – solution

From a deck on a building overlooking a road, provide a sloping berm to block the view of traffic on the road.

Step 1

Add 5' to the spot elevation on the deck, 97.5' + 5' = 102.5'. This is the adjusted height for a person standing on the deck. Add 8' to each of the contours along the viewshed. This is the height of a truck travelling on the road, 94' + 8' = 102.0'; 97' + 8' = 105.0'.

Step 2

Examine the existing topography and find a point along each line that aligns with the existing ridge. For line AB I chose a point at 74' from point A; for line AC I chose a point at 82' from point A.

Step 3

Create line of sight sections for each of the reference lines for the viewshed using the adjusted heights for the deck and the road. Use the reference lines of the viewshed as the length for each section. Add the midpoint spot elevations to the sections and determine the elevation along the line of sight that will obscure the vehicle. Pick a spot elevation that is up to 0.5' higher as the top elevation for the berm, see inset.

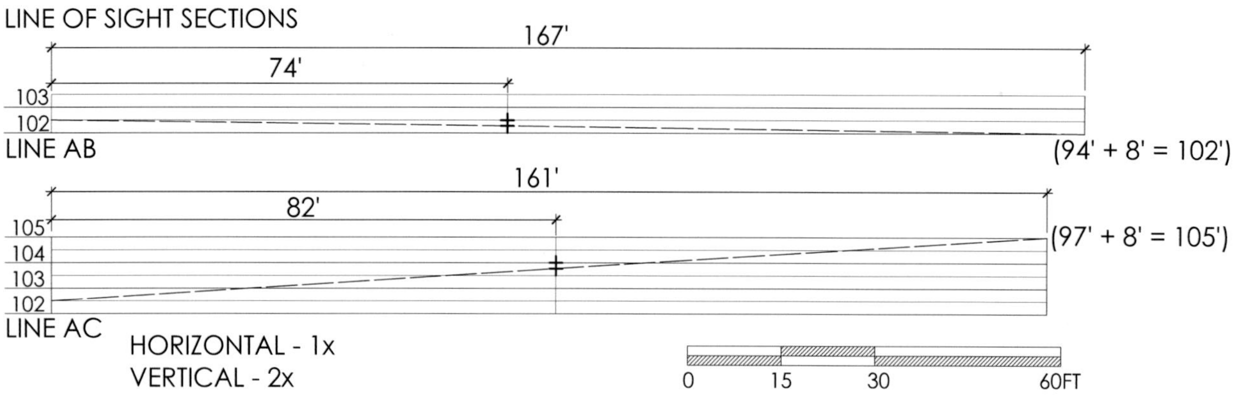

Step 4

Draw a reference line to connect the midpoint spot elevations on the plan. The line may be curved or straight. Determine the slope for the top of the berm, 104.0' – 102.5' = 1.5'; 1.5'/66' = 0.02272 or 2.27%. Using the slope for the top of the berm, locate the 103 spot elevation. Using the maximum slope on the side slopes create the closed contours around the spot elevations on the reference line.

Step 5

Create the berm using the maximum slope, 5:1. Label the proposed contour lines, the slope for the top of the berm and the maximum slope for the sides of the berm. Add any dimensions needed for clarity.

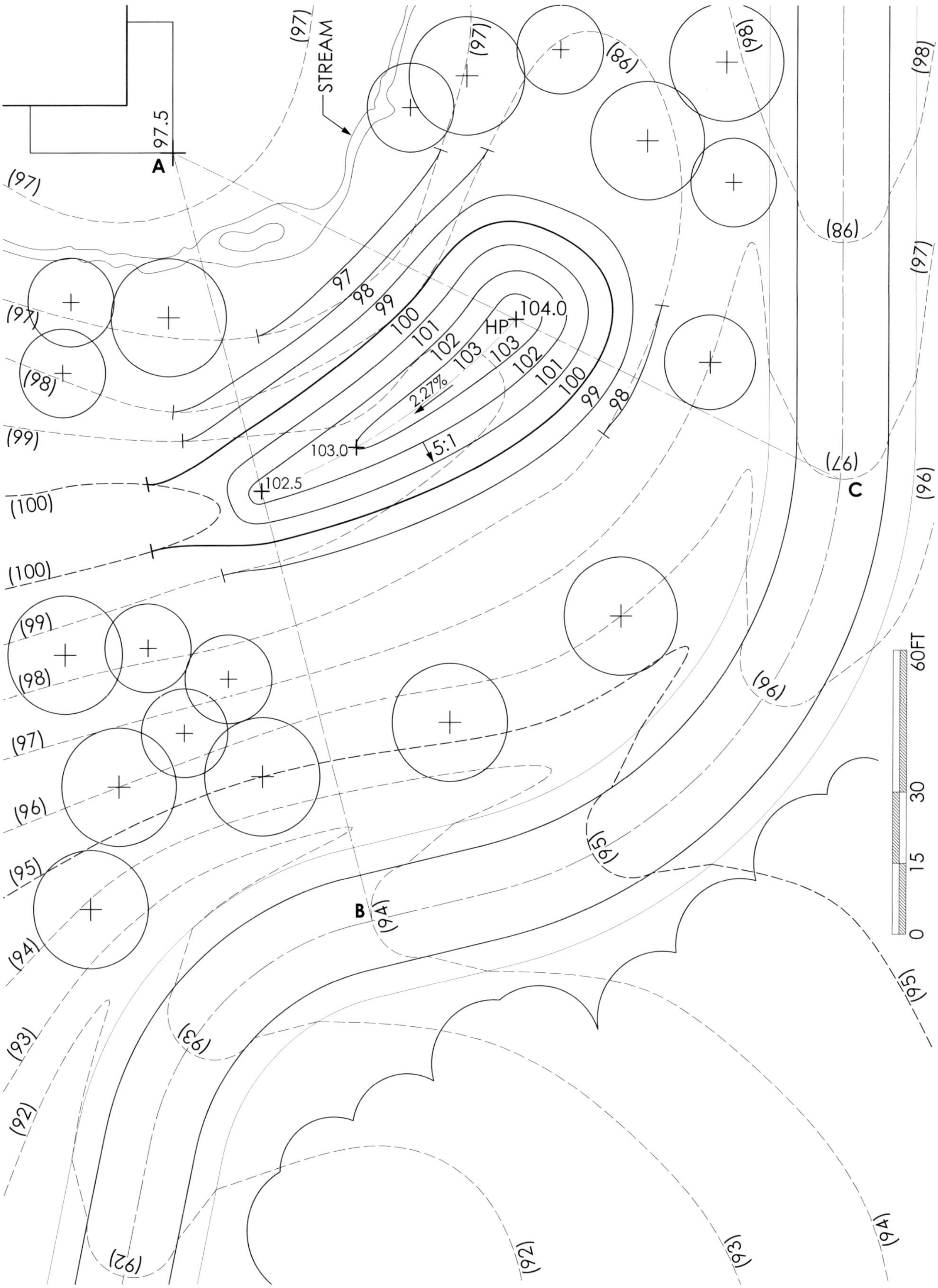

STREAM
97.5
A
B
C
HP
104.0
103.0
102.5
2.27%
5:1
97
98
99
100
101
102
103
(92)
(93)
(94)
(95)
(96)
(97)
(98)
(99)
(100)
0
15
30
60FT

Appendix
Metric standards in Canada and the United Kingdom

The first edition of this reference text was created in 2010 to satisfy a need for study material for the international Landscape Architecture Registration Examination's section E, the written site grading examination. The book expanded beyond the scope for which it was intended and traveled to countries outside of the United States, suggesting a need by international landscape designers for this information. The goal of the second edition was to incorporate metric examples for wider usability by an international audience. There were two challenges to overcome.

The obvious first challenge is that the United States is one of only three countries that still uses the imperial system of measurement, feet and inches. The majority of the world uses the metric system of measurement. This makes it difficult for designers outside of the US to easily translate minimum and maximum requirements for ramps and stairs into metric and for US designers to do the reverse.

The second challenge was that although the metric system is used globally, there is no universal standard for accessibility in building construction, each nation has its own building standards or regulations. A study published in 2006, *International Best Practices in Universal Design: A Global Review*, compared accessibility standards across several nations with established standards and the differences were numerous. The International Standards Organization (ISO) is developing a comprehensive set of standards for accessibility and the built environment, but these will have to be adopted by each nation. Therefore, providing a few metric examples might satisfy the requirements of one nation, but not another and would likely become outdated when the ISO standards are adopted.

Undeterred, I have included this appendix to provide a small comparison of the grading standards for stairs and ramps of three countries, Canada, the United Kingdom (UK) and the United States (US) as a way to help designers in the US and internationally apply some of the principles in this book to the metric system. I chose Canada because of proximity and because they use the LARE as their licensing examination, which allows for reciprocity in the United States. I chose the UK because of the closely related histories of the three countries.

The US and Canadian regulations are based on accessible or barrier-free design and the UK regulations are based on universal design. Accessible design, as promoted by the ADA requirements, provides specialized design that aids disabled persons in navigating the built environment. It covers the minimum requirements that a designer should incorporate into a site design and it focuses heavily on mobility rather than visual and hearing disabilities. Universal design is a term coined by architect Ronald L. Mace. It promotes the idea that all design solutions should be usable by the largest possible group of people, regardless of ability. It strips the idea of separate but equal from design and asks designers to create design for all regardless of ability. It addresses disabilities that affect mobility, hearing and sight into solutions that will work well for a majority of users. In many ways, universal design exceeds the requirements of the ADA. Requirements, whether accessible or universal, are not meant to hinder landscape architects' creativity but are given as a baseline from where to begin designing. A good designer should use them as a means of applying creative solutions to challenging topography.

A.1 Units of linear measurement

The meter, the standard unit of metric linear measurement, is divided into ten equal units, subsequent divisions are each divided into tens. The divisions are decimeters, centimeters and millimeters. In landscape design, the decimeter and centimeter designations are never used, fractions of a meter are labeled as 0.1 m, 0.2 m, etc. for large-scale plans or as millimeters 10 mm, 200 mm, etc. for construction details. A typical meter stick or ruler is divided into 1000 millimeters, with demarcations at 10 mm and 10 cm increments.

The most commonly used metric linear measurements in landscape design are the kilometer, the meter, and the millimeter and the most commonly used square measurement for measuring area is the hectare. Units used are based on scale, roads are shown in kilometers, and intricate detailing of small items in millimeters.

The foot, the standard unit of imperial linear measurement, is divided into 12 equal units or inches. Subsequent standard divisions are 1/2, 1/4, 1/8, 1/16 and 1/32 of an inch. The most commonly used linear measurements in topographic grading are given as decimal equivalents of a foot to correspond with civil engineering standards, see Figure 1.2–1. Units used are based on scale, roads are shown in miles, and intricate detailing of small items in fractions of an inch. The most commonly used linear measurements in construction detailing are in feet and inches. A comparison of the different measurement systems is highlighted in Appendix Figure 1.

All topographic plans must have a clearly labelled scale. Scales for large-scale master plans are 1:500 and up, 1:200–1:50 for site-specific plans and sections, 1:20–1:10 for enlargements and 1:10–1:5 for construction details. Units can be whole feet (US) or meters (Canada, UK) or half units, but they cannot switch between half or whole units within any given survey or plan. Always check the unit of measure on any topographic map or survey prior to grading.

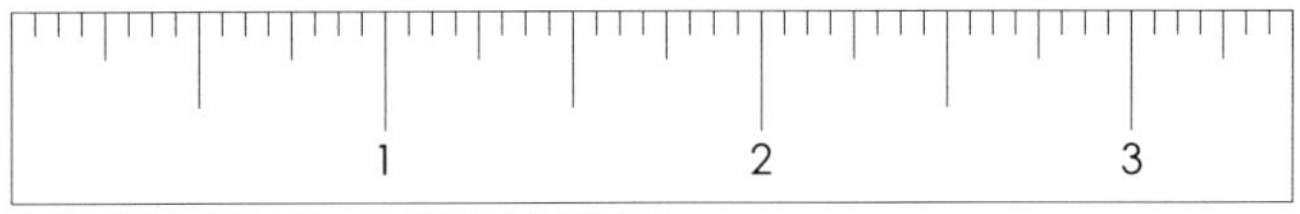

IMPERIAL RULER - UNITS IN INCHES

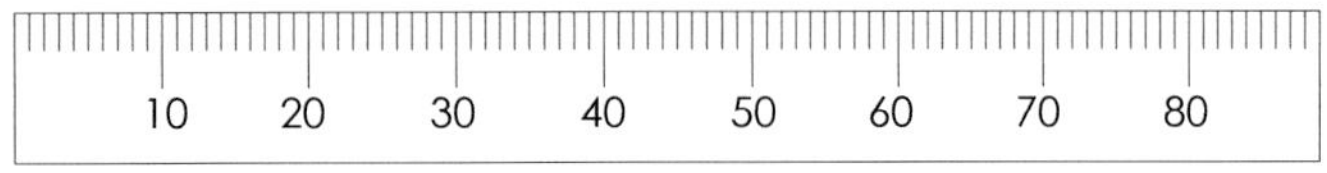

METRIC RULER - UNITS IN MILLIMETERS

NOTE:
1 METER = 3'– 3-3/8"
1 INCH = 25.4 mm

LINEAR MEASUREMENT EQUIVALENTS

PRIMARY UNIT OF LINEAR MEASUREMENT	FEET (1')	METER (1 m)
COMMONLY USED SUB MEASUREMENTS	INCHES (12")	MILLIMETER (1000 mm)
PRIMARY UNIT OF LONG DISTANCE MEASURE	MILE (5280')	KILOMETER (1 km or 1000 m)
PRIMARY UNIT OF AREA MEASUREMENT	ACRE (43,560 ft^2)	HECTARE (10,000 m^2)

A.1
Units of linear measurement

A.2 Comparison of accessible building regulations

The compiled regulations were summarized from the National Building Code of Canada (2015), The Building Regulations – Access To and Use of Buildings Approved Document M Volume 1: Dwellings and Volume 2: Buildings other than Dwellings (2015) and the 2010 ADA Standards for Accessible Design. See the literature cited section for the full citations. Note that the regulations cover all aspects of building for accessibility within structures and this brief comparative study focuses on only those elements that affect the grading of the exterior environment. Many local municipalities will have regulations that exceed the minimum requirements of national standards and they should be consulted prior to designing stairs, ramps and pathways. The table in Appendix Figure 2 summarizes the most pertinent national standards for external built features and the following discussion provides an opportunity for greater understanding of how to apply the regulations. Some of the measurements are given as ranges and each country may have a preferred measurement that they use in construction. For example, steps in the US are typically 6" tall because it is an easily divisible unit of a whole foot. Refer to the individual building regulations for each country for clarity.

Appendix Figure 2: Comparison Table of Building Regulations	**UNITED STATES**
Associated International Building Code or Standard	2010 ADA Standards for Accessible Design
PATHWAYS	
Path width, multi-use accessible	Minimum width 60" (1525 mm)
Path cross slope	1:48 max., 2.08$\overline{33}$%
RAMPS	
Ramp slope maximum	1:12 max.
Ramp slope minimum	1:20 min.
Ramp cross slope	1:48 max., 2.0833%
Ramp width minimum	Clear width 36" (915 mm)
Ramp landing (beginning and end)	60" long x width of ramp (1525 mm)
Ramp landing (intermediate)	60" x 60" for right turns (1525 mm x 1525 mm)
Ramp run length (going) maximum	30' (9150 mm)
Ramp landing cross slope	1:48 max., 2.0833%
HANDRAILS AND GUARDRAILS	
Handrails (continuous)	Yes, both sides
Handrail clearance	1-1/2" min. (38 mm) from vertical surface
Handrails height above ramp	34" min. (865 mm) – 38" max. (965 mm); height must be consistent
Handrail height above stair	34" min. (865 mm) – 38" max. (965 mm)
Handrail extension (ramps)	12" (305 mm) minimum at top and bottom of ramps
Handrail extension (stairs)	12" (305 mm) minimum at top and one tread length at the bottom of he stairs
Guardrails	42" required when grade difference is 30"
STAIRS	
Stair width	44" (1118 mm) min. width clear
Stair riser heights	4"(100 mm) – 7" (180 mm), preferred 6"
Stair tread (going) lengths	11"(280 mm) minimum (2R + T = 24" - 26")
Stair tread slope	1:48 max., 2.0833%
Stair landing length	Landing width same as stair width, length 48" (1219 mm) minimum
Stair landing slope	1:48 max., 2.0833%

A.2
Comparison Table of Building Regulations

UNITED KINGDOM	CANADA
The Building Regulations 2010 Access To and Use of Buildings Approved Document M Volumes 2 (2015)	National Building Code of Canada (2015)
PATHWAYS	
Minimum width 1500 mm for 1 wheelchair + 1 pedestrian; 1800 mm preferred	Minimum width 1100 mm
1:40 max. cross slope, 1:60 max. longitudinal slope	1:50 max.
RAMPS	
1:12 max.	1:10 max.,1:12 max. for barrier-free travel
1:20 min.	1:20 min.
1:40 max. cross slope,1:60 max. longitudinal slope	1:50 max.
Clear width 1500 mm	Clear width 1100 mm
1200 mm long x width of ramp	1500 mm long x 1500 mm wide
1500 mm x width of ramp; 1800 mm x 1800 mm for right turns	1200 mm x width of ramp
Length varies by slope: 1:12 max. slope – 2000 mm length; 1:15 slope – 5000 mm length; 1:20 min. slope – 10 m length	9000 mm
1:40 max. cross slope; 1:60 max. longitudinal slope	1:50 max.
HANDRAILS AND GUARDRAILS	
Yes, both sides. Double handrails at 1,000 and 750 mm; required for three or more steps	Yes, both sides; required for three or more steps
50 mm – 75 mm from vertical surface	50 mm – 60 mm from vertical surface
900 mm min. – 1000 mm max.	860 mm min. – 965 mm max.
900 mm min. – 1000 mm max.	865 mm min. – 1070 mm max.
300 mm top and bottom of ramps	300 mm at top and bottom of ramps
300 mm top and bottom of stairs	300 mm at top and bottom of the stairs
1100 mm required when grade difference is 600 mm	1070 mm required when grade difference is 600 mm
STAIRS	
1000 mm min. width clear	900 mm min. width clear
150 mm –170 mm	125 mm –180 mm
280 mm – 425 mm (2R + G = 550 mm – 700 mm)	280 mm minimum
n/a	1:50 max.
Landing width same as stair width, landing length 1200 mm minimum	Landng width same as stair width, landing length 1100 mm minimum
1:40 max. cross slope; 1:60 max. longitudinal slope	1:50 max.

A.2
continued

A.3 Stairs

Similarities: in all three countries, steps should be solid and uniformly consistent for an individual flight of stairs. All three countries have required minimums and maximums for riser heights and tread lengths. Landings with a maximum slope of 1:50 are required at the top and bottom of the flight of stairs. Intermediate landings should be the same width as the width of the stair and if the stair width exceeds 2 m or 6' then intermediate handrails are required.

Differences: the differences between the three countries are subtle, but numerous. While all three countries require continuous handrails on both sides of the stairs with a minimum width of 1 m or 44", the UK requires double handrails to accommodate people of different heights and abilities. The UK and Canada are more stringent than the US in accommodating people with visual disabilities. The UK requires 600 mm textured strips at the top and bottom of the stairs to signal to pedestrians that a vertical change is imminent. Canada only requires the textured striping at the top of the stairs. Both the UK and Canada require stair nosings that are color-contrasted to the stairs to increase visibility. All three require well-lit staircases.

The example in Appendix Figure 3 satisfies the requirements for stairs in both Canada and the UK. In addition to the requirements shown, you should add textured

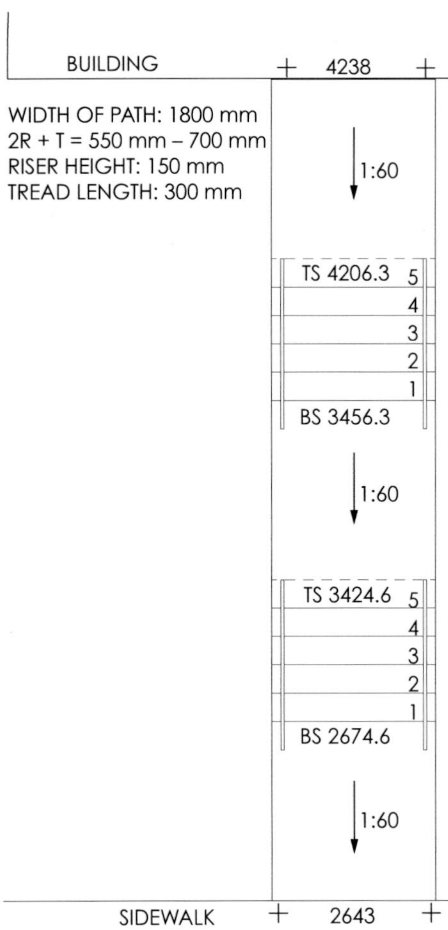

DISTANCE IN MILLIMETERS	STAIRS	LANDINGS	TOTAL
VERTICAL	1500 mm	95 mm	1595 mm
HORIZONTAL	3000 mm	5700 mm	8700 mm

TOTAL NUMBER OF STEPS: 10
SLOPE OF LANDINGS: 1:60 max.
NUMBER OF FLIGHTS OF STAIRS: 2, 5 steps per flight
NUMBER OF LANDINGS: 3, 1900 mm per landing

	STEPS PER STAIR	LENGTH IN MILLIMETERS	TOTAL
VERTICAL	5	150	750 mm
HORIZONTAL	5	300	1500 mm

TYPICAL LANDING	LENGTH IN MILLIMETERS
VERTICAL	31.7
HORIZONTAL	1900
SLOPE	0.016

MIN. LENGTH OF LANDING:
1100 mm (Canada)
1200 mm (UK)

HANDRAIL LENGTH:
3000 mm top and bottom (UK)
3000 mm top and 3000 mm + one tread bottom (Canada)

detectable warnings on the landings at the top and bottom of each flight of stairs to increase visibility. Note that the UK has a recommendation of 2R+T=550 mm–700 mm, this is similar to the US recommendation of 2R+T=24"–26". There are several combinations of riser heights to tread lengths that will result in evenly graded stairs and landings that don't exceed the maximum landing slope requirement. Aim for a landing slope minimum of 1.5% and the maximum, in the UK 1:60 (1.6%), in Canada 1:50 (2%) and in the US 1:48 (2.083%) to ensure positive drainage of stormwater. Once calculated, divide the total number of steps into flights of stairs to fit into the existing slope of the surrounding topography. Commonly used riser heights and tread lengths are 150 mm and 280 mm respectively.

A.4 Ramps

Similarities: in all three countries, the minimum slope for a ramp is 1:20 and the maximum slope is 1:12. All regulations require landings at the top and bottom of a ramp with continuous handrails along both sides that extend 300 mm (Canada and UK) or 12" (US) past the beginning and end of the ramp. Landings should be clear of obstructions.

Differences: all three countries require intermediate landings; however, the UK standards are the most stringent. The UK requires landings at a maximum of 2 m horizontally if the slope is 1:12 with longer ramp lengths between landings as the slope gets shallower. The illustration in Appendix Figure 4 taken from The Building Regulations

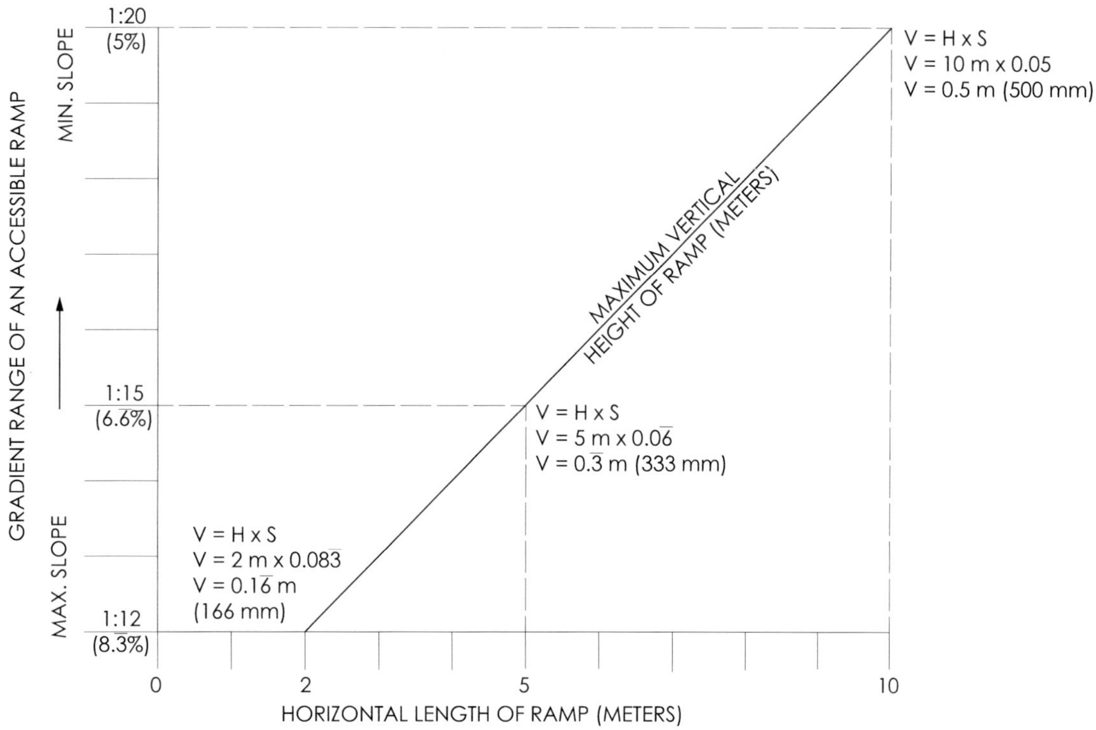

A.4
Location of intermediate landing based on ramp slope (UK building standard) – The Building Regulations Approved Document M Volume 2 (2015)

Approved Document M (2015) shows the sliding scale of horizontal ramp length to maximum ramp slope. The United States and Canada require intermediate landings at a maximum ramp length of 30' or 9 m, respectively, regardless of the steepness of the ramp. Ramp cross slope maximums are 1:50 and intermediate landing lengths vary by country with sizes changing for straight runs as well as ramps that turn 90°. Maximum slope on landings is 1:48, 1:50 and 1:60 in the US, Canada and the UK respectively.

An example of a stair/ramp combination using Canadian standards is shown in Appendix Figure 5. Start grading the stair/ramp combination at the top stair landing first. Decide the width for the ramp; exceed the required minimums if space allows to ensure two-way travel. Remember to add 300 mm for the handrail extension areas before drawing the ramp and intermediate landings. When interpolating make sure to carry the slope decimal out to five digits to ensure accuracy and round spot elevations only at the end of calculations.

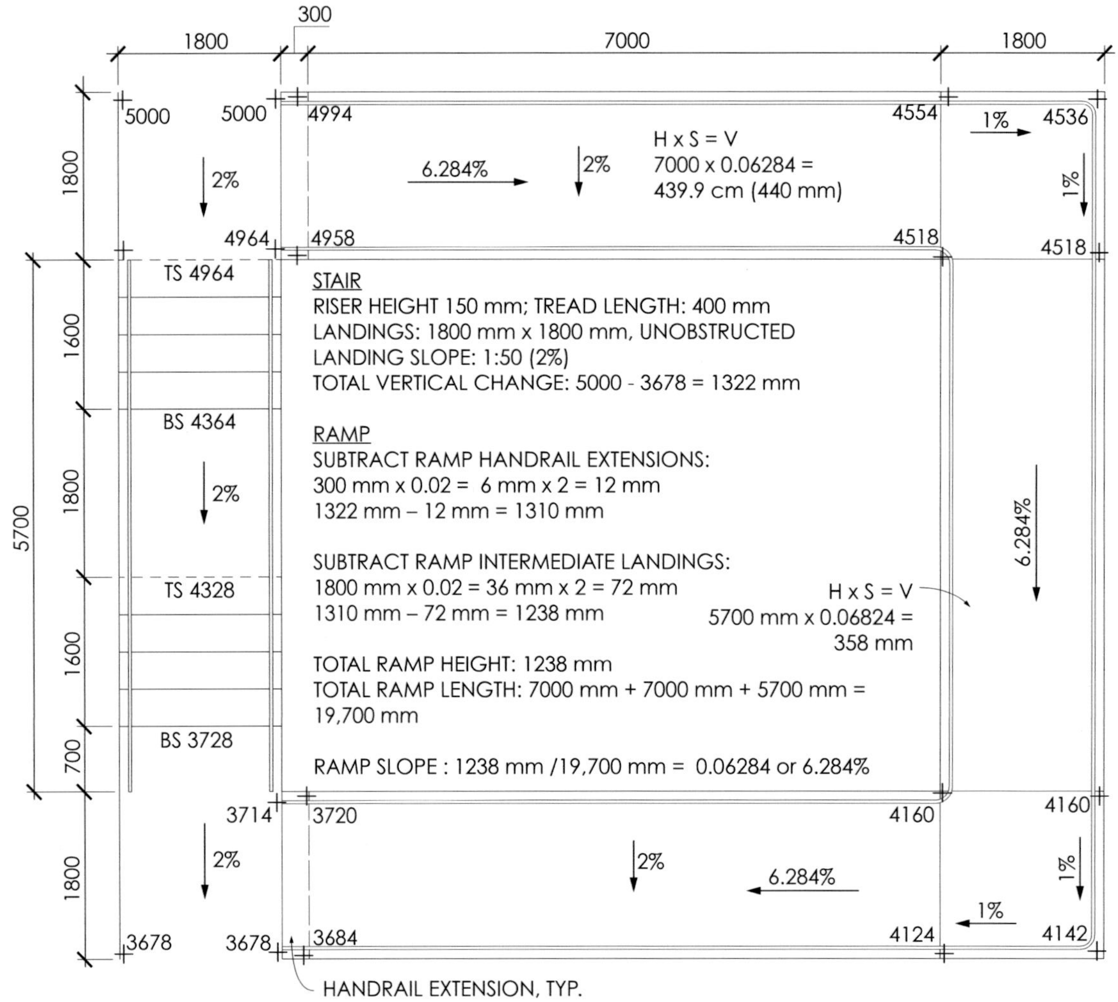

A.5
Stair/ramp combination (Canada)

When choosing horizontal lengths for ramps, round up to the nearest whole centimeter for ease of construction. For example, round 2232 mm to 2240 mm or 2300 mm. Remember to calculate the slope for the entire ramp based on the length of all the ramp runs.

For ramps in the UK, calculate the total vertical change for the ramp and then determine a ramp slope based on the chart in Appendix Figure 4. You may have to choose two different ramp slopes to accommodate any extra landings that are required. In Appendix Figure 6, the example has been updated to UK requirements. Note that the handrails have been omitted from the image to improve clarity.

1 The clearance for the handrails at the bottom of the stairs is 300 mm only; this changes the ramp run that is parallel to the stairs to 5300 mm. With the two perpendicular ramp runs at 7000 mm, the total horizontal length of ramp is 19,300 mm.
2 The total vertical change for the entire ramp is 5000 mm−3705 mm = 1295 mm as calculated from the stair elevations.

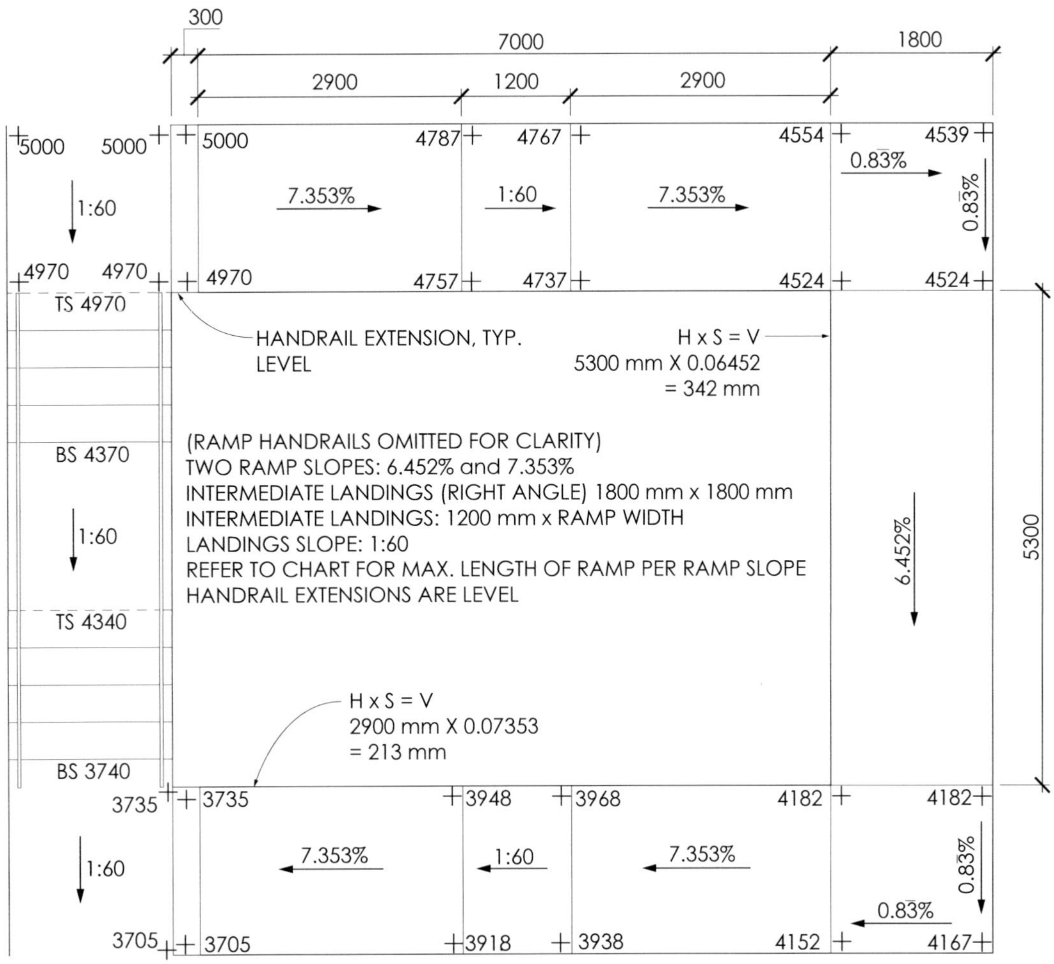

A.6
Stair/ramp combination (UK)

3 Recalculate the slope on the landings based on a maximum slope of 1:60 (1.66%). Subtract two landings from the overall vertical change, 1800 mm × 0.0166 = 30 mm; 30 mm × 2 = 60 mm; 1295 mm − 60 mm = 1235 mm.
4 Calculate the slope needed for the ramp, 1235 mm/19,300 mm = 6.39%. Based on the chart in Appendix Figure 4, the maximum length of a ramp run at 6.39% slope is between 5000 mm and 6000 mm. The maximum allowable slope for a 7000 mm long ramp run is 1:17 or 5.88% which is too shallow to meet grade.
5 Because the ramp run that is parallel to the stairs is a fixed horizontal length, the slope for this ramp run should be calculated first. At 1:15 slope the maximum ramp run length is 5000 mm and at 1:16 slope the maximum ramp run length is 6000 mm. The parallel ramp run is 5300 mm so the maximum slope must fall between 1:15 and 1:16. Using interpolation, a ramp slope of 1:15.5 (6.452%) has a maximum horizontal ramp length of 5500 mm.
6 Calculate the vertical change for the parallel ramp run using 6.452%. Because the ramp run is shorter than 5500 mm it will not exceed the maximum required horizontal length, 5300 mm × 0.06452 = 342 mm.
7 Subtract the vertical change of the ramp run from the adjusted ramp height, 1235 mm − 342 mm = 893 mm.
8 Next, because the two ramp runs of 7000 mm are too long, add an intermediate landing in the middle of each ramp run. The landing should be minimum 1200 mm long. There should be four short ramps that are equal at 2900 mm each, 2900 mm ramp + 1200 mm landing + 2900 mm ramp = 7000 mm. Note that the length encompassed by the ramp, 7000 mm is not increased, but rather a portion, 1200 mm, has been converted to a landing allowing the remaining length, 5800 mm, to be steeper.
9 Grade each intermediate landing with the maximum slope for landings, 1:60, and subtract the vertical change from the adjusted ramp vertical change, 1200 mm × 0.0166 = 20 mm; 20 mm × 2 = 40 mm; 893 mm − 40 mm = 853 mm.
10 Calculate the ramp slope required for the remaining ramp length, 2900 mm × 4 = 11,600 mm; 853mm/11,600 mm = 0.07353 or 7.35% slope.
11 Check the new slope against the chart. At 1:13 slope, the maximum ramp length is 3000 mm and at 1:14 slope, the maximum ramp length is 4000 mm. The calculated slope is 1:13.6 (7.35%) the maximum length of ramp is between 3000 mm and 4000 mm. Each remaining ramp run is only 2900 mm, so the ramps are shorter than the required maximum lengths.
12 Calculate all the spot elevations for each ramp run and landing to make sure they meet the required elevations as set by the stairs.

A.5 Summary

Constructability

Because the metric system of linear measurement is divisible by tens, it eliminates fractional division of whole units as used in the imperial system. For improved constructability, round up to whole centimeters rather than millimeters. Always use decimals to denote parts of a meter. For example, 2.3 meters = 2300 millimeters, or 2 meters + 300 millimeters.

Although it can be intuitive, for example, when looking at a plan of a sidewalk and the dimension is listed as 2, it is understood that the unit is in meters not millimeters; for improved clarity, and as good practice, add the unit designation to the drawing or to each measurement. Avoid mixing units in the same drawing, be consistent.

The minimum stair tread length in Canada and the UK is 280 mm. If space allows, opt for longer tread lengths outside, this makes steps easier to maneuver. Additionally, add intermediate landings to a flight of stairs to enable pedestrians to rest, this is similar to the need for intermediate landings on ramps. Aim for between 10–12 steps maximum before a landing. While it may seem preferable to remove stairs from an accessible entrance, some people with limited mobility prefer stairs to ramps because ramps are uneven ground which can be destabilizing. To accommodate the greatest number of people, include stairs in your site plan. The shift should be from stairs with ramps on the sides, to ramps with stairs on the sides. Making the primary access ramped accommodates the greatest number of people. Although not required by the ADA, it is good practice, and may be required by local standards, to incorporate added features that improve visibility such as detectable warnings, and contrasting colors for risers, treads, and nosings.

In ramp/stair combinations start with the stair elevations first, calculate and subtract values for 90° intermediate landings and handrail extensions based on maximum slope, 1:60 in the UK and 1:50 in Canada. Calculate the slope for the ramp based on the adjusted vertical distance. In the UK, check the ramp lengths chart to determine if additional landings are required. If they are, subtract the height for additional intermediate landings and recalculate the slope for the ramp. Check the chart again to confirm conformance to the required standards and divide the ramp lengths evenly between landings.

Calculations

While compiling this metric appendix, the greatest challenge was applying the UK building regulations to calculate the correct number of steps (riser height, tread length) and still find a landing that would not exceed the maximum landing slope, 1:60. In the imperial system, although the riser height is limited to 4"–7" and the tread length minimum is 11", there are fewer options for step riser heights and tread lengths. Because of constructability, whole inches are used for risers and treads, avoiding the need for ½" measurements. Also, because a foot is easily divisible by 3, most tread lengths are a whole foot or a foot + 3" or 6". The recommendation of 2R + T = 24" – 26" also implies that certain riser tread combinations are typical, thus, in imperial units 4" riser/18" tread, 5" riser/15" tread and 6" riser/12" tread are commonly used variations. This makes calculating the number of steps easier and because the maximum landing slope is 1:48 there is a greater likelihood that a suitable landing slope between 1.5% and 1:48 can be found.

In the UK regulations, the viable combinations are more numerous, and the landing slope maximum is shallower so finding riser heights and tread lengths that work with the landing slope are more difficult. In the UK, there are three whole centimeter riser heights, but 16 viable tread lengths, or 48 possible riser-tread combinations, and many won't meet the landing slope requirement. Although I was unable to find an easy calculation method for stairs and landings the following recommendations highlighted in Appendix Figure 7 minimized the number of calculations.

1 Start with the minimum riser height, 150 mm and choose a tread length in the middle of the recommended tread lengths. Calculate the number of steps required to meet the vertical change, Step A.
2 Multiply the number of steps by the riser height and tread length, Step B.
3 Subtract the total vertical change of the risers from the overall vertical change; this is the vertical change for the landing. Subtract the total horizontal length of the treads from the overall horizontal length; this is the horizontal length for the landing, Step C.
4 Calculate the slope for the landing using the landing vertical change and the landing horizontal length, Step D.

EXAMPLE 1:
OVERALL VERTICAL CHANGE: **2353 mm**
OVERALL HORIZONTAL DISTANCE: **11,205 mm**

INITIAL RISER HEIGHT: **150 mm**
INITIAL TREAD LENGTH: **350 mm**

EXISTING LANDSCAPE SLOPE: **21%**

UK REGULATIONS

RISER HEIGHTS (mm)
150
160
170

TREAD LENGTHS (mm)
280
<u>290</u>
<u>300</u>
<u>310</u>
<u>320</u>
<u>330</u>
340
<u>350</u>
360
370
380
390
400
410
420
425

STEPS TO CALCULATING STAIRS AND LANDINGS:

Ⓐ 2353 mm/150 mm = 15.69 or **15 STEPS**

Ⓑ 150 mm x 15 = 2250 mm RISER HEIGHT
350 mm x 15 = 5250 mm TREAD LENGTH

Ⓒ 2353 mm – 2250 mm = 103 mm LANDING HEIGHT
11,205 mm – 5250 mm = 5955 mm LANDING LENGTH

Ⓓ 103 mm/5955 mm = **1.7% LANDING SLOPE, TOO STEEP**

Ⓔ 103 mm/$0.01\overline{6}$ = 6180 mm 103 mm/0.015 = 6867mm LANDING **LENGTH NEEDS TO BE BETWEEN 6180 mm AND 6867 mm TO MEET SLOPE REQUIREMENTS (1:60)**

Ⓕ SHORTER TREAD LENGTH = LONGER LANDING LENGTH AND A SHALLOWER SLOPE

Ⓖ BECAUSE THE SLOPE IS SO CLOSE TO THE MAXIMUM, CHOOSE TREADS LENGTHS THAT ARE CLOSE TO THE INITIAL TREAD LENGTH:
340 mm x 15 = 5100 mm; 11,205 mm – 5100 mm = 6105 mm, TREAD IS TOO LONG
330 mm x 15 = 4950 mm; 11,205 mm – 4950 mm = 6255 mm, **MEETS REQUIREMENT**

Ⓗ CONTINUE CHECKING TREAD LENGTHS, THIS WILL GIVE YOU MORE OPTIONS TREAD LENGTHS **290, 300, 310,** AND **320 mm** ALSO FALL WITHIN THE RANGE OF 1.5% AND $1.\overline{6}$% SLOPE.

Ⓘ 15 STEPS: **150 mm** FINAL RISER HEIGHT
290, 300, 310, 320 AND **330** mm FINAL TREAD LENGTHS

A.7
Calculating steps using the UK building regulations by adjusting tread lengths

5 Using the vertical change for the landings, calculate the range of horizontal distance you will need for the landing to meet the required slope. Divide the vertical change by 1.5% slope to find the shallow limit and 1.667% slope to find the maximum, Step E.

6 If the resultant slope is shallow, less than 1.5%, increase the tread length. Increasing the tread length decreases the horizontal length of the landing making the landing slope steeper.

7 If the resultant slope exceeds the maximum slope, 1:60 (1.667%) but is close, 1.7%–1.8%, shorten the tread length, Step G. Decreasing the tread length increases the horizontal length of the landing, making the landing slope shallower. Recalculate the landing slope.

8 If you cannot get the landing slope to fall between 1.5%–1.667%, because it is too shallow, e.g. between 0.2%–1.0%, remove a step from the total number of steps and add the values to the height of the landing and the length of landing respectively, see Appendix Figure 8.

Sometimes none of these steps works. The best option is to choose a slope that is shallower than 1.5% but greater than 1.0% to maintain positive drainage on landings. Note that I did not check each tread length when I was trying to increase or decrease the landing slope, I chose tread lengths that were about 30 mm apart first and went up or down in tread length based on the calculation.

This challenge presented by the UK regulations is not a concern with the Canadian regulations. The Canadian regulations have more riser height variability, 125–180 mm, no tread length maximum and a landing slope maximum of 1:50 (2%). Most riser-tread combinations will meet the Canadian requirements.

EXAMPLE 2:
OVERALL VERTICAL CHANGE: **2353 mm**
OVERALL HORIZONTAL DISTANCE: **29,415 mm**

INITIAL RISER HEIGHT: **150 mm**
INITIAL TREAD LENGTH: **350 mm**

EXISTING LANDSCAPE SLOPE: **8.0%**

UK REGULATIONS

RISER HEIGHTS (mm)
150
160
170

TREAD LENGTHS (mm)
280
290
300
310
320
330
340
350
360
370
380
390
400
410
420
425

STEPS TO CALCULATING STAIRS AND LANDINGS:

(A) 2353 mm/ 150 mm = 15.69 or **15 STEPS**

(B) 150 mm x 15 = 2250 mm RISER HEIGHT
350 mm x 15 = 5250 mm TREAD LENGTH

(C) 2353 mm - 2250 mm = 103 mm LANDING HEIGHT
29,415 mm - 5250 mm = 24,165 mm LANDING LENGTH

(D) 103 mm/24,165 mm = **0.4% LANDING SLOPE IS TOO SHALLOW**

(E) SUBTRACT ONE STEP (RISER HEIGHT AND TREAD LENGTH) = **14 STEPS**

(F) 150 mm x 14 = 2100 mm RISER HEIGHT
350 mm x 14 = 4900 mm TREAD LENGTH

(G) 2353 mm – 2100 mm = 253 mm LANDING HEIGHT
29,415 mm – 4900 mm = 24,515 mm LANDING LENGTH

(H) 253 mm/24,515 mm = **1.03% LANDING SLOPE IS TOO SHALLOW**

(I) SUBTRACT ONE STEP (RISER HEIGHT AND TREAD LENGTH) = **13 STEPS**

(J) 150 mm x 13 = 1950 mm RISER HEIGHT
350 mm x 13 = 4550 mm TREAD LENGTH

(K) 2353 mm – 1950 mm = 403 mm LANDING HEIGHT
29,415 mm – 4550 mm = 24,865 mm LANDING LENGTH

(L) 403 mm/24,865 mm = **1.62% MEETS REQUIREMENT**

(M) 403 mm/$0.01\bar{6}$ = 24,180 mm; 403 mm/0.015 = 26,867 mm
LANDING LENGTH NEEDS TO BE BETWEEN **24,180 mm** AND **26,867 mm** TO MEET SLOPE REQUIREMENTS (1:60)

(N) CONTINUE CHECKING TREAD LENGTHS, THIS WILL GIVE YOU MORE OPTIONS
TREAD LENGTHS **280, 290, 300, 310, 320, 330, 340, 350, 360, 370, 380, 390 and 400 mm** ALSO FALL WITHIN THE RANGE OF 1.5% AND $1.\bar{6}$% SLOPE.

(O) 13 STEPS: **150 mm** FINAL RISER HEIGHT
280 mm – 400 mm FINAL TREAD LENGTHS

A.8
Calculating steps using the UK building regulations by changing the number of steps

A.6 Literature cited

Canadian Human Rights Commission. 2007. *International Best Practices in Universal Design: A Global Review, Revised Edition.* Ottawa, Canada: Betty Dion Enterprises, Ltd. www.chrc-ccdp.gc.ca/eng/content/international-best-practices-universal-design-global-review.

Ministry of Housing, Communities and Local Government. 2013. *The Building Regulations 2010, Approved Document K Protection from falling, collision and impact.* London, UK: HM Government Crown Copyright. www.gov.uk/government/publications/protection-from-falling-collision-and-impact-approved-document-k.

Ministry of Housing, Communities and Local Government. 2015. *The Building Regulations 2010, Approved Document M Access To and Use of Buildings, Volume 1: Dwellings.* London, UK: HM Government Crown Copyright. www.gov.uk/government/publications/access-to-and-use-of-buildings-approved-document-m.

Ministry of Housing, Communities and Local Government. 2015. *The Building Regulations 2010, Approved Document M Access To and Use of Buildings, Volume 2: Buildings other than Dwellings.* London, UK: HM Government Crown Copyright. www.gov.uk/government/publications/access-to-and-use-of-buildings-approved-document-m.

National Research Council of Canada. 2018. *National Building Code of Canada 2015 Volume 1, 14th Edition.* Ottawa, Canada: NRC Canada.

Standards Council of Canada. 2018. *B651-18 National Standard of Canada – Accessible Design for the Built Environment.* Toronto, Canada: CSA Group. www.csagroup.org/wp-content/uploads/B651-18EN.pdf.

United States Department of Justice. 2010. *2010 ADA Standards for Accessible Design.* Washington, DC: US Department of Justice Civil Rights Division. www.ada.gov/2010ADAstandards_index.htm.

Index

Page numbers in *italics* denote figures.

Index

For Product Safety Concerns and Information please contact our EU representative GPSR@taylorandfrancis.com Taylor & Francis Verlag GmbH, Kaufingerstraße 24, 80331 München, Germany

Batch number: 10733469

Printed by Printforce, the Netherlands